Agricultural Development and Sustainable Intensification

Sustainable Intensification (SI) has recently emerged as a key concept for agricultural development, recognising that yields must increase to feed a growing world population, but it must be achieved without damage to the environment, on finite land resources and while preserving social and natural capital. It also recognises that all initiatives must cope with the challenges of climate change to agricultural production, food security and livelihoods.

This multidisciplinary book presents state-of-the-art reviews of current SI approaches to promote major food crops, challenges and advances made in technology, and the institutional and policy measures necessary to overcome the constraints faced by smallholder farmers. Addressing the UN's Sustainable Development Goal 2, the various chapters based on evidence and experiences of reputed researchers show how these innovations, if properly nurtured and implemented, can make a difference to food and nutrition security outcomes. Case studies from around the world are included, with a particular emphasis on Asia and sub-Saharan Africa. The focus is not only on scientific aspects such as climate-smart agriculture, agroecology and improving input use efficiency and management, but also on institutional and policy challenges that must be met to increase the net societal benefits of sustainable agriculture intensification. The book is aimed at advanced students and researchers in sustainable agriculture and policy, development practitioners, policy makers and non-governmental and farmer organisations.

Udaya Sekhar Nagothu is Research Professor and Director (Centre for International Development), Norwegian Institute of Bioeconomy Research (NIBIO), Ås, Norway. He is editor of the books *Food Security: Country Case Studies* (Routledge, 2014) and *Climate Change and Agricultural Development: Improving Resilience through Climate Smart Agriculture, Agroecology and Conservation* (Routledge, 2016).

Other books in the Earthscan Food and Agriculture Series

Resistance to the Neoliberal Agri-Food Regime
A Critical Analysis
Edited by Alessandro Bonanno and Steven A. Wolf

Food Policy in the United States
An Introduction
(Second Edition)
Parke Wilde

The Commons, Plant Breeding and Agricultural Research
Challenges for Food Security and Agrobiodiversity
Edited by Fabien Girard and Christine Frison

Agri-environmental Governance as an Assemblage
Multiplicity, Power, and Transformation
Edited by Jérémie Forney, Chris Rosin and Hugh Campbell

Agricultural Development and Sustainable Intensification
Technology and Policy Challenges in the Face of Climate Change
Edited by Udaya Sekhar Nagothu

Environmental Justice and Soy Agribusiness
By Robert Hafner

The New Peasantries
Rural Development in Times of Globalization
(Second Edition)
Jan Douwe van der Ploeg

Farmers' Cooperatives and Sustainable Food Systems in Europe
Raquel Ajates Gonzalez

For further details please visit the series page on the Routledge website: http://www.routledge.com/books/series/ECEFA/

"In an epoch of a changing climate – and aware that we have not inherited our planet's natural resources from our ancestors but rather borrowed them from our posterity – this book highlights indispensable innovations needed to sustainably increase productivity in some of the world's most relevant crops to nourish humanity in the coming decades."

Hans Dreyer, Food and Agriculture Organization of the United Nations, Italy

"We have now entered the UN Sustainable Development Era where Goal No. 2 aims to promote sustainable agriculture and nutrition security through ecotechnologies. This book on agricultural development and sustainable intensification by Dr. Udaya Sekhar Nagothu is therefore a timely one. I hope it will be read and used widely."

M. S. Swaminathan, UNESCO Chair in Ecotechnology and Founder Chairman, M. S. Swaminathan Research Foundation, India

"A superb addition to the rapidly expanding evidence base on the sustainable intensification of agriculture. We can develop processes and agroecosystems where productivity is increased without adverse environmental impact and without the cultivation of more land. Sustainable intensification offers novel pathways to the reduction of food insecurity that will also help meet some of the vital targets set out in the Sustainable Development Goals. This timely book shows that significant innovations are emerging in many contexts and systems."

Jules Pretty, University of Essex, UK

Agricultural Development and Sustainable Intensification

Technology and Policy Challenges in the Face of Climate Change

Edited by Udaya Sekhar Nagothu

First published 2018
by Routledge
2 Park Square, Milton Park, Abingdon, Oxon OX14 4RN

and by Routledge
711 Third Avenue, New York, NY 10017

Routledge is an imprint of the Taylor & Francis Group, an informa business

British Library Cataloguing-in-Publication Data
A catalogue record for this book is available from the British Library

Library of Congress Cataloging-in-Publication Data
Names: Nagothu, Udaya Sekhar, editor.
Title: Agricultural development and sustainable intensification : technology and policy challenges in the face of climate change / edited by Udaya Sekhar Nagothu.
Description: New York, NY : Routledge, 2018. | Includes bibliographical references and index.
Identifiers: LCCN 2017053993| ISBN 9781138300590 (hardback) | ISBN 9780203733301 (ebook)
Subjects: LCSH: Sustainable agriculture. | Agricultural innovations. | Agriculture and state. | Food security.
Classification: LCC S494.5.S86 A366 2018 | DDC 338.1—dc23
LC record available at https://lccn.loc.gov/2017053993

ISBN: 978-1-138-30059-0 (hbk)
ISBN: 978-0-203-73330-1 (ebk)

Typeset in Sabon
by Swales & Willis Ltd, Exeter, Devon, UK

Contents

Figures

Tables

Boxes

Contributors

Maricelis Acevedo is Associate Director for Science, Delivering Genetic Gain in Wheat Project-International Programs, College of Agriculture and Life Sciences at Cornell University, USA.

Asfaw Adugna is Sorghum Breeder in Africa and works in Advanta Seed International (a UPL Group Company) based in Addis Ababa, Ethiopia.

Mansab Ali is a Senior Scientist in Crop Physiology, Production and Management – Vegetable Project Leader at The World Vegetable Center in Pakistan.

Ahmed Amri is Head of Genetic Resources, Deputy Director of Biodiversity and Integrated Gene Management at International Center for Agricultural Research ICARDA, Rabat, Morocco.

Meriem Aoun is Postdoctoral Research Assistant-durum breeding and rust pathology, Department of Plant Sciences, North Dakota State University, USA.

Jesse Binamira is FAO Consultant and Coordinator of the Philippine National Save and Grow Program, based at the Department of Agriculture in Manila, Philippines.

Esther Bloem is scientist in the Norwegian Institute of Bioeconomy Research, Ås, Norway.

Harold Bockleman is Supervisory Agronomist/Curator National Small Grains Collection, USDA-ARS, Small Grains and Potato Germplasm Research Unit, USA.

J. Michael Bonman is Supervisory Research Plant Pathologist, USDA-ARS, Small Grains and Potato Germplasm Research Unit, Aberdeen, USA.

Andrew Borrell is crop physiologist at the university of Queensland, Australia, and Centre Leader of the Queensland Government's Hermitage Research Facility.

Allison M. Chatrchyan is the Director of the Cornell Institute for Climate Smart Solutions, and a Senior Research Associate, at Cornell University in Ithaca, New York.

Ronnie Coffman is Director of International Programs, College of Agriculture and Life Sciences, Cornell University, USA.

Nga Dao is the Director, Center for Water Resources Conservation and Development (WARECOD), in Hanoi, Vietnam.

Hugo Despretz is a Research Agronomist specialized in protected cultivation of vegetables in the (sub) tropics – and currently works as independent Consultant.

Pham Van Du worked as Delivery Manager of FAO's Regional Rice Initiative and earlier with Ministry of Agriculture and Rural Development, Vietnam.

Ngo Tien Dung worked as Deputy Director of Vietnam's Plant Protection Department, Ministry of Agriculture and Rural Development based in Hanoi, Vietnam.

Warwick Easdown is Senior Scientist with expertise in Communication, Agricultural Education and Extension, and the Regional Director, South Asia office, The World Vegetable Centre.

Mahmoud El-Sohl is independent consultant, and was Director General (retired), ICARDA, Beirut, Lebanon.

Sita Ghimire is Senior Scientist at Biosciences eastern and central Africa – International Livestock Research Institute (BecA-ILRI) Hub.

Elias M. Gichangi is Senior Research Scientist at the Kenya Agricultural and Livestock Organization (KALRO).

Xinyao He is Associate Scientist, Wheat Pathology, CIMMYT, Mexico.

Chu Thai Hoanh is Emeritus Scientist and a former Principal Researcher at the International Water Management Institute (IWMI).

Elizabeth Humphreys is currently a Co-Editor-in-Chief of *Field Crops Research*. She was a Senior Scientist (Water) at the International Rice Research Institute (IRRI) until 2015.

Philomin Juliana is PostDoctorate fellow, Genomic selection, Global Wheat Program, CIMMYT, Mexico.

James K Karanja is Senior Principal Research Scientist at the Kenya Agricultural and Livestock Organization (KALRO), Kabete, Kenya.

Jan Willem Ketelaar is Chief Technical Advisor of the FAO Regional Integrated Pest Management/Pesticide Risk Reduction Programme, at FAO's Regional Office for Asia-Pacific, Bangkok.

Linda McCandless is Assistant, Director of Communications, International Programs, College of Agriculture and Life Sciences, Cornell University, Ithaca, USA.

Connie Madembo is a Master student at University of Zimbabwe doing part time studies at CIMMYT-Zimbabwe.

Douglas J. Merrey is former Deputy Director General and founding Director for Africa at the International Water Management Institute (IWMI). He is currently an independent consultant.

Gemma Molero is Associate Scientist, Wheat Physiologist at CIMMYT, Texcoco Mexico.

Manoranjan K. Mondal is the Coordinator, SIIL-Polder Project at International Rice Research Institute (IRRI), and was Project Leader, Polder Community Water Management Project.

Alma Linda Morales-Abubakar is Programme Development Officer of FAO Regional Integrated Pest Management/Pesticide Risk Reduction Programme, at FAO's Regional Office for Asia-Pacific, Bangkok.

Munyaradzi Mutenje is a Socio-economist at CIMMYT, Zimbabwe.

Mupenzi Mutimura is Senior Research Fellow at Rwanda Agriculture Board, Rwanda.

Udaya Sekhar Nagothu is Research Professor and Director (International Department) at the Norwegian Institute of Bioeconomy Research, Ås, Norway.

Donald M.G. Njarui is Senior Principal Research Scientist at the Kenya Agricultural and Livestock Organization (KALRO).

Isaiah Nyagumbo is the Regional Cropping Systems Agronomist at CIMMYT-Zimbabwe.

Avakat Phasouysaingam is with Faculty of Agriculture, National University of Laos, and Principal Investigator, FAO, Save and Grow Program in Laos.

Peter Selkwena Setimela is a Senior Seed Systems Scientist at CIMMYT, Zimbabwe.

Pawan Singh is a Senior Wheat Pathologist at CIMMYT, Mexico.

Mehreteab Tesfai is scientist at the Norwegian Institute of Bioeconomy Research, Ås, Norway.

Emmanuel Torquebiau is senior scientist in agroforestry, and climate change correspondent with CIRAD, the French Agricultural Research Centre for International Development.

Cahyana Widyastama is IPM and FFS Facilitator, and Director of Field Indonesia (TFA) that provides technical assistance to Governments, CSOs, farmers and local communities.

Jason D. Zurn is Research Geneticist (plants), USDA-ARS, National Clonal Germplasm Repository, Corvalis, USA.

Preface

Achieving and maintaining food and nutrition security is essential to meet the Sustainable Development Goals, most notably the SDG-2 that aims to eradicate hunger. 'Business as usual' cannot achieve the goal, especially when smallholders are constrained due to climate change, lack of suitable technology and policy options. One promising approach is sustainable intensification of agriculture, the focus of this book. Sustainable intensification offers innovative pathways to increase crop productivity, income of smallholders and at the same time minimize impacts on the environment. The approach emphasizes on integrating gender and addressing equity issues that are important for reducing food insecurity.

A main message of the book is to use diverse and integrated approaches in agriculture development that will enable to address the challenges of climate change and food security. Exploiting the diverse farming systems, crops and genetic resources, and integrating suitable measures for efficient management of soil, water, pests and diseases becomes necessary for promoting sustainable production on smallholder farms. On the post-production side, support services including efficient extension services, post-harvest processing and marketing are to be strengthened. Sustainable intensification thus enables a holistic value chain approach.

This interdisciplinary book highlights the multi-functional benefits of sustainable intensification, including crop diversification through crop rotations and intercropping of cereals or millets with legumes, conservation agriculture, save and grow method, no-tillage and similar climate smart approaches suited for different crops in diverse agro-ecological settings. The various chapters in the book drafted by 43 experienced researchers and consultants from several disciplines representing 27 agencies worldwide bring together a mosaic of sustainable agriculture intensification approaches. The disparate book chapters provide technology, institutional and policy recommendations that will be relevant for future agricultural development through sustainable intensification. The book will be useful to a wide range of audience including scientists, research fellows, practitioners and policy makers.

I would like to sincerely thank all the contributors of the various chapters, in particular the lead authors who have spent time in coordinating and drafting the chapters. Thanks to Sankalp Nagothu who has helped me in editing and formatting some of the chapters in the book. This book would not have been possible without the support from the Norwegian Institute of Bioeconomy Research.

1 Agricultural development and sustainable intensification

Technology and policy innovations in the face of climate change

Udaya Sekhar Nagothu, Esther Bloem and Andrew Borrell

Introduction

The main challenges the world faces today, and will continue to face in the coming decades, are climate change, rapid shift in population demographics and unequal economic growth (FAO, 2009; Nelson *et al.*, 2010; FAO, 2016). These challenges will not only have a direct impact on food and nutrition security (FNS) but also become a source of conflict in many countries (Beddington *et al.*, 2012). Climate change, resulting in frequent extreme weather events in the form of droughts and floods will make it even more difficult to address the FNS challenges, in particular, for the poor and vulnerable communities. Despite the global advances made in agricultural development, hunger and malnutrition remain as perennial problems in many parts of the world, especially in sub-Saharan Africa (SSA) and South Asia (FAO, 2015). Agriculture is one sector that is highly vulnerable to the ongoing changes and needs immediate attention. To address the major challenges linked to FNS as well as improve our adaptation and mitigation to climate change we need sustainable and innovative approaches. Ensuring FNS and securing rural livelihoods for millions of poor and smallholders therefore is crucial to reduce the mass migrations across continents that we are witnessing today. The aging population and migration of people is leading to a dramatic change in cultural landscapes and loss of productive workforce. In the long run, it will be detrimental to the development of the countries from where migrations are taking place.

To summarize, main drivers impacting FNS include environmental (climate change and variability, extreme weather events, loss of biodiversity and land degradation), technological (soil and water management, seed quality, pest and disease management, post-harvest losses, poor extension services) and socio-economic and policy-related issues (poverty, economic inequality, lack of appropriate policies and implementation) (NEPAD/CAADP, 2003; FAO, 2015; Nagothu, 2015; EU, 2016). These drivers, either separately or in combination, influence food production as well as access to food in general. New theories mostly attribute socio-economic constraints for poor nutrition and irregular diets, causing hidden hunger and obesity. The type of crops cultivated, the variety of food produced, the access to diverse food types, the socio-economic constraints and the type of food consumed, all have an impact on FNS.

To begin with, early humans were hunters and gatherers, foraging for food that gave them access to a rich diversity of diet. In fact, foraging provided early humans diverse nutrition compared to smallholders of the present day. After the first Agricultural Revolution that started approximately 12,000 years ago, humans started to manipulate a few plant and animal species to meet their food needs. The actual transition to agriculture started between 9500 and 8500 BC and this wave continued till 3500 BC (Harari, 2011). Wheat and goats were the first to be domesticated in the Middle East regions and it then spread to other parts of the world. Domestication and cultivation of a few crops led to the production of more food, but the food was less diverse and poorer in terms of minerals, vitamins and other nutritive elements in the diets. Harari (2011) writes that in fact it was the crops that domesticated humans, and not vice versa. Humans stopped foraging and settled down around the land where crops were cultivated in order to protect their crops from wild animals and other communities. Domestication was definitely a big step for mankind, but its long-term impacts on the environment, human nutrition and health is questionable. The large environmental and deforestation impacts began when humans started to use fire, followed by the domestication of crops. This also meant that over the years humans gradually started to rely on a few crops to meet their major demand for food over the years. In the 1960s, 'The Green Revolution' further narrowed the range of crops cultivated, focusing largely on a few high yielding varieties of wheat and rice.

During the Green Revolution, conventional agricultural intensification and increase in food production was accomplished by introducing high-yielding crop varieties, increasing use of chemical fertilizers and pesticides, and developing irrigation and farm mechanization. In the process, beneficial functions of biodiversity (e.g. natural pest control and pollination) and environmental quality have been largely disrupted, threatening overall sustainability (Tscharntke *et al.*, 2012; IPES-Food, 2016). Due to increased specialization in the production process, farmers have reduced the use of genetic diversity within the crop types and livestock species. For example, even within rice or wheat, only a few high yielding varieties currently dominate the agricultural landscapes, overlooking the diversity of local types. Although the 'planned biodiversity' of cropping systems has resulted in higher yields and productivity, it has led to monocultures dominated by a few cereal crops such as rice, wheat and maize. According to the Food and Agriculture Organization (FAO), more than half of the calories consumed globally come from just three crops: maize, wheat and rice (Thornton, 2012).

Monocultures not only contribute to malnutrition, but also affect the environment negatively, due to excess use of inputs leading to land and water pollution (IPCC, 2007). Monocultures have a strong influence on the composition and abundance of the associated biota, including pest complexes, soil invertebrates and microorganisms, which in turn affect plant and soil processes (Swift and Ingram, 1996). As a result, we also see a reduction in soil organic matter, besides changes in water and nutrient composition of soils. This is mainly due to increased use of irrigation water and fertilizers, resulting in degradation of soils in irrigated agricultural lands. For example, serious problems of salinity arising within arid and semi-arid areas due to excess use of irrigation water that makes the land uncultivable after some years. High concentrations of soluble salts containing sodium (Na^{+}) and calcium (Ca^{2+}) ions in agricultural soils and/or irrigation water increases the osmotic potential of the soil solution, making it difficult for plants to take up water and nutrients (Khang *et al.*, 2008). Consequently, plant growth is suppressed and ultimately crop yield is significantly reduced (Lamsal *et al.*, 1999; Ologundudu *et al.*, 2014). Increasing use of fertilizers, pesticides and irrigation does not necessarily mean increased production beyond a certain limit, as supported by the law of variable proportion, but rather adds to the costs of cultivation and negative environmental impact. This situation is quite dominant in the rice-growing regions of South Asia, especially the semi-arid regions where irrigation has been introduced. Millets that once dominant in these regions have been replaced largely by rice and maize (Seetharam *et al.*, 1989). Millets are nutritious and drought resistant and can easily be grown under water scarce conditions with fewer inputs. Chapter 6 in this book provides some details about the advantages of growing millets. Thus major shifts in agricultural cropping patterns during the Green Revolution significantly contributed to the new FNS challenges.

Agricultural scientists worldwide are now faced with an intriguing question: *How to address the challenge of FNS without repeating the environmental damage of the mid-twentieth century*? To feed the growing population and achieve sustainable FNS, innovative processes in smallholder agriculture are needed that are ideally *low cost and low in terms of risk, and that can make significant contributions to productivity and income from the same parcel of land with minimal environmental impacts*. In this chapter, we first attempt to briefly analyse the relevance of sustainable agriculture, followed by a conceptual framework of sustainable intensification (SI) of agriculture or sustainable agricultural intensification (SAI), and a brief discussion on the critical perspectives of SAI. We then focus on the key factors such as technology, institutions and policy drivers necessary for SAI implementation. Towards the end, the chapter provides an outline of the book and a brief conclusion.

From agricultural intensification towards sustainable agriculture intensification

All along, agricultural intensification has been a prerequisite to the development of human civilization. Humans started cultivating food crops in order to be food secure (FAO, 2004). Human migration over the centuries, and subsequent colonization, has led to the spread of crops from their centres of origin to new areas. A classic example is the spread of potato from South America, its centre of origin, to Europe and Asia. Agricultural intensification also contributed to population growth, and further evolved over the years as a result of globalization and increasing world trade in agriculture (Hazell and Wood, 2008). This also led to spatial differences in agriculture dominated by cereal crops in certain regions of the world, whereas in other areas, crops such as coffee, tea, spices, cacao and sugarcane dominated to meet the demands of the international market. The last hundred years witnessed a significant change in agricultural development, with the introduction of chemical fertilizers, high yielding varieties and expansion of agriculture at the cost of forest clearance and loss of biodiversity (Grigg, 1987).

The widespread use of chemical fertilizers and high yielding varieties no doubt reduced the yield gap. This paradigm dominated agricultural research in the '60s and '70s. Nevertheless, the negative impacts associated with the Green Revolution were not adequately addressed in the process. The SAI approach claims that its objective is to increase the agriculture efficiency and productivity and at the same time reduce environmental impacts. It also expected to contribute to the increase in smallholder income through a comprehensive transformation of both the input and output markets. It is obvious that smallholder farming will succeed only when it is aligned with the markets, increasing market demand for food that provides opportunities to earn more income in a sustainable manner.

A transformation to SI is thus justified both by necessity (to safeguard global sustainability, a precondition for long-term agricultural viability) and by opportunity (to use sustainable practices as a vehicle for a second green revolution). The Montpellier Panel report, 2013 (IIED, 2016 and the publication by Garnett *et al.*, 2013) represents one of the most recent attempts by a large number of scholars to conceptualize sustainable intensification. SI (or SAI) is an essential means of adapting to climate change and resulting in lower emissions per unit of output produced. A paradigm shift towards the SAI approach is seen as one of the promising alternatives proposed in recent years to address the serious challenges in the agriculture sector in order to achieve FNS in the future.

Rockström *et al.* (2016) argue that sustainable agriculture is the only strategy that delivers productivity enhancement to meet rising food demands and enables an Earth system operating within planetary boundaries, given the decisive role of world agriculture on human development and on Earth system processes. Malnutrition and hunger are linked to poverty and inequitable food access, more than to inadequate food production per se according to some studies (FAO, 2013). According to Oborn *et al.* (2017), sustainable agriculture includes ecological, economic and social dimensions, where food and nutrition security, gender and equity are crucial components. However, the problems associated with lack of self-sufficiency in food production are serious in SSA, whereas in south Asia, it is mainly the lack of crop diversity and hidden hunger that are the major challenges. Thus, SAI remains a major target of research and development to address these challenges. Future SAI initiatives consider to: i) sustainably increase agricultural productivity and incomes; ii) adapt and build resilience to climate change; and iii) reduce and/or stop greenhouse gas emissions, where possible (FAO, 2013). Any new agricultural technology introduction should have minimum impacts on the environment, increase crop diversity, require low investments and use site-specific assessments to identify suitable agricultural production technologies and practices (FAO, 2013).

The ratification of the Paris Agreement in 2015, and other global initiatives if properly implemented, can positively contribute to FNS (UNFCCC, 2015). However, the agriculture sector, both at the national and international levels needs adequate investments in order to achieve FNS.

Sustainable Development Goal (SDG) number 2 clearly states the need to end hunger, achieve food security and improved nutrition, and promote sustainable agriculture (UN, 2015). The concept of sustainable agriculture, through its links to food security, nutrition, gender, health, employment, economic development and environment can contribute significantly to achieving a wide range of SDGs. This linkage between SDGs and sustainable agriculture has to be exploited by scientists and policy makers to achieve the goals of FNS. To strengthen these linkages, islands of knowledge (e.g. crop, discipline, scale, environment, organization) need to be better integrated and mapping is needed to identify gaps and make connections between these islands (Borrell and Reynolds, 2017).

Improvement in the agriculture sector can enhance livelihoods for the majority of the rural population in developing countries. We have experienced this in major parts of Asia that have gained significant pace over the last few years, but agricultural development in Africa has remained slow (Posthumus and Wongtschowski, 2014). This is the region where the next population expansion and economic growth will happen. Due to the lack of proper infrastructure development, lack of capacity and targeted investments, applied technologies remain ineffective in achieving FNS. State support in the SSA countries covering aspects of agricultural extension, knowledge and knowhow on new methodologies for implementation exist, but are scattered and inaccessible to end users, especially the rural poor in remote locations (Posthumus and Wongtschowski, 2014).

According to an IAASTD report (2009), social and economic inequalities, political uncertainties and changing environmental conditions demand a new approach to sustainable agricultural production and consumption at all levels. The report advocates integrated innovation processes in smallholder agriculture that are implementable, sustainable and contribute to higher productivity and income. Innovations in agricultural technologies, knowledge transfer to farmers and institutional development should go hand in hand in order to provide multiple benefits (ecological, nutritional and socio-economic): i) using multiple approaches (technological, extension and institutional); ii) with multiple actors (e.g. public and private sectors, civil society, NGOs and other relevant actors); and iii) at multiple scales from field/farm clusters to village, district and national levels (Gatzweiler and von Braun, 2016). Translational research, spanning from the molecular to farm scale, should be the focus of international crop research aimed at addressing global challenges such as food security and climate change (Borrell and Reynolds, 2017). The key word is translational – taking scientific discoveries and translating them to useful yield outcomes in the field.

Sustainable agricultural intensification – conceptual framework

Sustainable agricultural intensification integrates three main goals – environmental health, economic profitability, and social and economic equity. *SAI is a concept to increase yields without adverse environmental effects and without cultivation of more land* (Petersen and Snapp, 2015). This concept was explained in more detail by Pretty *et al.* (2011), where SI of agriculture is expected *to produce more outputs from the same area of land with a more efficient use of agricultural inputs (seeds, fertilizers, water and pesticides) including innovative technology and knowledge interventions on a durable basis, while building resilience and the social and natural capitals, and reducing environmental damage and improving the flow of environmental services*. This is achievable using techniques such as *precision farming* which is better-suited to the needs of small-scale farmers and reduces the negative environmental effects of food production. Precision farming does not need to be technology intensive. It could include a move towards more need-based input application. This could be highly relevant for smallholders who are constrained by credit and inputs. Any savings on input could be useful to the smallholder economy, as well as to the environment. Promoted since the 1990s, SAI has received increasing attention, particularly in the arena of international agricultural development (Pretty, 1997). The Food and Agriculture Organization (FAO, 2012), the CGIAR (Beddington *et al.*, 2012) and other organizations have promoted SAI as a necessary approach to achieve FNS in the twenty-first century. They have highlighted SAI as a means to meet the rising global food demand, yet the technologies under SAI, and the organizing principles for the approach need to be specified and improved further to meet the different agroecological conditions.

Associated closely with SAI is the aspect of integrated water management, a prerequisite, especially in the context of irrigation, and a necessary condition for agricultural development, as well as improving smallholder adaptation in the face of climate change. As water is becoming a critical resource, improving water productivity of major food crops will be necessary (Evans and Sadler, 2008). Practices leading to sustainable water management will be an integral component of SAI development.

In addition to technological advancements, institutional and policy innovations are crucial for promoting SAI. This should be accompanied by adequate investments in the agriculture sector for developing and scaling up high impact SAI systems that are suitable for smallholders. The most important measures include: i) diverse climate-resilient cropping systems with a focus on major crops/agri-food systems (rice, maize, wheat, potato, legumes, sorghum, millets and vegetables); ii) integrated crop–livestock systems; iii) sustainable water management; iv) innovative institutions; and v) agricultural information services and inclusive policies.

SAI requires proven integrated approaches in smallholder agriculture that are easily implementable, low cost and low risk, and can make significant contributions to higher productivity of the major food crops and minimize impact on the environment. (NEPAD/CAADP, 2003; FAO, 2015; Nagothu, 2015; EU, 2016).

Sustainable agriculture intensification – a critical perspective

SAI is an evolving concept and its meaning and objectives are subject to debate and contest (Rockström *et al.*, 2016). There are concepts such as sustainable intensification (The Montpellier Panel, 2013), ecological intensification (Tittonell, 2014), eco-efficient agriculture or agroecological intensification, that are similar to SAI (Gliessman, 2007). Tittonell (2014) examines the conceptual differences between sustainable and ecological intensification as used in research, development and policy, particularly with respect to the balance between agriculture and nature. According to Campbell *et al.* (2014), the 'sustainable intensification' (SI) approach and 'climate-smart agriculture' (CSA) are highly complementary. Their study shows that CSA in fact provides the foundations for incentivizing and enabling SI, with its emphasis on improving risk management, information flows and local institutions to support adaptive capacity. As SAI is often loosely defined, several approaches mentioned above can be labelled under it, making it challenging for scientists to differentiate. All these approaches are expected to sustain ecosystem services while minimizing environmental costs and maintaining biodiversity (Tscharntke *et al.*, 2012). The argument is relevant, since agricultural production is highly dependent on ecosystem services such as pest control, pollination by bees, and soil fertility among others.

The opportunity costs borne by smallholders practising SAI in terms of reduced production to ensure ecosystem services is debatable. Some argue that SAI focuses too much on the production side and too little on the post-harvest aspects of food systems, including food access, processing and marketing (Silici *et al.*, 2015). When dealing with smallholders, it is important to address the entire product value chain (VC), especially the extension and market aspects that contribute to the positive socio-economic impacts of smallholders. Unless smallholders see the benefits in the short term from SAI, they are not likely to adopt them. There are also critics of SAI that challenge the ability of anything to be sustainable in a world with increasing human population through to 2050, coupled with economic theory that is focused on 'growth' as the ideal (Heinberg, 2015).

There is no single approach of SAI that works under all conditions. Vanlauwe *et al.* (2014) analysed the ability of SAI to address a very diverse group of smallholders that dominate sub-Saharan agriculture. Farmers in this region are characterized by large variation in socio-technical conditions, production objectives, and biophysical environments that potentially demands a multitude of SAI pathways. According to Cook *et al.* (2015), SAI is only one aspect of the larger 'food system', and there are other issues such as food consumption and waste, property rights, gender and income that need to be adequately addressed in the process.

Sustainability is not just about environment, but the interaction between environment and agriculture production, including how they respond to each other. According to Cleary (2014), the concept of resilience is central to agriculture landscapes when addressing climate change and needs to be discussed more adequately in the SAI debate. Thus resilience of agro-ecosystems is important, needs more emphasis, and also demonstrates the close link between agriculture production and environment. Given the looming climate change threats to agriculture, considerable attention is required to build resilient farming systems (Elagib, 2015). And within a broader perspective, the socio-economic resilience of the smallholders is also crucial for adapting to climate change and variability. Resilience suffers from a lack of attention to power dynamics within households and communities (Nelson and Stathers, 2009).

A practical challenge associated with SAI is the measurement of sustainability indicators that could be locality specific and vary from one situation to another (Cleary, 2014). Hence, it can be difficult to establish and measure the sustainability index, especially when it is linked to the issue of scale and geographical context. In a macro-level context, environmental and agricultural outcomes may not necessarily be interdependent. The technical specifications and measurement of intensification or extensive adjustment in any given case are actually quite complex. Changes in the productivity are likely to be accompanied by adjustments for inputs used in the process and thereby cost implications for smallholders (FAO, 2004). Moreover, there could be other reasons influencing the ecosystem, rather than the impacts from input use. Such complexities can make it difficult to develop a broader political consensus, incentives at farm level for reducing environmental impacts, and a general framework for promoting SAI at the policy level. Despite the success of SAI, challenges remain in scaling out the SAI practices in most cases (Pretty, 1997). A fundamental shift, therefore, in the institutions and policy will be required for a wider adoption of SAI practices. Such changes will be challenging in many developing countries where it is actually needed (Tilman *et al.*, 2002).

Sustainable agriculture intensification – the role of technology

Technology has played an important role in agricultural development throughout history. It has been the basis for the Green Revolution where new cultivars, use of fertilizers and irrigation development primarily contributed to increases in food production (Tilman *et al.*, 2002). Yet, yields of important crops such as rice and wheat have now stopped increasing in some intensively farmed parts of the world, a phenomenon called 'yield plateauing' (*The Economist*, 2016). The spread of existing best practices should enable yields to approach these plateaus. But to go beyond them, without negative environmental effects will require research and development leading to improved technologies that are sustainable. This will be a significant challenge for scientists. Farmers are risk averse, since the cost of getting things wrong (messing up an entire season's harvest) will have considerable consequences, particularly for smallholders in the developing world. Nevertheless, we have to move forward and the use of innovative technologies that will be important for advancement of SAI. Sustainability of technology advancement has been at the centre of debate in recent years. New paradigms of agricultural development suggest new technologies that are sustainable and simultaneously increase the net societal benefits. A survey of 89 projects by Pretty *et al.* (2003), demonstrated an average per project increase in per hectare food production significantly due to use of sustainable approaches such as crop rotations with legumes and integrated pest management, and new and locally appropriate crop varieties. Besides, other interventions such as improved water use efficiency, soil health and fertility, and pest control with minimal or zero pesticide use also played a substantial role. In the following part of this section, some of the promising sustainable agriculture techniques that have the potential to contribute to higher productivity and at the same time ensure ecosystem services are discussed.

Soil management and nutrient use efficiency

Soil fertility is fundamental to crop production and can be maintained and enhanced in many ways, including leaving crop residues in the field after harvest, ploughing under cover crops, adding composted plant material, or animal manure, need-based fertilizer application, or, in some cases, even leaving the land fallow (UCS, 2017). Many farmers also take advantage of the benefits of having plants growing in the soil at all times, rather than leaving the ground bare between cropping periods, which may lead to unintended problems such as soil erosion. The planting of cover crops such as hairy vetch, clover or oats helps farmers to achieve the basic goals of preventing soil erosion, suppressing weeds and enhancing soil quality. Using appropriate cover crops, including tree crops in agro-forestry is beneficial because it reduces the need for chemical inputs such as herbicides, insecticides and fertilizers. It also contributes to additional income for smallholders at times of climate vulnerability and crop failures due to extreme climates.

It is necessary to increase nutrient-use efficiency by aligning the temporal and spatial nutrient supply with plant demand (Matson *et al.*, 1998). Need-based fertilizer application has the potential to reduce losses while maintaining or improving yields, and at the same time reduce the costs of cultivation for smallholders.

Integrated pest management

Understanding a farm as an ecosystem rather than a factory offers opportunities for effective pest control. For example, many insects, spiders, birds and other species are natural predators of agricultural pests. Managing farms so that they harbour populations of pest predators is a sustainable and effective pest-control technique (UCS, 2017). This will help in reducing the use of pesticides and the negative impacts on the environment, which has reached serious levels in most of the agriculture intensive areas. One of the unfortunate consequences of the intensive use of chemical pesticides is the indiscriminate killing of biota and loss of biodiversity. Reintroduction of natural, disease-fighting microbes into plants and/or soil, and release of beneficial organisms that prey on the pests is necessary and sustainable in the long run (Hokkanen and Lynch, 1995; Bale *et al.*, 2008). One problem associated with biocontrol agents is that they are more specific in their action as compared to chemical pesticides that act on a broader spectrum of pests or diseases. Climate also plays a significant role in the success of biocontrol agents, as they are easily affected by changes in temperature or humidity. Hence, a combination of methods including physical, chemical and biological is recommended for effective control of pests and diseases. Integrated Pest Management (IPM) approaches that result in lower pesticide use will benefit not only farmers, but also wider environments and human health (Pretty and Bharucha, 2015). Future success of biological control is strongly dependent on the investment in research and development by governments and other organizations that are committed to reduce the use of chemicals, in addition to the use of new tools and technologies to monitor and manage the pest and disease outbreaks.

Sustainable irrigation water management

With global freshwater resources under extreme stress and overuse, sustainable irrigation water management, including both surface and groundwater, becomes a pre-requisite for future agricultural development and sustainable intensification, as well as improving smallholder adaptation in the face of climate change. Therefore, a dedicated chapter on water and ecosystems is included in the book to show the importance of this subject. Often this is a key gap, since commodity driven approaches focusing on major food crops miss the larger watershed or ecosystem dimension. In fact, commodity based work should be complemented by ecosystem sustainability at scale. From farm to basin scales, there are several examples of water management interventions used to manage rainfall efficiently in smallholder farms that can achieve significant long-term impacts.

McCartney *et al.* (2016) highlight how different approaches to improve water management can be used to ameliorate some of the major impacts of climate change on agricultural production in Southeast Asia and in some cases generate significant mitigation benefits by reducing greenhouse gas (GHG) emissions. These approaches will be more relevant for crops such as rice that are water demanding and also a major source of GHG emissions. Smart water management in rice can save a significant amount of freshwater resources. Groundwater use will be a key component of adaptation strategies in many areas, and dependence on groundwater as a buffer to climate variability is likely to increase under climate change. According to Barron (2009), improved water management in rain-fed systems is necessary to make it cost-effective for increasing food production and building resilience. Increasing overall water productivity in irrigated agriculture, especially for crops such as rice, wheat and potato will be important in the future. Cofie and Amede (2015) demonstrate in their study that both technical and institutional innovations in water management are necessary for creating and sustaining resilient agrarian communities in SSA. Such innovations are best developed and implemented in active consultations involving researchers, households, private sector, and other stakeholders with a management or regulatory responsibility.

Crop rotations and integrated systems

Monocultures dominated by cereal production with two or three cycles of the same crop in a year may become progressively susceptible to diseases and insect pests because of insufficient crop diversity (Tilman *et al.*, 2002). Crop rotation is an age-old technique that has been shown to benefit both soil and environment, and increase crop diversity. Rotating a variety of crops improves soil fertility and reduces pesticide use, especially in monocultures with cereals. It is a key element of the permanent and effective solution to pest problems because many pests have preferences for specific crops, and continuous growth of the same crop guarantees them a steady food supply, so that populations increase (UCS, 2017).

In rotations, farmers can plant crops such as soybeans and other legumes alternating with cereals that replenish plant nutrients, thereby, reducing the need for chemical fertilizers, e.g. maize or rice followed by legumes in the same plot. Multi-year, multi-crop rotations produce high yields for each crop in the rotation, control pests and weeds with less reliance on chemical pesticides and enhance soil fertility with less need for synthetic fertilizers (Davis *et al.*, 2012). Rotations that include nitrogen-producing legumes such as peas, beans and alfalfa improve soil fertility. Recent research shows that nitrogen from legumes remains in the soil longer than the nitrogen from synthetic fertilizers, leaving less to leach into groundwater or run off fields and pollute streams (Gardner and Drinkwater, 2009). Leaving land fallow without cultivating any crop also helps soils to replenish the nutrients. Farmers have traditionally followed this practice to give some rest to the soils. Similarly, rice integrated with growing fish in south and Southeast Asian countries has shown to be more productive than monocultures. The following are examples of some of the crop rotations that can simultaneously contribute to soil fertility improvement, human nutrition and climate adaptation.

Maize–legume cropping systems with improved maize and legume varieties have been the focus of regional projects such as the SI of Maize–Legume cropping systems for food security in Eastern and Southern Africa (SIMLESA), showing their potential to contribute to improved productivity and ultimately food security in ESA, while at the same time ensuring environmental sustainability (Thierfelder *et al.*, 2013). These systems, if properly managed, contribute to increased yields, improved soil fertility, reduced GHG emissions, better nutrition (Vitamin A, more proteins), lower fertilizer costs, increased income and reduced vulnerability to climate change.

Diversified millet–legume cropping systems (pearl/finger millet/pigeon pea/beans/peanuts): integrating a greater diversity of legumes and other crops (including millet, sorghum, cassava and sweet potatoes), depending on the agroecosystem, rotations and intercropping associations, provides opportunities for soil fertility improvement and increased productivity, leading to improved nutrition and better market value. Millets and sorghum show resilience and tolerance to droughts when used in combination with CA systems and will be environmentally sustainable in semi-arid and arid agroecologies of the region. Greater climate resilience, improved soil quality, increased yields, increased agrobiodiversity, better nutrition (iron, Vitamin A, more proteins) with a more diverse diet, lower fertilizer costs, increased income, increased equality within households, and increased farmer knowledge and capacity to apply SAI principles should result from the adoption of diversified millet–legume cropping systems (Bezner Kerr *et al.*, 2012; Snapp *et al.*, 2014).

Integrated crop–livestock systems interact to create a synergy whereby recycling maximizes the use of available resources and overall productivity. Crop residues from the cultivation of legumes can be used for animal feed, while livestock and livestock by-products and their processing can enhance farm productivity by providing nutrients that improve soil fertility and at the same time reduce the use of chemical fertilizers (Rota, 2008; IFAD, 2008). For small farmers, this provides an optimal allocation of scarce resources between crop and livestock production. However, smallholders need to plan properly and should have ready access to knowledge, assets and markets to take advantage of the system. Crop–livestock integration can ensure sustainability in the long term, both on the economic and environmental fronts.

New crop varieties and crop diversity

New cultivars that are high yielding, with tolerance to abiotic (drought, floods, salinity) and biotic (pests, diseases) factors, and that are rich in minerals and vitamins will be important components of future agricultural development (Ronald, 2011). High yielding varieties have largely contributed to yield increases in rice, wheat, maize and other major food crops. In the future, crop diversity needs to be conserved, explored and utilized, thereby increasing the diversity of crops cultivated. A number of local crops and crop varieties (landraces) of the food crops are richer in nutrients and more tolerant to pests and diseases than the new crop varieties developed and cultivated. Capturing such crop diversity should be an integral part of the SAI initiatives in the future.

Sustainable agriculture techniques – the role of institutions and policy innovations

Institutions: Technology is important, but efficient extension, market services and policy support are essential for scaling out SAI practices. One way to reinforce policy advocacy for SAI is by validating and consolidating evidence of the technology benefits, in contrast with the negative impacts of monocultures (Cook *et al.*, 2015). If policy makers are convinced about the potential of SAI, it becomes easier to develop and implement the science-based policies on a larger scale. Politically motivated policies such as the fertilizer subsidy policies introduced in many countries during the '80s and '90s has led to overuse of fertilizers and, subsequently, negative impacts on the environment. Local stakeholders can play an important role by pursuing the changes that are needed in agricultural policy and practice. Policy impact can be increased by fostering partnerships between various sectors and avoiding duplication of efforts, including pooling resources to scale up SAI technologies. Creating and raising awareness of the benefits of a technology is a commonly acknowledged prerequisite for farmers to decide whether to adopt a technology or not (Odendo *et al.*, 2009).

Enabling policies: The conditions that create policy are important for providing an enabling environment for men and women farmers to change their practices and start to use new technology, as well as accepting institutional innovations related to how knowledge is being delivered and the opportunity for demand-driven services (Haug *et al.*, 2016). According to Vanlauwe *et al.* (2014), the institutional context needs to be appropriate for delivering the necessary goods and services underlying SI, ensuring inclusiveness across different household types and facilitating local innovation. Evidence suggests that up-front investment costs can significantly improve adoption of certain innovative practices (McCarthy *et al.*, 2011; Vanlauwe *et al.*, 2014). Policies are also important when determining the type of Extension and Advisory Services (EASs) that a country should establish. With proper policies and incentives, SI could both promote transitions towards greener economies as well as benefit from progress in other sectors (Pretty and Bharucha, 2014). Issues that should be considered include public and/or private sector involvement, level of decentralization, financing, partnership and linkages, as well as capacity of staff and their motivation and organizational culture. One important outcome of SAI is that farmers are able to minimize their use of pesticides and fertilizers, thereby saving money and protecting future productivity, as well as the environment. However, the various proven technology options available within the scientific domain have to be validated in real time conditions and made accessible to smallholders. The validated technologies should be given adequate institutional and policy support for scaling up involving the public and private sectors.

Extension and training: Technologies may be available, but often are not accessible to smallholders due to poor extension and training services. Providing farmers with the right knowledge, training and investments are a prerequisite to put SAI into practice and further scaling out to other farmers through continuing adaptation and innovation (Cook *et al.*, 2015). Chapter 10 in this book highlights the importance of extension services and strengthening market value chains to ensure successful implementation of SAI practices by smallholders. Any future investments in the agriculture (including livestock) sector should be linked to developing and scaling up innovations that will significantly promote FNS, and at the same time adapt to climate change. Scaling up requires adequate investments, as well as institutional and policy backing that can address specific needs of smallholders, including gender mainstreaming.

Market improvements: Woelcke (2003) showed that focusing solely on increasing output prices revealed negligible impacts on technology adaptation and nutrient balances. Woelcke (2006) in a case study in Uganda, showed that in order to pursue SAI the market value needs to be improved. That is, transaction and transportation of costs have to be reduced, innovative credit schemes for smallholders have to be introduced, and alternative forms of labour acquisition have to be promoted, to provide sufficient economic incentives for the adaptation of environmentally sound production methods. Efforts to increase yields should be accompanied by improving post-harvest services to smallholders in order to contribute to positive net benefits of technology adoption (Woelcke, 2006). Market improvements have the potential to simultaneously increase household welfare and nutrient balances. Policy options focusing only on input market improvements would probably not be a promising strategy, considering the extreme fertilizer price reduction needed to induce farmers to switch to more SAI practices. The overall impact of subsidies through fertilizer price reduction on income is very modest, reaching 5.5 per cent increase when fertilizer prices are reduced to 5 per cent of the current price (Woelcke, 2006). With new structural changes in the agriculture sector across countries, subsidies will be gradually erased. Governments should therefore try to develop partnerships with the private sector to address FNS challenges in the future.

Public private partnerships: Public private partnerships are crucial to scale out SAI in the current situation where governments are constrained with funds to invest in the agriculture sector. Partnerships among several relevant actors – public, private, NGOs and farmers will be essential to promote SAI on a wider scale. Smallholders should have a key role in such partnerships, and this is possible if farmer collectives are well organized. In Rwanda, dairy cooperatives are functioning well due to the fact that smallholders are given their due share of profit (FAO, 2014). Integrated crop–livestock systems are becoming important SAI initiatives that are able to increase productivity and at the same time generate profits to smallholders.

Gender integration and mainstreaming: Women play an important role in agriculture in the developing world, and any future success of smallholder farms using the SAI approach will need a stronger gender integration. Gender has not been given adequate importance in agricultural development, despite the significant role of women in irrigated agriculture and the increasing burden they are facing due to climate risks (Raj and Nagothu, 2016). Several studies show that women play a significant role in ensuring household food and nutritional needs in most developing countries (Turral *et al.*, 2011). At the farm level, both men and women complement each other to ensure agriculture production and income. With increased migration of men to urban areas, as observed in SSA and other regions of the world, the role of women in agriculture will become even more important. Thus, while promoting SAI, the role of women, their specific vulnerabilities and capabilities for adapting in the context of changing economic, climate and agrarian systems should be taken into consideration.

Mainstreaming adaptation strategies with socially inclusive and gender sensitive approaches in the agricultural development sector have the potential to reduce the gender gap (Raj and Nagothu, 2016). However, in practice, it depends upon how it is being conceptualized and facilitated. In many contexts, these 'gender role changes may not be sufficient to bring uncontested positive changes in gender relations' (Tatlonghari and Paris, 2013). In their study, Dirutu *et al.* (2015) found gender differences in the adoption patterns for some of the SAI practices. According to them, women farmers are less likely to adopt minimum tillage and animal manure in crop production than men, indicating the existence of certain socio-economic inequalities and barriers. However, the study finds no gender differences in the adoption of soil and water conservation measures, improved seed varieties, chemical fertilizers, maize–legume intercropping and maize–legume rotations. Whereas, a study by Theriault *et al.* (2017) shows that women cultivating cereal crops are less likely than their male counterparts to adopt yield-enhancing and soil-restoring strategies, although no differential is apparent for yield-protecting strategies. It clearly demonstrates that while introducing SAI practices, institutions and policy measures should take the gender differences into consideration, as this determines the success of SAI practices (Oborn *et al.*, 2017). At the same time, a better understanding of gender differences is crucial for designing effective policies to close the gender gap while sustainably enhancing farm productivity (Theriault *et al.*, 2017).

Book outline

This book consists of 12 chapters focusing on major food crops (including rice, maize, wheat, potato, legumes, millets, vegetables), integrated crop–livestock systems, water management, innovative agricultural extension services, market value chains and inclusive policies. This multidisciplinary book will bring together state-of-the-art technologies with institutional and policy innovations addressing key FNS priorities based on long-term experiences. The different chapters are based on evidences and experiences of reputed researchers from various countries in their respective fields and will show how the innovative technologies, institutional and policy options, if properly validated and implemented, can make a significant difference to FNS outcomes and adapt to climate change in the future. The book chapters will focus on the latest developments in research and development of major food crops and sustainable practices that contribute to productivity and ecosystem services. The emphasis within various chapters will be on innovations in soil, water and fertilizer management, new crop varieties and integrated cropping systems that are likely to contribute to the increase in yields and, at the same time, ensure sustainable management of natural resources. Any new interventions must be low cost and low risk for smallholders and demonstrate the positive outcomes.

The importance of innovative extension services to provide timely and correct information to farmers, the need for strengthening market value chains, and for enabling policy support to scale out SAI practices is analysed in later parts of the book. Though the chapters may not be addressing all the principles and challenges of SAI, they do attempt to throw some light on the key SAI issues in terms of improving crop/product value chains and addressing FNS. The varied experience of authors from different disciplinary backgrounds and geographical regions brought together in this book is one of the major strengths. This book attempts to address the question, *whether SI or SAI is the right approach to the growing FNS challenges or not?*

Conclusion

By 2050, agricultural scientists and policy makers have the responsibility to address the key goals of FNS amidst the challenges of climate change and increasing population. This needs to be accomplished through increasing the efficiency of agricultural inputs used and at the same time minimizing the negative environmental impacts. Governments have to realize that capacity building of smallholders, institutional and policy options are crucial for the long-term sustainability of agricultural development. Separate chapters in the book focusing on major food crops (rice, maize, wheat, potato, millets–legumes, vegetables) discuss in detail the challenges and potential of enhancing productivity through SAI.

The pursuit of SAI will require significant increases not only in technology intensification (for example, new crop varieties, soil and water management, integrated pest management) but also addressing challenges related to knowledge sharing and improving markets, storage facilities and extension services. The chapter shows that a VC approach integrated with SAI is an effective way to address FNS in the future, especially in the climate and economic vulnerable zones. The chapter also acknowledges the critics of SAI and emphasizes that care should be taken to avoid pitfalls while promoting the concept for the benefit of smallholders. The ultimate goal of SAI while promoting future agricultural development is to simultaneously maximize the net societal, economic and environmental benefits while promoting agricultural development.

References

Bale, J.S., van Lenteren, J.C. and Bigler, F. (2008) Biological control and sustainable agricultural production. Available at: www.ncbi.nlm.nih.gov/pmc/articles/PMC2610108/ (accessed 01 August 2017).

Barron, J. (ed.) (2009) *Rainwater Harvesting: A lifeline for human wellbeing*. United Nations Environment Programme, Nairobi/Stockholm Environment Institute, Stockholm.

Beddington, J., Asaduzzaman, M., Clark, M., Fernandez, A., Guillou, M., Jahn, M., Erda, L., Mamo T, van Bo, N., Nobre, C.A., Scholes, R., Sharma, R. and Wakhungu, J. (2012) Achieving food security in the face of climate change: Final report from the Commission on Sustainable Agriculture and Climate Change. Available at: https://cgspace.cgiar.org/bitstream/handle/10568/35589/climate_food_commission-final-mar2012.pdf (accessed 01 August 2017).

Bezner Kerr, R., Msachi, R., Dakishoni, L., Shumba, L., Nkhonya, Z., Berti, P.R., Bonatsos, C., Chione, E., Mithi, M., Chitaya, A., Maona, E. and Pachanya, S. (2012) Growing healthy communities: Farmer participatory research to improve child nutrition, food security and soils in Ekwendeni, Malawi. In: *Ecohealth Research in Practice*, pp. 37–46. Available at: http://linkinghub.elsevier.com/retrieve/pii/S1877343514000359 (accessed 01 August 2017).

Borrell, A.K. and Reynolds, M. (2017) Integrating islands of knowledge for greater synergy and efficiency in crop research. *Food and Energy Security*, 6(1), pp. 26–32.

Campbell, B., Thomton, P., Zougmore, R., van Asten, P. and Lipper, L. (2014) Sustainable intensification: What is its role in climate smart agriculture? *Current Opinion in Environmental Sustainability*, 8, pp. 39–43.

Cleary, D. (2014) Deconstructing sustainable intensification and issues around sustainability metrics. Available at: www.nature.org/science-in-action/science-features/david-cleary-critique.pdf (accessed 13 July 2017).

Cofie, O. and Amede, T. (2015) Water management for sustainable agriculture intensification. *Water Resources and Rural Development*, 6, pp. 3–11.

Cook, S., Silici, L., Adolph, B. and Walker, S. (2015) Sustainable agricultural intensification revisited. Available at: http://pubs.iied.org/pdfs/14651IIED.pdf (accessed 13 July 2017).

Davis, A.S., Hill, J.D., Chase, C.A., Johanns, A.M. and Liebman, M. (2012) Increasing cropping system diversity balances productivity, profitability and environmental health. *PLoS ONE*, 7(10), pp. e47149.

Dirutu, S.W., Kassie, M. and Shiferaw, B. (2015) Are there systematic gender differences in the adoption of sustainable agricultural intensification practices? Evidence from Kenya. Available at: www.sciencedirect.com/science/article/pii/S0306919214001109 (accessed 29 July 2017).

Elagib, N.A. (2015). Drought risk during the early growing season in Sahelian Sudan. *Natural Hazards*, 79(3), pp. 1549–1566.

European Union (EU) (2016) Designing the path: a strategic approach to EU agricultural research and innovation (draft paper). Available at: https://ec.europa.eu/programmes/horizon2020/en/news/designing-path-strategic-approach-eu-agricultural-research-and-innovation (accessed 10 August 2016).

Evans, G.R. and Sadler, E.J. (2008) Methods and technologies to improve water use efficiency. Available at: http://onlinelibrary.wiley.com/doi/10.1029/2007WR006200/full (accessed 25 October 2017).

Food and Agricultural Organization (FAO) (2004) The ethics of sustainable agricultural intensification. Available at: www.fao.org/docrep/007/j0902e/j0902e03.htm (accessed 13 July 2017).

Food and Agricultural Organization (FAO) (2009) The special challenge for sub-Saharan Africa. *High level expert forum: how to feed the world 2050*. Rome: Available at: www.fao.org/fileadmin/templates/wsfs/docs/Issues_papers/HLEF2050_Africa.pdf.

Food and Agricultural Organization (FAO) (2012) Towards the future we want: End hunger and make the transition to sustainable agricultural and food systems. Available at: www.fao.org/docrep/015/an894e/an894e00.pdf (accessed 10 August 2016).

Food and Agricultural Organization (FAO) (2013) *Climate-Smart Agriculture Sourcebook*. Food and Agriculture Organization of the United Nations, Rome, Italy.

Food and Agricultural Organization (FAO) (2014) FAO supports dairy farmers to bring quality milk to the market. Available at: www.fao.org/africa/news/detail-news/en/c/238284/ (accessed 15 June 2017).

Food and Agricultural Organization (FAO) (2015) The state of food insecurity in the world 2015. Meeting the 2015 international hunger targets: taking stock of uneven progress. Available at: www.fao.org/3/a-i4646e/i4646e00.pdf (accessed 16 August 2017).

Food and Agriculture Organization (FAO) (2016) The future of food and agriculture: Trends and challenges. Rome: FAO. Available at: www.fao.org/3/a-i6583e.pdf (accessed 18 January 2018).

Gardner, J.B. and Drinkwater, L.E. (2009) The fate of nitrogen in grain cropping systems: a meta-analysis of 15N field experiments, *Ecological Applications*, 19(8), pp. 2167–2184.

Garnett, T., Appleby, M.C., Balmford, A., Bateman, I.J., Benton, T.G., Bloomer, P., Burlingame, B., Dawkins, M., Dolan, L., Fraser, D. *et al.* (2013) Sustainable intensification in agriculture: premises and policies. *Science*, 341, pp. 33–34.

Gatzweiler, F.W. and von Braun, J. (eds) (2016) Technological and institutional innovations for marginalized smallholders in agricultural development. ISBN 978-3-319-25716-7: DOI 10.1007/978-3-319-25718-1.

Gliessman, S. (2007) *Agroecology: The ecology of sustainable food systems*, 2nd edn. Boca Raton, FL: CRC Press/Taylor & Francis.

Grigg, D. (1987) The Industrial Revolution and land transformation. Available at: https://dge.carnegiescience.edu/SCOPE/SCOPE_32/SCOPE_32_1.4_Chapter4_79-109.pdf (accessed 25 October 2017).

Harari, Y.N. (2011) *Sapiens: A brief history of humankind*. Harmondsworth: Penguin, pp. 87–91.

Haug, R., Hella, J.P., Nchimbi-Msolla, S., Mwaseba, D.L. and Gry, S. (2016) If technology is the answer, what does it take? *Development in Practice*, 26 (3), pp. 375–386.

Hazell, P. and Wood, S. (2008) Drivers of change in global agriculture. Available at: www.ncbi.nlm.nih.gov/pmc/articles/PMC2610166/ (accessed 21 July 2017).

Heinberg, R. (2015) The end of growth. Available at: http://richardheinberg.com/bookshelf/the-end-of-growth-book (accessed 25 October, 2017).

Hokkanen, H.M.T. and Lynch J.M. (1995). Available at: www.sadrabiotech.com/catalog/Biological%20Control%20Benefits%20and%20Risks%20BOOK.pdf (accessed 01 August 2017).

IAASTD (International assessment of agricultural knowledge, science and technology for development) The Synthesis Report (2009) In: McIntyre, B.D., Herren, H.R., Wakhungu, J. and Watson, R.T. (eds). Washington, DC: Island Press.

Intergovernmental Panel on Climate Change (IPCC) (2007) *Climate Change 2007: Synthesis Report*. International Panel on Climate Change. Cambridge, UK and New York, NY: Cambridge University Press.

International Institute for Environment and Development (IIED) (2016) New paradigm for African agriculture sees sustainable intensification in a new light. Available at: www.iied.org/new-paradigm-for-african-agriculture-sees-sustainable-intensification-new-light (accessed 27 October 2017).

International Fund for Agricultural Development (IFAD) (2008) Improving crop–livestock productivity through efficient nutrient management in mixed farming systems of semi-arid West Africa. Online, www.ifad.org/lrkm/tags/384.htm (accessed 29 July, 2017).

IPES-Food (2016) From uniformity to diversity: A paradigm shift from industrial agriculture to diversified agro-ecological systems, International Panel of Experts on Sustainable food systems. Available at: www.ipes-food.org (accessed on 15 November 2016).

Khang, N.D., Kotera, A., Sakamoto, T. and Yokozawa, M. (2008) Sensitivity of salinity intrusion to sea level rise and river flow change in Vietnamese Mekong Delta: Impacts on availability of irrigation water for rice cropping. *Journal of Agriculture and Meteorology*, 64 (3), pp. 167–176.

Lamsal, K., Paudyal, G.N. and Saeed, M. (1999) Model for assessing impact of salinity on soil water availability and crop yield. *Agricultural Water Management*, 41, pp. 57–70.

Matson, P.A., Naylor, R. and Ortiz-Monasterio, I. (1998) Integration of environmental, agronomic, and economic aspects of fertilizer management. *Science*, 280, pp. 112–115.

McCarthy, N., Lipper L. and Branca, G. (2011) Climate-smart agriculture: Smallholder adoption and implications for climate change adaptation and mitigation. *Mitigation of Climate Change in Agriculture Series 4*. Rome, Italy: Food and Agriculture Organization of the United Nations (FAO).

McCartney, M.P., Johnston, R. and Lacombee, G. (2016) Building climate resilience through smart water and irrigation management systems: Experiences from Southeast Asia. In: Nagothu, U.S. (ed.) *Climate Change and Agriculture Development*. London: Routledge, pp. 41–65.

Nagothu, U.S. (2015) The future of food security: Summary and recommendations. In: Nagothu, U.S. (ed.) *Food Security and Development: Country case studies*. London: Routledge, pp. 274.

Nelson, G.C., Rosegrant, M.W., Palazzo, A., Gray, I., Ingersoll, C., Robertson, R., Tokgoz, S., Zhu, T., Sulser, T., Ringler, C., Msangi, S. and You, L. (2010) *Food Security, Farming, and Climate Change to 2050: Scenarios, Results, Policy Options*. Washington, DC: IFPRI.

Nelson, V. and Stathers, T. (2009) Resilience, power, culture, and climate: a case study from semi-arid Tanzania, and new research directions. *Gender and Development*, 17 (1), pp. 81–94.

NEPAD (New Partnerships For Africa's Development (2003) Comprehensive Africa Agriculture Development Program (CAADP), Midrand, South Africa: NEPAD.

Oborn, I., Vanlauwe, B., Phillips, M., Thomas, R., Brooijmans, W. and Atta-Krah, K. (2017) Sustainable intensification in smallholder agriculture: An integrated systems research approach. Available at: www.routledge.com/Sustainable-Intensification-in-Smallholder-Agriculture-An-integrated-systems/Oborn-Vanlauwe-Phillips-Thomas-Brooijmans-Atta-Krah/p/book/9781138668089 (accessed 21 November, 2017).

Odendo, M., Obare, G. and Salasya, B. (2009) Factors responsible for differences in uptake of integrated soil fertility management practices amongst smallholders in western Kenya. *African Journal of Agricultural Research*, 4 (11), pp. 1303–1311.

Ologundudu, A.F., Adelusi, A.A. and Akinwale, R.O. (2014) Effect of salt stress on germination parameters of rice (Oryza sativa, L.). *Notulae Scientia Biologicae*, 6 (2), pp. 237–243.

Petersen, B. and Snapp, S. (2015) What is sustainable intensification? Views from experts. *Land Use Policy*, 46, pp. 1–10.

Posthumus, H. and Wongtschowski, M. (2014) *Innovation Platforms. Note 1. GFRAS Good Practice Notes for Extension and Advisory Services*. Lindau, Switzerland: GFRAS.

Pretty, J.N. (1997) The sustainable intensification of agriculture. *Natural Resources Forum*, 21 (4), pp. 247–256.

Pretty, J.N., Morison, J.I.L. and Hine, R.E. (2003) Reducing food poverty by increasing agricultural sustainability in developing countries. *Agriculture, Ecosystems and Environment*, 95, pp. 217–234.

Pretty, J.N., Toulmin, C. and Williams, S. (2011) Sustainable intensification in African agriculture. *International Journal of Agricultural Sustainability*, 9 (1), pp. 5–24.

Pretty, J and Bharucha, Z.P. (2014) Sustainable intensification in agricultural systems. *Annals of Botany*, p. 1 of 26. Available at: www.aob.oxfordjournals.org.

Pretty, J. and Bharucha, Z.P. (2015) Integrated pest management for sustainable intensification of agriculture in Asia and Africa. *Insects*, 6, pp. 152–182.

Raj, R. and Nagothu, U.S. (2016) Gendered adaptation to climate change in canal-irrigated agro-ecosystems. In: Nagothu, U.S. (ed.) *Climate Change and Agriculture Development*. London: Routledge, pp. 259–278.

Rockström, J., Williams, J. and Daily, G. (2016) Sustainable intensification of agriculture for human prosperity and global sustainability. *Ambio*, Available at: http://link.springer.com/article/10.1007/s13280-016-0793-6.

Ronald, P. (2011) Plant genetics, sustainable agriculture and global food security. *Genetics*, 188 (1), pp. 11–20.

Rota, A. (2008) Integrated crop-livestock farming systems. Available at: www.ifad.org/documents/10180/31bb4140-1a73-4507-83ed-6fc8cb64bed7 (accessed 29 July 2017).

Seetharam, A., Riley, K.W. and Harinarayana. G. (1989) *Small Millets in Global Agriculture*. New Delhi, India: Oxford and IBH.

Silici, L., Bias, C. and Cavane, E. (2015) Sustainable agriculture for small-scale farmers in Mozambique: A scoping report. Available at: http://pubs.iied.org/14654IIED/ (accessed 11 July 2017).

Snapp, S.S., Blackie, M.J., Gilbert, R.A., Bezner-Kerr, R. and Kanyama-Phiri, G.Y. (2014) Modeling and participatory, farmer-led approaches to food security in a changing world: a case study from Malawi. 2014. *Secheresse*, 24, pp. 350–358.

Swift, M.J. and Ingram, J.S.O. (1996) *Global Change and Terrestrial Ecosystems*, GCTE Report No. 13 Wallingford, UK.

Tatlonghari, G.T. and Paris, T.R. (2013) Gendered adaptations to climate change: a case study from the Philippines. In: Alston, M. and Whittenbury, K. (eds) *Research, Actio and Policy: Addressing the gendered impacts of climate change*. Amsterdam: Springer.

The Montpellier Panel (2013) *Sustainable Intensification: A new paradigm for African agriculture*. London. Available at: http://ag4impact.org/publications/montpellier-panel-report2013/ (accessed 18 January 2018).

Theriault, V., Smale, M. and Haider, H. (2017) How does gender affect sustainable intensification of cereal production in the West African Sahel? Evidence from Burkina Faso. *World Development*, 92, pp. 177–191.

The Economist (2016) The Future of Agriculture. Available at: www.economist.com/technology-quarterly/2016-06-09/factory-fresh (accessed 25 October, 2017).

Thierfelder, C., Mwila, M. and Rusinamhodzi, L. (2013) Conservation agriculture in eastern and southern provinces of Zambia: Long-term effects on soil quality and maize productivity. *Soil and Tillage Research*, 126, pp. 246–258.

Thornton, P. (2012) Recalibrating food production in the developing world: Global warming will change more than just the climate. CCAFS Policy Brief no. 6. CGIAR Research Program on Climate Change, Agriculture and Food Security (CCAFS). Available online at www.ccafs.cgiar.org.

Tilman, D., Cassman, K.G., Matson, P.A., Naylor, R. and Polasky, S. (2002) Agricultural sustainability and intensive production practices. Available at: www.nature.com/nature/journal/v418/n6898/full/nature01014.html (accessed 17 July 2017).

Tittonell, P. (2014) Ecological intensification of agriculture: Sustainable by nature. *Current Opinion in Environmental Sustainability*, 8, pp. 53–61.

Tscharntke, T., Clough, Y., Wanger, T.C., Jackson, L., Motzke, I., Perfecto, I., Vandermeer, J. and Whitbread, A. (2012) Global food security, biodiversity conservation and the future of agricultural intensification. *Biological Conservation*, 151, pp. 53–59.

Turral, H., Burke, J. and Faures, J-M. (2011) *Climate Change, Water and Food Security*, FAO Water Reports 36. Food and Agriculture Organization of the United Nations, Rome. Available at: www.fao.org/docrep/014/i2096e/i2096e.pdf (accessed 16 July 2017).

Union of Concerned Scientists (UCS), (2017) Solutions: Advance sustainable agriculture. Available at: www.ucsusa.org/our-work/food-agriculture/solutions/advance-sustainable-agriculture#.WWSKrWYUlsM (accessed 11 July 2017).

United Nations (UN) (2015) Sustainable Development Goals. Available at: https://sustainabledevelopment.un.org/?page=view&nr=164&type=230 (accessed 21 July 2017).

UNFCCC (2015) Adoption of the Paris Agreement. Available at: https://unfccc.int/resource/docs/2015/cop21/eng/l09.pdf (Accessed 25 October 2017).

Vanlauwe, B., Coyne, D., Gockowski, J., Hauser, S., Huising, J., Masso, C., Nziguheba, G., Schut, M. and Van Asten, P. (2014) Sustainable intensification and the African smallholder. Available at: http://humidtropics.cgiar.org/wp-content/uploads/downloads/2014/12/Sustainable-intensification-and-the-African-smallholder.pdf (accessed 15 August, 2017).

Woelcke, J. (2003) Bio-economics of sustainable land management in Uganda. In: (Series edited by Heidhues, F. and von Braun, J.), Lang, P. (ed.), *Developing Economics and Policy*. Frankfurt am Main, Berlin, Bern, Bruxelles, New York, Oxford, Wien.

Woelcke, J. (2006) Technological and policy options for sustainable agricultural intensification in eastern Uganda. *Agricultural Economics*, 34, pp. 129–139.

2 Save and grow

Translating policy advice into field action for sustainable intensification of rice production

Jan Willem Ketelaar, Alma Linda Morales-Abubakar, Pham Van Du, Cahyana Widyastama, Avakat Phasouysaingam, Jesse Binamira and Ngo Tien Dung

Introduction

The Food and Agriculture Organization (FAO) estimates that the global population will reach about 9.7 billion by the year 2050. Nearly all the increase in population will be in developing countries where about 815 million people remain chronically undernourished (FAO, 2017). Global urbanization, which is at 54 percent today (Statista, 2017), is expected to continue and reach about 66 percent by 2050 (UN DESA, 2015). As cities and income level grow, consumption patterns change, particularly in urban areas. FAO estimates that food production will have to be increased (in addition to food crops for biofuels) by about 70–100 percent so as to feed the world's population. Eighty percent of production increases in the developing countries can only come from increased land productivity and cropping intensity, given reduction in agricultural land area resulting from changed land-use patterns (FAO, 2011). Farmers will also have to contend with other yield reducing challenges such as drought, floods, pest and disease induced by climate change (FAO, 2016 a, b, c).

Every day, humanity consumes millions of tons of cereals – predominantly maize, rice and wheat – in an almost endless variety of familiar forms – from steaming bowls of rice to tortillas, *naan*, pasta, pizza, pies and pastries. Millions of tons more reach us indirectly having been fed first to cattle, pigs and poultry supplying our ever increasing demands for meat, milk and egg products.

Cereals are the single most important item in the human diet, accounting for an estimated 43 percent of the world's food calorie supply. Globally, cereals also are a main source of proteins – around 37 percent – a close second to fish and livestock products. Cereals also supply 6 percent of the fat in our diets. These three cereals are truly important for global food security (FAO, 2016 a, b, c).

By 2015, world annual demand for maize, rice and wheat is expected to reach some 3.3 billion tons, or 800 million tons more than 2014's record combined harvest. Much of the increased cereal production will need to come from existing farmland. But one-third of that land is degraded (e.g., due to salinization, erosion and chemical pollution) and farmers' share of water is under growing pressure from other sectors. Yield growth rates of major cereal crops have been declining steadily from 3.2 percent in 1960 to 1.5 percent in 2000 (FAO, 2009).

Rice plays a critical role in global food security and is a major staple crop for 800 million of the world's poorest peoples. Rice is grown by 150 million smallholders, mostly in Asia, where 90 percent of the world's rice is produced and consumed. Asia is also home to nearly two-thirds (64.3 percent) of the world's food insecure population. The majority of them eat rice as a staple food and are dependent on rice production for their livelihoods (FAO, 2015 a, b).

Global rice production will have to grow to meet projected global supply shortfall by 2050, when demand will reach about 600 million metric tons, leading to an increasing annual requirement of 100 million metric tons of rice for every 1 billion people added to the global population. Average global rice yield in 2010 was estimated to be 4.5 tons per hectare (FAO, 2016 a, b, c), with the highest yields achieved by smallholders in the irrigated rice production areas in Asia. However, the current growth rate of rice has slipped to 1 percent, too low to meet future global rice demands. While the global per capita rice consumption has stagnated, population is still rising drastically. According to the International Rice Research Institute (IRRI), the global rice demand is estimated to increase from 439 million tons in 2010 to 555 million tons in 2035 (Regino, 2017).

Climate change, the most serious environmental challenge facing humanity, is expected to have far-reaching impacts on production of maize, rice and wheat. Without adaptation and adoption of climate smart practices, rice production in Southeast Asia is forecast to decline by 10 percent by 2050. The rice sector is a major contributor of greenhouse gas emissions in the agriculture sector of producing countries, amounting to an estimated 500 million tons of emissions CO_2e/year, representing up to 50 percent of agricultural emissions in some rice-producing countries and 10 percent of all agriculture sector emissions worldwide. Mitigation efforts, including through wide-scale adoption of climate smart practices, are urgently needed (FAO, 2016 a, b, c).

Today's high-yielding varieties are intolerant to major biotic stresses worsened by abiotic stress due to climate change, such as higher temperatures, drought, flooding and salinity. Rising sea levels and increased frequency of major storms (e.g., typhoons in the Philippines, cyclones in South Asia) will pose a particular threat to rice-based landscapes in coastal areas. Since river deltas in Bangladesh, Myanmar and Vietnam have been responsible for half of rice production increases over the past 25 years, a serious loss of their production capacity could negatively impact on global food security (FAO, 2016 a, b, c).

If rice production is to grow and meet global demands in the future, smallholders will have to intensify production on available production land and make more efficient use of natural resources while at the same time adapting to climate change. Use of good quality seeds and novel improved varieties with better tolerance biotic (e.g., better pest and disease protection) and abiotic stress (e.g., submergence-tolerant varieties in flood-prone areas) should be part of this equation. These smallholder producers are mostly resource-poor farmers, many with less than one hectare of landholdings. For the foreseeable future, smallholder agriculture will remain the mainstay of production of humanity's global food needs, particularly in Asia. It is on the relatively small plots of rice land in Asia where smallholders will continue to produce most of our daily rice needs. These farmers also hold a key role in protecting and managing vitally important rice-based landscapes, including humanity's most expansive man-made wetlands where most of our rice is produced. Smallholder rice farmers are also the custodians of vital ecosystem services provided by natural biological processes, e.g., biological control, pollination and nutrient cycles. If smallholder rice farmers are to intensify rice production, they will have to learn how to do so in a sustainable manner (FAO, 2010).

Unfortunately, the push for intensification of agricultural production has driven farmers to increase the use of chemical inputs such as fertilizers and pesticides in efforts to increase crop yields. The intensive use of pesticides harms vital ecosystem services, particularly aquatic fauna such as fish and waterfowl and continuing patterns of pesticide use are damaging biodiversity in wetlands, especially in rice production landscapes (Ramsar Convention, 2012 and further illustrated in Box 2.1).

Box 2.1 Impact of pesticide use and practices on biodiversity

A study carried out in 2014 in Central Thailand documented the impact of synthetic and natural pesticide applications in rice fields. Despite wetland conversion to intensive rice production, Central Thailand remains a center for wetland biodiversity in Southeast Asia. However, the study found that abundance and diversity of aquatic fauna was generally higher on fields treated with natural rather than synthetic insecticides. Fishes and waterfowl tended to be less abundant on fields exposed to synthetic insecticides and herbicides. The study provides insights into how current pesticide use and practices influence the aquatic faunal communities in the rice fields – especially non-target organisms – degrading some of the essential ecosystem services they support. These services include the turnover of carbon

(continued)

(continued)

and nutrients, and the control of potentially noxious herbivores and disease vectors by predators. None of these services can be fully substituted by technological means, and the decline in rice field biodiversity is therefore of concern, especially on a longer-term outlook (Cochard *et al.*, 2014).

Source: Adapted from Cochard *et al.* (2014).

Pesticide use also raises food safety concerns and jeopardizes export potential of agriculture produce, causes frequent poisoning and chronic health problems and exposes the most vulnerable – women and children – directly or indirectly to toxic substances. Other global phenomena such as climate change, the "greying" and "feminization" of agriculture, vanishing rice cultures and heritage varieties, are directing governments to examine and redesign policies and strategies to support a more *sustainable* rice production intensification.

These concerns and renewed government interest in boosting productivity have led FAO to work with its member governments to promote *sustainable* intensification of agricultural production. The new paradigm emphasizes the need for more efficient use of diminishing production resources and better management and use of agro-ecological processes, promoted by FAO under the banner of *Save and Grow* (FAO, 2011). In 2013, FAO initiated a Regional Rice Initiative (RRI) for the purpose of working with selected member countries to apply the *Save and Grow* concepts, principles and associated good practices in rice-based production landscapes in Asia. In 2016, FAO published policy guidance for – and practical examples of – *sustainable* intensification of cereal crops, in particular maize, rice and wheat (FAO, 2016 a, b, c).

This book chapter, illustrated with several case studies presenting results from *Save and Grow* interventions in rice-based landscapes in Indonesia, Lao PDR, the Philippines and Vietnam during the 2013–17 period, will outline RRI implementation progress and results to date and argue the case for increased policy support for – and investments in – sustainable intensification of crop production education programs for smallholder farmers. It is argued that such investments are essential for ensuring food and nutrition security for future rural and urban generations in Asia. Lessons from such cases can provide important information to other parts of the world where rice is becoming an important part of the diet.

The chapter started with a background section. This is followed by a description of the (historical) background and the context for FAO's RRI in response to the request from member countries to promote more efficient use of diminishing resources and better management and use of agro-ecological processes for *sustainable* intensification of agricultural production. Subsequently, case studies are presented that briefly describe the *Save and Grow* farmer education interventions and results thereof in Indonesia, Lao PDR, Philippines and Vietnam. Building on case study evidence, results of the RRI initiative to date are discussed with explanations on how these contribute to local, national and global policy and development goals. Finally, some conclusions and recommendations are presented, forward-looking and relevant for how Asian countries can develop, implement, invest and scale out efforts for long-term realization of *sustainable* crop intensification.

The Regional Rice Initiative

The Regional Rice Initiative – a response to the emergent call from member countries

Farmers are faced with the major challenge of intensifying agricultural production to feed the world's growing population in the face of declining availability of labor, water and agricultural land, lower productivity and changing consumer patterns (FAO, 2011). It was against this background that government representatives, at the 31st Session of the FAO Regional Conference for Asia and the Pacific (APRC) held in Hanoi in March 2012, requested FAO to: i) strengthen the capacities of the Member Countries on rice production; and ii) develop a Regional Rice Strategy to harmonize diverse rice-related issues. The Regional Rice Strategy, published by FAO in 2014, was intended to provide evidence-based strategic guidelines to member nations to help them: i) develop and adjust their rice sector strategies in the light of broader regional and global trends and national priorities; and ii) choose among key strategic options while considering the implied trade-offs (or consequences) with the end view of assisting countries to achieve sustainable food security (FAO, 2014a). In December 2012, the 145th Session of the FAO Council endorsed the formulation and implementation of a regional initiative for Asia to strengthen rice-based production systems as part of its new Strategic Objective 2: "*Make agriculture, forestry and fisheries more productive and sustainable.*" Since 2013, the RRI has supported three pilot countries – Indonesia, Lao PDR and the Philippines – by: i) focusing on goods and services produced by and available from rice ecosystems and landscapes; and ii) identifying and undertaking sustainable rice production practices to enhance resilience and increase efficiencies in rice production to improve food and nutrition security.

Indonesia and the *Philippines* were chosen as the pilot countries because they are major rice importing countries but are strongly committed to improving food security by reducing their dependency on rice imports and increasing food production. *Lao PDR* was selected as it suffered from the second highest prevalence of undernourished in the region in terms of the proportion of undernourished in total population at that time (FAO/WFP, 2012). Sustainable increase of rice production in Lao PDR is considered critical for reducing hunger and malnutrition and improving livelihoods of impoverished farmers, especially smallholders. In response to the interest of other countries and as a way to geographically expand the sharing and utilization of experiences gained from the Regional Rice Initiative, pilot Save and Grow activities were implemented in *Vietnam* starting in 2014.

Concepts and approaches of the Regional Rice Initiative

The RRI foresees long-term impact on: i) improved food and nutrition security through effective provision and utilization of ecosystem services and goods derived from rice-based farming systems and landscapes; and ii) poverty reduction through increased productivity and income generating opportunities, and improved access to market in the rice sector. These can be achieved through: i) adopting innovative and sustainable rice farming practices, i.e. *Save and Grow* based Sustainable Intensification of Rice Production, which allows rice farmers to increase productivity and improve rice quality in a sustainable way despite less agricultural and labor input; ii) generating more knowledge and evidence on the sustainability and resource-use efficiency; and iii) formulating and implementing national rice policies or strategies drawing on the vision and strategic options laid out by the Regional Rice Strategy for Sustainable Food Security in Asia and the Pacific, while contributing to regional and global policy processes such as the Convention on Biological Diversity.

A two-pronged approach is employed with practical work at field level that provides evidences for policy reform and work at the policy level to support the implementation of innovative and sustainable rice farming practices at field level. At the same time, the Initiative contributes to global policy processes and facilitates member country implementation of international treaties such as the Convention on Biological Diversity and the Ramsar Wetlands Convention.

The RRI associates with farmer education interventions implemented through National IPM Programs through Save and Grow Farmers Field Schools (Box 2.2). In these field-based educational interventions farmers are exposed to an array of practical options and good ecological practices that are more productive, sustainable and efficient in resource use while promoting better agroecology-based intensification and diversification of rice-based farming systems. These include: rice–fish, rice–livestock and rice–vegetables diversified production systems, Integrated Pest Management (IPM), Agroforestry and management of Trees Outside of Forest (TOF), improved canal operation techniques based on “MASSCOTE,” simple mechanization options to reduce labor and drudgery from manual transplanting, weeding and harvesting and promotion of Climate-Smart Agriculture. The RRI supports relevant studies with local and international research and development organizations that have led to several innovative publications aimed at awareness raising, practical guidance and policy support for appreciation and responsible management of the multiple goods and services provided by rice-based production landscapes (FAO, 2014b; FAO, 2017). Finally, RRI also works with national and local policy makers to promote the development of better enabling policy and legal frameworks to support effective ecosystem-based management of insect pests and diseases. Through a global initiative led by FAO, the RRI engages with the Globally Important Agricultural Heritage Systems (GIAHS) to encourage and formally recognize the importance of the preservation of rice heritage and culture. All these approaches are integrated in the overarching *Save and Grow* paradigm (Figure 2.1) that espouses agro-ecological approaches that move agricultural development away from a focus on singular focal areas – e.g., improved seed, pest control, water management – to more holistic and agro-ecology based solutions that integrate all components of the farming system (FAO, 2014b).

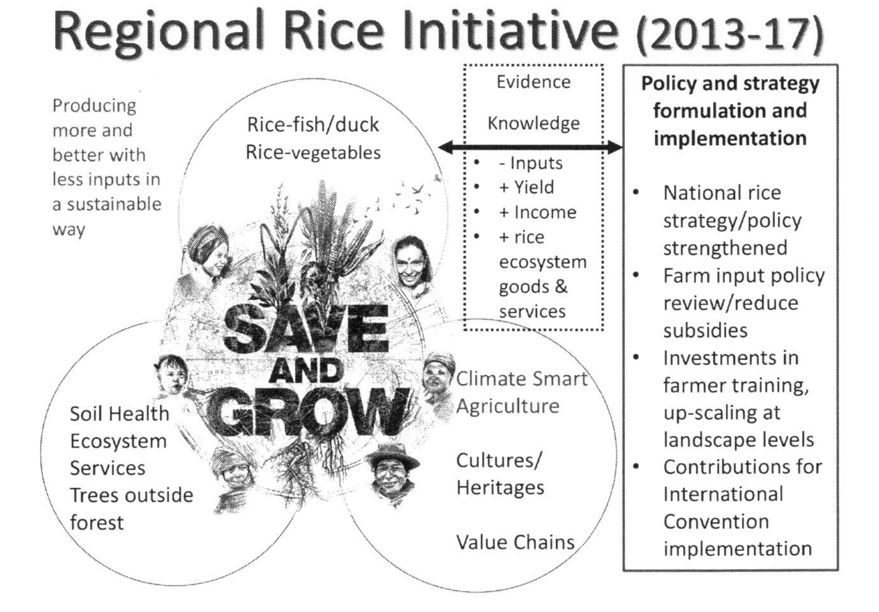

Figure 2.1 FAO's Regional Rice Initiative: Save and Grow: from field to policy

Source: Authors' own compilation.

Box 2.2 Farmer Field Schools

Save and Grow Farmer Field Schools (FFS) build knowledge and skills of farmer groups and communities on application of good agricultural practices. The FFS employs discovery-based group learning processes. Usually, a group of 25–30 farmers meet one morning weekly for an entire crop growing season and engage in experiential learning activities to gain an ecological perspective of managing ecosystems and skills in informed decision making based on location-specific conditions. The learning process is facilitated by extension workers or trained farmers. Non-formal education methods are employed and the field is used as the primary resource for discovery-based learning (FAO, 2013). Starting with FAO-supported FFS in rice IPM in Indonesia in 1989, FFS are now used in 90 countries in different regions around the world to facilitate learning about various topics including, genetic resource management, managing water, soils and fertility, crop nutrition, system of rice intensification (SRI), conservation agriculture, fisheries and animal husbandry, health, nutrition, child care, climate smart agriculture, postharvest, Farmer Life Schools, etc. (FAO, 2016 a, b, c). The FFS is recognized as one of the best educational and capacity building tools for training farmers on complex skills such as natural resource management, diversifying production and accessing markets to increase rural incomes (Swanson and Rajalahti, 2010). Implemented through a nation-wide national IPM program, the Philippine experience over a 15 year period (1993–2007) has shown that IPM-FFS interventions have resulted in sustainable reductions in insecticide use while farmers adopted IPM, maintaining growth in national rice production output (Ketelaar and Abubakar, 2012).

Case studies on Save and Grow Farmer Field Schools (FFS)

Save and Grow FFS in Indonesia

In Indonesia, 75 rice farmers in three Save and Grow FFS in the West Java district of Indramayu established field studies to compare Save and Grow practices and conventional farmers' practices. A summary of the practices that were applied is provided in Table 2.1. Many of the practices that were applied in the Save and Grow fields are not new (e.g., transplanting single seedling per hill) but were not commonly practiced in the community prior to the FFS intervention (e.g., introduction of fish into the rice field). The natural biological control employed under the Save and Grow practices included augmenting the functional role of generalist predators (such as spiders), the dominant controlling agent for many rice pests, including Brown Plant Hopper (*Nilaparvata lugens*). The populations of these generalist predators are boosted early in the season by the availability of alternative prey resources that are semi-aquatic (e.g., chironomid midges and ephydrid flies that have an aquatic larval stage and terrestrial adult stage) and that also provide food for the fish in the paddy. The strong generalist predator populations already established by the time the first potential rice pest species arrive reduce the chance of survival of subsequent pests and are also key to the unfolding of the seasonal pest dynamics (Settle *et al.*, 1996). This hypothesized flow of energy in tropical rice ecosystems, based on empirical field research in Indonesia during the 1990s, is clearly demonstrated in Figure 2.2 (Gallagher *et al.*, 2005).

Following rice harvest in October 2016, farmers reported rice yields ranging from 7.2–7.7mt/ha in Save and Grow plots compared with 8.0–8.3mt/ha in Farmers' Practice plots. However, the produce from Save and Grow plots – organically grown but not certified – was sold at IDRp 2,000 (US$ 0.15) higher per kg or IDRp 6,300 (US$ 0.48) compared to the IDRp 4,300 (US$0.33) price of rice produced using conventional farmers' practices. This gave higher benefits of about US$507/ha even from growing rice alone using Save and Grow practices (Figure 2.3).

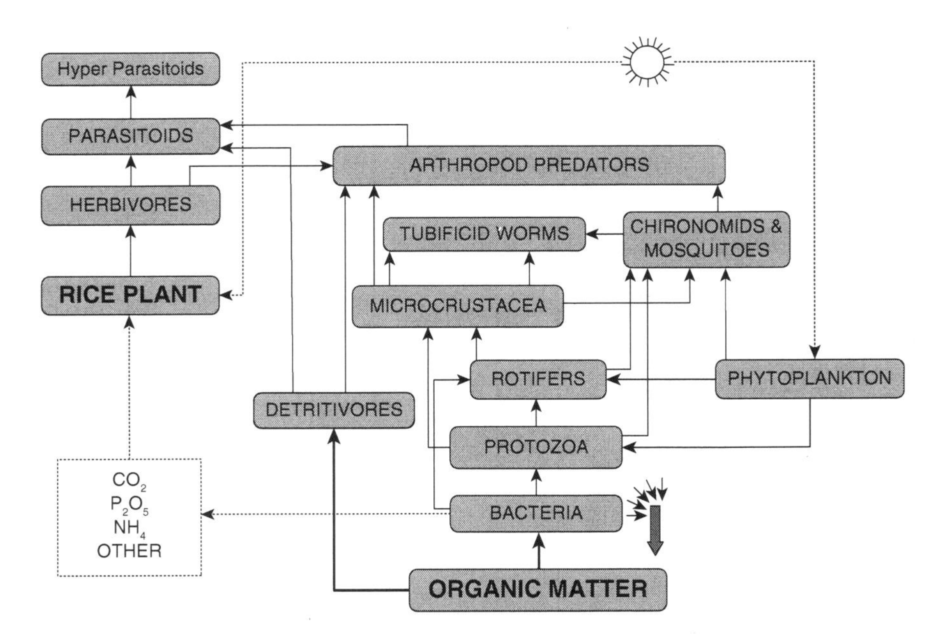

Figure 2.2 Flow of energy in tropical rice ecosystems

Source: Adapted from Gallagher *et al.* (2005) and based on published original field research in rice paddies in Indonesia (Settle *et al.*, 1996).

Table 2.1 Save and Grow practices versus conventional farmers' practices, Indramayu, West Java, Indonesia

Save and Grow practices	*Conventional farmers' practices*
Soil tillage using tractor + Soil levelling/ flattening + fixing the dykes	Soil tillage using tractor + Soil levelling/flattening + fixing the dykes
Government-released variety (Sri Putih and Ciherang) + farmer-bred variety (Idaman Suka Slamet); different varieties for broader genetic base	Government-released variety (Sri Putih)
Farmer selected seeds	Village-subsidized seeds (certified seeds)
Manual selection: winnowing and floating test applied to select heavy seeds	No seed selection process; seeds sown directly
Planting distance of 30x30 cm	Planting distance of 30x30 cm
2–3 seedlings/hill	3–5 seedlings/hill
Age of seedling 20 days	Age of seedling 23–25 days
Irrigation 35 cm constantly flooded	Irrigation 5 cm constantly flooded
Chemical + organic fertilizers	Chemical fertilizers only
Handweeding	Herbicides
Integrated pest management including use of natural biological control such as predators and parasitoids; botanicals	Chemical pesticides, 6–15 applications during the rainy season
Planting of vegetables on the bunds and rice fields inc. luffa, winged bean, choy sum, chili, eggplant, yellow velvet, land water spinach, bitter gourd, yardlong bean, maize	Not applied
Introduction of fish to enhance biodiversity and provision of food	Not applied

Source: Authors' own compilation

When the benefits from the introduction of fish into the paddy and planting vegetables on the bunds were considered, it resulted in a difference of about US$757/ha (Figure 2.4) compared with growing rice only using conventional farmers' practices (FIELD Indonesia, 2016). This calculation does not include the value of pest population regulatory ecosystem services provided by fish. Neither does this reflect the value of the negative impact of the excessive chemical pesticide use on the environment and ecosystem services. In Indramayu, for example, the spraying of pesticides for pest control by conventional farmers during the rainy season ranged from 6–15 times per season (Fox and Winarto, 2016).

The initial investment in restructuring the field (i.e., labor for digging ditches) to accommodate the introduction of fish increased the overall cost

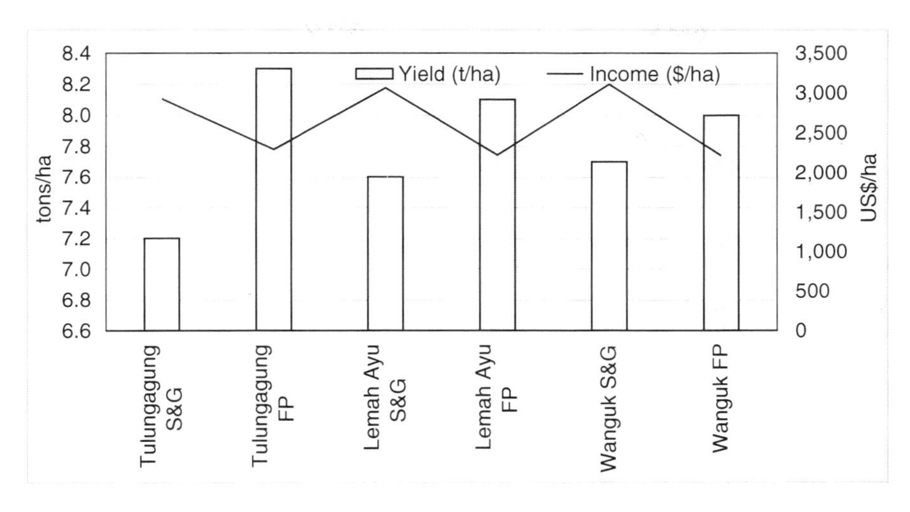

Figure 2.3 Rice yields and income from Save and Grow (S&G) and conventional farmers' practices (FP) experiments in FFSs, Indramayu, Indonesia, 2016 (n=3)

Source: Authors' own compilation from project data.

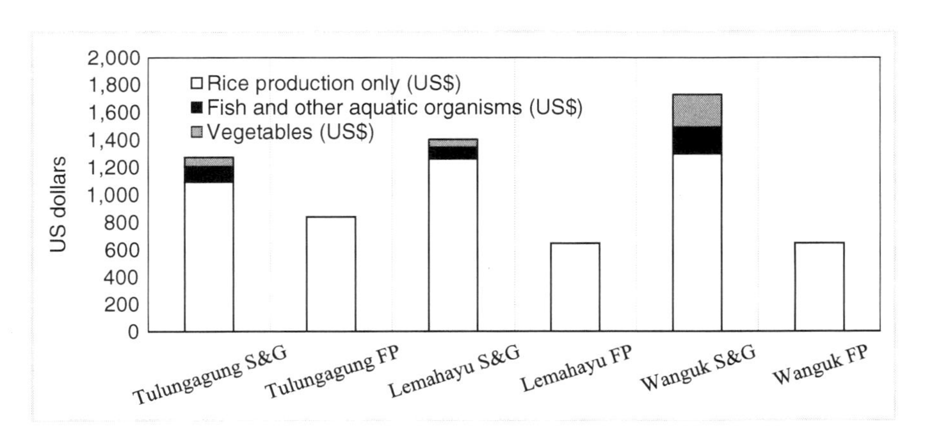

Figure 2.4 Comparison of benefits (in US$/ha) from Save and Grow (S&G) and conventional farmers' practices (FP) experiments in FFSs, Indramayu, Indonesia, 2016 (n=3)

Source: Authors' own compilation from project data.

of production otherwise reduced by adoption of Save and Grow practices (e.g., reduced seed input as a result of use of younger seedlings, singly planted, wider spacing and reduced use of agrochemicals, both pesticides and fertilizers). The cost of restructuring the field would no longer occur in subsequent seasons. Additionally, planting vegetables on the bunds maximizes utilization of the production area and at the same time provides habitat for *natural enemies* enhancing natural biological control. In the Indramayu example (Table 2.2), however, while the means of the three villages were highly significantly different between S&G and FP, the increases varied highly between the villages so that the paired analysis showed a non-significant result. This was mainly due to the four times higher vegetable benefits in Wangkuk. Without the vegetable data, the benefits from only rice and fish were significantly different in the paired analysis.

Save and Grow FFS in the Philippines

Rice is the Philippines' most important food staple. It influences the livelihood and economies of several million Filipinos. It contributes around 20 percent of the Gross Value Added (GVA) of the country's agriculture and employs some 2.5 million households. It is the country's major food staple and accounts for about 25 percent of food expenditures of the poorest 30 percent of the population. Among the low-income households, rice provides the largest single source of calories and protein in the diet. However, farming households are food insecure and net farm incomes are not enough to cover their poverty thresholds.

The Philippines' rice productivity growth has been low since the years of the Green Revolution as public investment in the modernization of the rice sector falls behind and the country contends with threats such as

Table 2.2 Comparison of benefits from Save and Grow and conventional farmers' practices, Indramayu, West Java, Indonesia, 2016 (n=3)
Exchange rate 1US$: 13,000IDRp.

	Save & Grow + Rice–fish				*Farmer's practice*
	From rice (US$)	*From fish (US$)*	*From vegetables (US$)*	*Sub-total (US$)*	*Only from rice (US$)*
Tulungagung	1,092	113	69	1,274	839
Lemahayu	1,262	82	58	1,402	645
Wangkuk	1,297	190	238	1,726	646
Mean/ha	1,217	129	122	1,467	710
Std	110	56	101	233	112
Diff to FP					
T-test	−5.61**			−5.08**	
Paired t-test	3.98 ns			4.07 ns	

Source: Authors' own compilation from project data.

urbanization, increasing cost of agricultural inputs and farm labor (that has significantly decreased farm incomes) and disruptions resulting from environmental changes such as weather disturbances. This has resulted in an increase in the volume of imported rice to address domestic consumption requirements.

The price of rice in the Philippines is the highest among ASEAN countries due to the cost of production. In view of the lifting of quantitative restrictions in rice imports in 2017, there is a need for drastic changes in the way farmers produce so that domestic rice production can be globally competitive and at the same time give good profits and returns on investments (FAO, 2015 a, b).

It was within the context of achieving rice self-sufficiency and making domestic rice production more competitive, given relatively high rice production costs, that the Philippines – and its Department of Agriculture in particular – embarked on the FAO-supported Save and Grow RRI pilot activities in 2013. The Save and Grow pilot project was designed to support the overarching goals of the agriculture sector under the 2011–2016 Philippine Development Plan (PDP) which are focused on achieving food security and raising farmer incomes. This was in line with the government's rice targets under the Food Staples Sufficient Programme. Through field experiments conducted as part of FFS-based farmer education interventions, the Philippines aimed to gather evidence of the potential for farmers to "*Save and Grow*" resulting from adoption of sustainable rice production intensification good practices. In particular, the DA was keen to demonstrate that adoption of Save and Grow crop production strategies, the application of improved agricultural practices and ecologically sound technologies, could: (1) increase rice productivity by at least 10 percent; (2) increase cost efficiencies by reducing cost of production by at least 15 percent; and (3) increase farm incomes by at least 15 percent.

In the Philippines, participants in Save and Grow FFS established and implemented field studies to compare Save and Grow practices and conventional farmers' practices. A summary of the practices that were applied are provided in Table 2.3.

Many of the practices that were applied in the Save and Grow fields are not new (e.g., transplanting fewer seedlings at 1–2 seedlings per hill) but were not commonly practiced in the community prior to the FFS intervention (e.g., planting vegetables on the bunds). Some of the interventions were new and making use of recent technological advances (e.g., the use of IRRI's Site Specific Nutrient Management (SSNM). Save and Grow practices included natural biological control such as augmenting the functional role of generalist predators (such as spiders) for the control of Brown Plant Hopper (*Nilaparvata lugens*) and the introduction of ducks into the paddy field at the early stage of the crop that preyed on both semi-aquatic neutrals as well as early plant feeders. Save and Grow practices also included the planting of vegetables (e.g., eggplant) and climbing herbaceous legume crop

Table 2.3 Save and Grow practices and conventional farmers' practices

Save and Grow practices	*Conventional farmers' practices*
Thorough land preparation (plowing, first and second harrowing and final leveling)	Farmers' local practice – not well leveled
Best inbred seeds in the location either purified or certified/hybrid	Good seeds or certified seeds
Planting distance of 20 x 20 cm, 25 x 25 cm and 30 x 30 cm	Planting distance of 20 x 20 cm
Transplanting 1–2 seedlings/hill	Transplanting 4–7 seedlings/hill
Age of seedling 12–14 days	20–25 days
Soil analysis and Site Specific Nutrient Management (SSNM)	No soil analysis; no basal fertilizer application
Use of IRRI phone application Nutrient Manager	Farmers' local practice based on experience
Application of organic fertilizer (vermicast)	No organic fertilizer
Alternate wet and dry water management (1–2 cm)	Constantly flooded fields (3–5 cm)
Timely use of rotary weeder	Hand weeding
Integrated pest management including use of natural biological control such as predators and parasitoids	Integrated pest management
Planting of vegetables (e.g., eggplant) and climbing herbaceous legume crop (*pole sitaw*) on the bunds	Not applied
Introduction of ducks to enhance biodiversity and provide food	Not applied

Source: Authors' own compilation.

(*pole sitaw*) on the bunds to maximize productivity and in view of providing additional food for family nutrition.

At the end of one season, 634 farmers (354 female) reported an average rice yield of 6.7mt/ha from 20 Save and Grow field experiment plots compared with 5.27mt/ha in Farmers' Practice plots or a difference of 1.43mt/ha, an average yield increase of 27.2 percent. In Save and Grow FFS plots, the average reduction in cost of production was 17 percent or actual money value saved of US$132 per ha. Savings resulted in increases in net income of about US$803. The average production cost using Farmers Practices was US$769 while in Save and Grow plots it was only US$638 (Figure 2.5). The reduction in the cost of production was attributed to the application of improved management practices (e.g., use of younger seedlings, wider spacing, organic fertilizer and alternate wet and dry water management). In addition, informed management decision making based on agro-ecosystem analysis, the use of ecosystem services such as natural pest control to avoid unnecessary use of chemical pesticides and the use of biological control agents also contributed to effective and sustainable pest management.

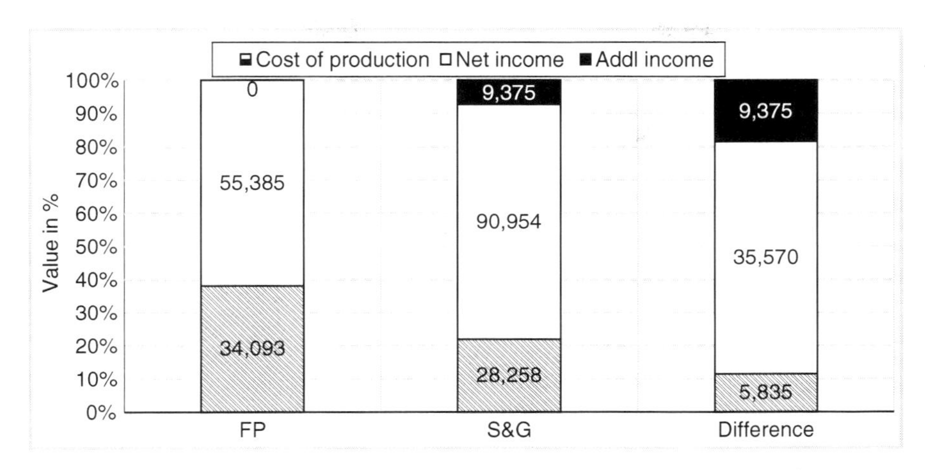

Figure 2.5 Comparison of cost of production, net income and additional income using Save and Grow and conventional farmers' practices, Philippines

Source: Authors' own compilation from project data.

The 72 percent increase in net income generated by the application of Save and Grow practices was attributed to higher yields, benefit derived from smart use of enhanced agro-biodiversity through multiple cropping as well as introducing ducks in rice production, and utilizing existing aquatic biodiversity. Savings from reduced cost of production resulting from improved management practices (e.g., reduced number of seedlings per hill) also contributed to the increased net income.

Planting vegetables on the bunds maximizes utilization of the production area and at the same time provides habitat for *natural enemies* enhancing natural biological control. Where ducks were introduced in the rice fields these provided additional regulatory ecosystem services by eating the insect pests found at the stem and base of rice plants.

Enhancing aquatic biodiversity in rice based cropping systems provided an average additional income of US$208.56 per hectare. For example, duck egg production alone became a source of additional income and nutrition. Information on existing aquatic biodiversity was not available until local agricultural schools – as part of the project – carried out assessments to determine existing species of inland fish (e.g., cat fish and mud fish) and molluscs (e.g., *kuhol* and another local snail species known as *ponggok*). These aquatic inland species had been lost as a result of high chemical use in rice production and their return is a result of reduced chemical use by farmers trained in IPM-FFS under the Philippine National IPM Programme that started in 1994 (FAO, 2015 a, b). The process of involving agricultural secondary schools and colleges was necessary to raise awareness among children (of farmers) and other stakeholders in the community (e.g., teachers and local leaders) on the importance of aquatic biodiversity to sustainable rice production and its contribution to food security.

Save and Grow FFS in Lao PDR

In Lao PDR the majority of its population is engaged in rice farming, which remains overall low-input and low-output production in terms of rice yields, input costs and profitability. The cultural and economic importance of paddy rice production for households in the lowlands of Lao PDR cannot be overstated. Annual rice production is viewed by households and the Government alike as an indicator of poverty and food security. Over the past decade the adoption of new technologies has resulted in productivity improvements in lowland rice systems, yet further gains are being sought to maintain national rice self-sufficiency. The Government of Laos has established optimistic yield targets for both the lowland rainfed and irrigated rice production systems. However, survey evidence shows that, despite the adoption of improved technologies, most rainfed farmers remain subsistence-oriented and there is a significant yield gap between the current situation and the proposed targets (Newby *et al.*, 2013). Only 10 percent of the total rice production area is irrigated while the bulk of production (60 percent of annual rice harvest) takes place in the rainfed lowlands, intended primarily for domestic consumption. Following a significant expansion in production area and an increase in productivity of rice during the 1990s, the country has been producing a stable and increasing rice surplus over the last decade. In recent decades, most intensification efforts have taken place in the irrigated lowlands, with a focus on introduction of improved and high-yielding varieties and increased use of inorganic fertilizers. With expanding production, the Lao PDR has achieved self-sufficiency at the national level. However, substantial regional/provincial differences in rice surplus/deficits remain requiring more geographically focused interventions to achieve self-sufficiency at local levels at all times (FAO/WB, 2012).

Use of agro-chemical pesticides remains overall low, certainly in comparison with production practices in some of the neighboring rice producing countries (e.g., Thailand and Vietnam). That said – and as the Lao Government is keen on further intensifying production, including for export to China – farmers are increasingly encouraged to make use of modern seeds and chemical production inputs – mostly N-based fertilizers – in an effort to boost rice yields. Inefficiencies in input use, most notably seeds and inorganic fertilizers, are commonly observed and depress farmer profits. Lao rice production also remains overall labor-intensive, involving manual labor for transplanting, weeding and harvesting, often done by women. Lao rice farmers lack knowledge about best management practices for growing improved varieties and on what fertilizers to use, when, how much and how best to apply. Farmers' understanding of agroecosystems and ecosystem services, such as natural biological control, is limited. Equally important to mention is that rural communities derive much of their food and nutrition security from products collected and harvested from rice landscapes. During the wet season farmers collect on a daily basis a variety of aquatic organisms in rice-base wetlands for local consumption and cash income when sold on local markets. These harvested natural resources from rice landscapes provide for cash income and are a vitally important source of proteins and vitamins (e.g., Vitamin A) for rural communities in particular.

Within the context of the Regional Rice Initiative it therefore made sense to introduce the Save and Grow based concepts and good practices associated with Sustainable Intensification of Rice Production. It also made sense to build on existing farmer training implementation networks, available within the context of the National IPM Programme, and explore the potential of Farmers Field School-based education interventions. At the outset, the FFS was identified as a practical and effective educational tool to develop farmer knowledge and skills for sustainable management of rice-based farming systems. The FFS training was envisioned to help raise awareness about natural resources management and more rationale and efficient use of production inputs while raising land productivity and farmer profits in primarily the *irrigated* lowlands of Lao PDR.

Figure 2.6 Save and Grow-SIRP Farmer Field School in Xaybouly, Savannakhet, Lao PDR

Source: Photo by Jan Willem Ketelaar.

During the 2015–2017 period some 54 season-long Save and Grow-SIRP Farmers Field Schools were implemented in 17 districts in six provinces (Sayabouly (11), Xiengkhuang (10), Khammouane (1), Salavan (2), Savannakhet (17), Champassack (13)), mostly (50 out of 54) during *wet season* rice production cycles. Each FFS met on a weekly basis, 17 times throughout the full production cycle. Some 1,562 farmers (37 percent female) participated in these FFSs.

Two FFS, involving some 50 rice farmers, were implemented in Phaxay district, Xieng Khouang province, Lao PDR during the 2015 wet season. In the FFS, the farmers learned about efficient management, growing healthy, well-yielding crops with fewer and more sustainable production inputs. Farmers explored making optimal use of multiple goods and services of paddy-based farming systems – conservation and management of aquatic biodiversity (including the introduction of cultured fish species) – in combination with improved agronomic practices such as wider plant spacing/reduced seeding rates, improved water management, and reduced use of chemical pesticides through the application of ecologically-sound IPM and natural biological control. For details see Table 2.4.

A results assessment study was carried out by the National University of Laos during the 2016 wet season rice production cycle, one year after the original FFS interventions in Phaxay district, Xieng Khuang. University staff collected data from 27 FFS alumni farmers, intended to assess adoption of Save and Grow good practices and results thereof. Data was also collected from 16 control rice farmers (from a different village in the same district but

Table 2.4 Practices applied in their own rice fields by FFS S&G (n=27) and non-FFS Farmers (n=26), Xieng Khuang during the wet season 2016 (one year after the 2015 FFS intervention)

FFS S&G-SIRP farmers	*Non-FFS farmers*
Seed rate of 74kg/ha	Seed rate of 87kg/ha
Planting distance ranging from 20 x 20 cm to 25 x 30 cm	Planting distance ranging from 15 x 15 cm to 30 x 30 cm
Age of seedling 22 days	Age of seedling 28 days
Manure: 4,101kg/ha	Manure: 2,989 kg/ha
Use of 16-20-00 nitrogenous fertilizer – basal application: 92 kg/ha	Use of 16-20-00 nitrogenous fertilizer – basal application: 165 kg/ha
Use of 16-20-00 nitrogenous fertilizer at 20 DAT: 142 kg/ha	Use of 16-20-00 nitrogenous fertilizer at 20 DAT: 185 kg/ha
Frequent hand weeding 30 days after transplanting	1–2 times hand weeding 30 days after transplanting
51% of farmers kept 10–15 cm water in the field	35% of farmers kept water in the field

Source: Authors' own compilation from project data.

with no exposure to FFS training) and 26 non-FFS farmers, i.e., farmers from the same villages but who did not participate in the FFS S&G (Table 2.5). In addition to interviews with the control, FFS trained and non-FFS trained farmers, field crop observations were done and focus group discussions (FGDs) were held with village heads, representatives of FFS alumni and landless farmers (i.e., landless laborers) to support or verify information obtained from the formal interviews.

In terms of adoption of Save and Grow practices, FFS graduate farmers transplanted younger seedlings, wider-spaced and used substantially less seed inputs (74 kg/ha) compared to the non-FFS (87 kg/ha) and control (121 kg/ha) farmers resulting into substantial savings on seed inputs. The 74 kg/ha is the average seed rate of FFS graduate farmers who had decided to divide their 2016 rice crops on an average 1.4 ha landholdings to compare Save and Grow-SIRP with conventional crop management practices. These farmers did not prepare separate seed beds for Save and Grow-SIRP and conventional practice plots. However, some FFS graduates were already applying 30–50 kg/ha seed rates demonstrating further potential to reduce seed inputs if Save and Grow is practiced on entire farmer landholdings. The pooled results from both groups of farmers in the FFS villages showed the same trend (Table 2.5). The FFS-trained farmers differed not significantly

Table 2.5 Yields, income, production cost, gross margin and returns on investment of FFS and non-FFS farmers, Lao PDR, wet season 2016

Exchange rate 1US$:8,289.81Laos

	Yield (t/ha)	*Income (kip)*	*Total cost*	*Gross margin*	*RoI*
Control farmers (n=16)					
Mean	2.91	10,196,355	4,034,698	6,161,657	1.53
Std	0.50	1,758,165	497,515	1,415,507	0.28
FFS Graduates (n=27)					
Mean	4.25	14,881,873	4,073,303	10,808,570	2.73
Std	1.07	3,373,121	607,637	3,848,494	1.05
Diff to control	**	**	ns	**	**
Non-FFS farmers (n=26)					
Mean	3.84	13,429,754	4,443,714	8,986,041	2.10
Std	1.12	3,935,951	894,061	3,831,206	1.02
Diff to control	**	**	ns	**	**
Diff FFS to non-FFS	ns	ns	ns	ns	*

Source: Authors' own compilation from project data.

from the non-FFS farmers in the same village. However, compared to the farmers in the control village, there were highly significant differences for both FFS-trained and non-FFS farmers in all categories (productivity/yield, income, gross margins and returns on investment) except for total cost. This could be indicative of a strong diffusion effect and/or a village effect (Phasouysaingam, 2017).

Save and Grow Farmer Field Schools (FFS) in Vietnam

Starting in 2014, pilot activities in Vietnam were carried out by the Vietnam National IPM Programme and its FFS implementation network in partnership with civil society organizations (CSO) at regional and national levels. The regional CSO, Field Alliance/Thai Education Foundation and the local CSO, Center of Initiatives on Community Empowerment and Rural Development (ICERD) carried out the training on aquatic (including rice-fish) and agrobiodiversity and pesticide impact assessment for communities.

Initial surveys carried out by ICERD showed that farmers in selected sites in Bac Giang and Quang Binh provinces in northern Vietnam practice rearing fish in ponds close to rice fields. Farmers release fish from the ponds into the rice fields but only after harvesting. Pesticide use – about 5–7 applications per crop, not including herbicides – made raising fish in standing rice crops impossible. As part of the process of community education on pesticide risk reduction, the FAO-supported National IPM Programme assisted the formulation of Community Action Plans and trained farmers in FFS to reduce pesticide risk, including the use of biological control options for pest management. Among other agreements, various stakeholders in the commune – including local government, people's organizations (Farmers' Union, Women's Union, Youth Union, etc.) and farmers agreed to reduce use of pesticides for improved rice ecosystem health and to integrate rice–fish–aquatic biodiversity in the production system. A total of 105 FFS farmers (55 women) from Bac Giang and Quang Binh provinces applied knowledge and skills on improved rice intensification practices to an area of 34 hectares. Rice fields adjacent to each other were selected for the Save and Grow study plots. Following good principles and practices of Save and Grow, rice farmers applied efficient management, growing healthy, well-yielding crops with fewer and more sustainable production inputs.

Farmers explored making optimal use of multiple goods and services of paddy-based farming systems – including conservation and management of aquatic biodiversity (including both captured and cultured fish species) – in combination with improved agronomic practices such as wider plant spacing/reduced seeding rates, improved water management, and reduced chemical pesticides through the application of ecologically sound IPM. This included augmentation of natural biological control through applications of *Metarhizium anisopliae* as an alternative to chemicals for population regulation of the brown planthoppers (*Nilaparvata lugens*) (Table 2.6).

Table 2.6 Practices applied by FFS S&G and non-FFS Farmers, Bac Giang and Quang Binh, Vietnam

Farmer Field School Save & Grow	*FFS farmers*
• Seed rate of 22 kg/ha	• Seed rate of 28 kg/ha
• Planting distance of 25 x 25 cm	• Planting distance of 20 x 20 cm
• Age of seedling 7–10 days	• Age of seedling 15–20 days
• Manure: 8,000 kg/ha	• Manure: 8,000 kg/ha
• Nitrogenous fertilizer: 139 kg N, Phosphate (P_2O_5): 416 kg, Potassium: 139 kg/ha	• Nitrogenous fertilizer: 222 kg N, Phosphate (P_2O_5): 500 kg, Potassium: 278 kg/ha
• Basal fertilizer: 83 kg N; 416 kg P; 56 kg Kali and 8,000 kg manure	• Basal fertilizer: 56kg N; 500kg P and 8,000kg manure
• 1st additional fertilizer: 56 kg N (7–10 DAT)	• 1st additional fertilizer: 110kg N; 110kg Kali (10–15DAT)
• 2nd additional fertilizer: 83kg K (panicle initiation stage)	• 2nd additional fertilizer: 56kg N; 168kg Kali (panicle initiation stage)
• Weed management: hand-weeding 7–10 days after transplanting	• Weed management: spray herbicide 7–10 days after transplanting
• Use of *Metarhizium anisopliae* for BPH management	• Spray chemical pesticide for BPH management 2 times per season
• Water management: "Nông – Lộ – Phơi" which means putting water in at shallow level (2–3 cm), keeping water in until the soil is exposed and the field is dry, before putting in water again	• Water management: always keep water in the field at 2–3 cm
• Released fish	• Fish not released

Source: Authors' own compilation.

Table 2.7 summarizes the differences between yields and benefits farmers obtained if the farmers only grew rice and when aquatic biodiversity was integrated in the production system. The average gross income from integrated rice–fish–aquatic production is US$7,751 compared to US$1,892 obtained from producing only rice. Utilization of integrated rice–fish–aquatic biodiversity production practices resulted in average gross income ranging from 211–551 percent compared with producing rice only. Farmers' experience and the aquatic biodiversity species and numbers – especially fish – accounted for the big difference in benefits (Figure 2.6). Informed management decision making based on agro-ecosystem analysis, the use of ecosystem services such as natural pest control to avoid unnecessary use of chemical pesticides and the use of biological control agents also contributed to effective and sustainable pest management.

The Save and Grow FFS provided opportunities for farmers and other community stakeholders to field test rice intensification practices while making optimal use of the multiple goods and services of paddy-based farming systems.

Table 2.7 Yields and benefits from rice only and integrated rice–aquatic biodiversity production systems, Vietnam, 2014

Parameters	*Total Bac Giang*	*Total Quang Binh*	*Average*
A. Rice yields (kg/ha)	6,120	5,417	5,769
B. Gross income from rice production only (US$/ha)	2,215	1,569	1,892
C. Yields of fish and other aquatic organisms (kg/ha)	7,913	1,860	4,886
D. Gross income from fish and other aquatic organisms (US$/ha)	9,981	1,738	5,860
E. Gross income from rice, fish and other aquatic organisms [B + D] (US$/ha)	12,196	3,307	7,751
F. Input costs (US$/ha)	4,547	1,402	2,975
G. Profits [E – F] (US$/ha)	7,649	1,905	4,776
H. Difference in gross income between rice production only and integrated rice-fish-aquatic biodiversity production (% increased)	551	211	381

Exchange rate: US$1:VND21,405.

Source: Authors' own compilation from project data.

Figure 2.7 A female farmer harvesting fish from the rice field

Source: Photo by Ngo Tien Dung.

The results are expected to generate government support for scaling up the pilot activities to improve quality and efficiency in the rice value chain and facilitate the development of more resilient and *sustainable* rice production intensification at landscape level (FAO, 2015 a, b, c).

Discussion

As outlined in case studies above, FAO has supported field experiments and farmer training on development and promotion of *sustainable* intensification of rice production in Indonesia, Lao PDR, Philippines and Vietnam during the 2013–17 period. The results of the *Save and Grow* policy inspired field pilots in all countries have clearly shown the enormous potential of rice farmers making more efficient use of production inputs while raising land productivity and incomes. The Lao case study on *adoption* of Save and Grow practices by FFS graduates clearly illustrates that farmers can intensify production and make rice farming more profitable and climate smart while at the same time becoming better custodians of the multiple goods and services provided by rice-based farming systems and landscapes.

The positive results and shown impact of the Save and Grow capacity building interventions can be attributed to the education approach used in the *Farmers Field Schools*. This approach has been shown to be ideally suitable and effective for smallholder farmers to learn about knowledge-intensive innovations and sustainable management of natural resources (Swanson and Rajalahti, 2010). Ecosystem-literacy training for smallholder farmers, preferably through Farmers Field School-based educational interventions, is essential for smallholder farmers to acquire the required agro-ecological knowledge and skills needed for *sustainable* rice production intensification (Ketelaar and Abubakar, 2012).

The FAO supported Save and Grow pilots documented in this chapter have primarily focused on development and promotion of sustainable intensification of rice production in the *irrigated* lowlands. Various development initiatives – past and present – have shown that also farmers in *rainfed* production areas can raise productivity and intensify rice production in a sustainable manner. The innovative development work done on adaptation of *System of Rice Intensification (SRI)*, a recognized *Save and Grow*-based suit of best practices for sustainable rice intensification (FAO, 2016 a, b, c), and promoting SRI among thousands of *rainfed* rice farmers within the context of an EU funded and Asian Institute of Technology implemented regional action research project in the Lower Mekong sub region illustrates the case (www.sri-lmb.ait.asia).

Key development areas include the better understanding and management of soil health and development and promotion of climate smart practices, both relevant for mitigation (e.g., alternate wetting and drying) and adaption (as described in Chapter 5 of this book with the zero tillage potato case study from Vietnam). FAO is continuing development work with rice farmers on soil health in the Philippines as part of RRI work. Other related initiatives in RRI countries and beyond focus on identification, application, farmer training and development of *climate-smart* rice production and development of relevant resource materials.

For sustainable intensification of rice production to be scaled out beyond pilot intervention areas, RRI pilot and other Asian countries are now investing in FFS-based farmer education and ecosystem literacy training. In the *Philippines*, the RRI pilot was designed to support the overarching goals of the agriculture sector under the Philippine Development Plan (PDP) 2011–2016, which are achieving food security and raising incomes. The pilot project was in line with the Department of Agriculture's Food Staples Sufficiency Programme. The positive results of the pilot activities in 2013 were used to strengthen the curriculum for farmer training to achieve rice self-sufficiency by increasing labor and land productivities and production cost efficiencies through sustainable rice production intensification. The regions where the RRI were piloted have expanded the implementation of FFS Save and Grow region-wide with more municipalities implementing FFS Save and Grow and funded by their respective regional rice programs (e.g., Northern Mindanao, Central Mindanao and CARAGA Regions). The Government of the Philippines under its National Rice Programme is committed to further development and the expansion of Save and Grow FFS under the second phase of the Regional Rice Initiative to which FAO has also committed additional funding support from 2014–2017.

In *Lao PDR*, the National Policy of the Lao Government promotes rice production intensification for maintaining rice self-sufficiency and for export production. The Lao Government has also called for efforts to halve labor inputs in rice production. The RRI pilot project provided evidence that Lao rice farmers through field testing of sustainable rice intensification practices can obtain higher yields with reduced labor inputs while making rice farming more profitable and making optimal use of the multiple ecosystem good and services provided by rice landscapes. This has led to a growing commitment of the Ministry of Agriculture and Forestry (MAF) to engage its technical line agencies (i.e., Department of Agriculture, Department of Livestock and Fisheries and Department of Agriculture Extension and Cooperatives) to support the Save and Grow work and endorse the Farmers Field Schools to provide education opportunities for smallholder farmers to acquire knowledge and skills for *sustainable* intensification of rice production.

In *Indonesia*, the Government has embarked on a target of 1 million hectares of rice–fish production, partly attributed to the results demonstrated by the pilot RRI and supported by the country's ambitious and well-funded rice–fish national program under the *Jajar Legowo*[1] model. Some of the benefits from rice–fish production that have been documented are: i) increased productivity from the paddy field (e.g., rice production increase by 10–20 percent, 6–7.5 tons/ha/crop; additional production of fish/aquatic animals:

1.2–1.5 tonnes/ha); ii) symbiotic relationship between fish and rice (e.g., paddy field provides fish with free food; fish help in weeding and pest control in the paddy); iii) sound ecological and environmental benefits (e.g., less negative impact on the environment from reduced use of chemicals and contribution to food safety); and iv) social and economic benefits from sustainable management of natural resources and application of these in crop management (e.g., increased net income by about US$4,800–7,800/ha with reduced production costs for both rice and fish/freshwater prawn; addresses concerns about "greying" of agriculture by making rice cultivation an attractive livelihood for the younger generation) (Soetrino, 2015).

In *Vietnam*, the Save and Grow FFS interventions, with a focus on agro-aquatic biodiversity management, have been scaled out to 50 rural communities in five provinces. The capacity building activities support community ownership in planning, management and implementation of local *Save and Grow* programs in support of the Government's new rural area development program "*Agriculture, Rural Areas and Farmers*" (called *Tam Nong*). The farmer empowerment processes supported by the *Save and Grow* interventions are aligned with the pillars of this new government program that (1) puts people at the heart of development starting from local participation in planning; (2) supports rural livelihoods, sustainable natural resource management including biodiversity; and (3) provides access to and control over natural resources in agricultural production back in the hands of farmers especially women.

Conclusions, recommendations and policy implications

The results of the *Save and Grow* capacity building interventions in all countries have clearly shown the enormous potential of rice farmers to make more efficient use of production inputs while raising land productivity and incomes. Given that smallholder farmers around the world will continue to produce the bulk of our daily food well into the twenty-first century, it is important that farmers innovate, become climate-smart and have access to productive, affordable and ecologically sound production inputs and sources of information.

What constitutes *Save and Grow* for one specific group of smallholder farmers in one particular location could be quite different for others, farming in different locations and rice-ecosystems and under different socio-economic conditions. There is no one size fits all in terms of a *Save and Grow* package to be communicated to farmers on a large scale. The *Farmers Field School*-based participatory and discovery-based learning processed employed have clearly been instrumental in allowing farmers to experiment with, learn about and adapt combinations of Save and Grow practices, diversify farming systems and become confident in adopting a suit of location and situation specific best practices that are economically feasible as well as ecologically sound.

Equally important will be the development of a *more enabling policy and regulatory environment* for sustainable intensification of rice production. This will include a necessary review of agricultural input subsidy policies, most notably on agro-chemicals, both fertilizers and pesticides. As long as the public and private sector continue to promote the use of and make agro-chemicals available at below market prices, farmers will continue to overuse such chemicals, with detrimental impacts on the land productivity, farmer income and the environment. Innovative (dis)incentives can help drive the rice sector towards better sustainability and more responsible management of ecosystem goods and services at landscape level. In this, the private sector has an important role to play as is, for example, illustrated by the recent First Global Sustainable Rice Conference pledge by leading private sector actors to commit by 2025 to 100 percent sourcing of fair and sustainable rice products throughout their global supply chains, supporting smallholder farmers and using Sustainable Rice Platform standards (SRP, 2017). Countries should also explore diversification of rice-based farming systems meeting multiple objectives and development goals. This is equally important for achieving and maintaining local and national food and nutrition security, in line with globally agreed Sustainable Development Goals, most notably SDG-2 (Zero Hunger).

Key lessons learned and policy implications from the implementation of the Regional Rice Initiative for sustainable intensification of rice production landscapes –relevant for rice production in Asia and beyond – are summarized below:

- Sustainable rice production intensification is feasible, allowing farmers to both *Save and Grow*.
- Improved management practices, use of more productive and affordable production inputs and responsible management of agro-biodiversity brings significant increase in yields, reductions in costs and higher rural incomes. This could provide incentives for the younger generation to take on a renewed interest in rice production and securing livelihoods in rural areas.
- Adoption/adaptation of *Save and Grow* best practices by smallholder farmers is location and situation specific and requires farmers to actively engage, learn and experiment on-farm.
- Recognizing the knowledge-intensive nature of *Save and Grow*, smallholder farmers trained in season-long Farmer Field Schools have been shown to master the concepts and skills required for sustainable management of natural resources and apply these in crop management with good results.
- Supportive government policies, innovative (dis)incentives and a better regulatory environment with regards to encouraging more responsible and less environmentally damaging use of agro-chemicals

in rice production is vital for sustainable intensification to take place at landscape level and beyond.

- Investments in farmer education and ecosystem-literacy training, preferably through Farmers Field Schools, are vital for large-scape adoption and scaling out of the *Save and Grow* good results.

Note

1 *Jajar legowo* is a planting method practiced in Indonesia where the rice seedlings are planted with only 2–3 seedlings per hill (7–10 days after germination), in a layout where each two rows of seedlings (spaced at 25 cm × 12.5 cm) is separated by a 50 cm empty row. The empty row increases the rice production by enhancing the edge effects of higher air movement and sunlight, while also providing the fish with room to forage among the plants.

References

Cochard, R., Maneepitakb, S. and Kumar, P. (2014) Aquatic faunal abundance and diversity in relation to synthetic and natural pesticide applications in rice fields of Central Thailand. *International Journal of Biodiversity Science, Ecosystem Services and Management*, 10 (2), pp. 157–173. Available at: www.tandfonline.com/doi/abs/10.1080/21513732.2014.892029 (accessed 8 February 2017).

FIELD Indonesia (2016) *Farmer Field School on Save and Grow integrated with rice–fish and better management of aquatic biodiversity. Final report*, Jakarta.

Food and Agriculture Organization of the United Nations (FAO) (2009) *How to Feed the World in 2050*. Rome: FAO.

Food and Agriculture Organization of the United Nations (FAO) (2010) *Sustainable Crop Production Intensification through an Ecosystem Approach and An Enabling Environment: Capturing efficiency through ecosystem services and management*. Report of the 22nd session of the Committee on Agriculture, Rome.

Food and Agriculture Organization of the United Nations (FAO) (2011) Save and Grow: A policymaker's guide to the sustainable intensification of smallholder crop production. Available at: www.fao.org/docrep/014/i2215e/i2215e.pdf (accessed 27 September 2017).

Food and Agriculture Organization (FAO), World Bank (WB) (2012) *Lao People's Democratic Republic Rice Policy Study*. Rome: FAO.

Food and Agriculture Organization of the United Nations (FAO) (2013) Empowering farmers to reduce pesticide risks. Bangkok: FAO, 68 pp. Available at: http://v1.vegetableipmasia.org/docs/Empowering%20Farmers%20To%20Reduce%20Pesticide%20Risks%2028Oct.pdf (accessed 27 September 2017).

Food and Agriculture Organization of the United Nations (FAO) (2014a) A regional rice strategy for sustainable food security in Asia and the Pacific. Bangkok: FAO. Available at: www.fao.org/docrep/019/i3643e/i3643e00.htm (accessed 27 September 2017).

Food and Agriculture Organization of the United Nations (FAO) (2014b) The multiple goods and services of Asian rice production systems. Rome: FAO. Available at: www.fao.org/3/a-i3878e.pdf (accessed 27 September 2017).

Food and Agriculture Organization of the United Nations (FAO) (2015a) Regional Rice Initiative pilot project: Philippines. Available at: www.fao.org/fileadmin/templates/agphome/scpi/Document_pdfs_and_images/Presentation_RRI-Philippines.pdf (accessed 18 September 2017).

Food and Agriculture Organization of the United Nations (FAO) (2015b) *Stories from the Field: Philippines, Save and Grow*. Bangkok: FAO.

Food and Agriculture Organization of the United Nations (FAO) (2015c) *Stories from the Field: Vietnam, Save and Grow*. Bangkok: FAO.

Food and Agriculture Organization of the United Nations (FAO) (2016a) Farmer Field School guidance document: Planning for quality programmes. Rome: FAO. Available at: www.fao.org/3/a-i5296e.pdf (accessed 31 July 2017).

Food and Agriculture Organization of the United Nations (FAO) (2016b) Save and Grow in practice – maize, rice, wheat: A guide to sustainable cereal production. Rome: FAO. Available at: www.fao.org/publications/save-and-grow/maize-rice-wheat/en/ (accessed 27 September 2017).

Food and Agriculture Organization of the United Nations (FAO) (2016c) The state of food and agriculture: Climate change, agriculture and food security. Rome: FAO. Available at: www.fao.org/3/a-i6030e.pdf (accessed 27 September 2017).

Food and Agriculture Organization of the United Nations (FAO) (2017) The state of food and agriculture: Leveraging food systems for inclusive rural transformation. Rome: FAO. Available at: www.fao.org/3/a-I7658e.pdf (accessed 16 October 2017).

Food and Agriculture Organization of the United Nations (FAO), International Fund for Agricultural Development (IFAD) and World Food Programme (WFP) (2012) *The State of Food Insecurity in the World: Economic growth is necessary but not sufficient to accelerate reduction of hunger and malnutrition*. Rome: FAO.

Fox, J. and Winarto, Y.T. (2016) Farmers' use of pesticides in an intensive rice-growing village in Indramayu on Java. In: Winarto, Y.T. (ed.) *Krisis pangan dan 'sesat piker': Mengapa masih berlanjut?* Jakarta, Indonesia: Yayakan Pustaka Obor.

Gallagher, K., Ooi, P., Mew, T., Borromeo, E., Kenmore, P. and Ketelaar, J.W. (2005) Ecological basis for low-toxicity Integrated Pest Management (IPM) in rice and vegetables. In: Pretty, J. (ed.) *The Pesticide Detox*. London: Earthscan.

International Rice Research Institute (IRRI) (2010) *Rice in the Global Economy: Strategic research and policy issues for food security*, edited by S. Pandey, D. Byerlee, D. Dawe, A. Dobermann, S. Mohanty, S. Rozelle and B. Hardy. Los Banos: IRRI.

Ketelaar, J.W. and Abubakar, A.L. (2012) Sustainable intensification of agricultural production: The essential role of ecosystem-literacy education for smallholder farmers in Asia. In: Mijung, K. and Diong, C.H. (eds) *Biology Education for Social and Sustainable Development*. The Netherlands: Sense Publishers, pp. 173–182.

Newby, J.C., Manivong, V. and Cramb, R.A. (2013) Intensification of lowland rice-based farming systems in Laos in the context of diversified rural livelihoods. Contributed paper prepared for presentation at the *57th AARES Annual Conference, Sydney, Australia, 5–8 February, 2013*.

Phasouysaingam, A. (2017) Case study on FFS graduate farmer adoption of Save and Grow (S&G)-based sustainable intensification of crop production. *Unpublished, consultancy report submitted to FAO*. National University of Laos, Vientiane: FAO.

Ramsar Convention (2012) *Draft Resolution XI.15: Agriculture–Wetland Interactions: Rice paddy and pesticide usage.* [Prepared by the Scientific and Technical Review Panel, submitted by the Standing Committee] Bucharest: Ramsar.

Regino, A. (2017) Of rice and men: Cultivating the next green revolution. In: *The McGill International Review: Agriculture, Climate Change and Environment.* Available at: http://mironline.ca/rice-men-cultivating-next-green-revolution/ (accessed 02 October 2017).

Settle, W.H., Ariawan, H., Astuti, E.T., Cahyana, W., Hakim, A.L., Hindayana, D., Lestari, A.S. and Pajarningsih, S. (1996) Managing tropical rice pests through conservation of generalist natural enemies and alternative prey. *Ecology*, 77 (7), pp. 1975–1988.

Soetrino, I.C.K (2015) Availability and use of aquatic biodiversity in rice field and ecosystem: Case study in West Java [PowerPoint slides].

Statista: The Statistics Portal (2017) Urbanization worldwide by continent. Available at: www.statista.com/statistics/270860/urbanization-by-continent/ (accessed 16 October 2017).

Sustainable Rice Platform (2017) Declaration of the First Global Sustainable Rice Conference, Bangkok, 4–5 November. Available at: www.sustainablerice.org/ (accessed 17 October 2017).

Swanson, B.E. and Rajalahti, R. (2010) Strengthening agricultural extension and advisory systems: Procedures for assessing, transforming, and evaluating extension systems. *Agriculture and Rural Development Discussion Paper 45*, Washington DC: The International Bank for Reconstruction and Development and World Bank.

United Nations Department of Economics and Social Affairs (UN DESA), Population Division (2015) *World Urbanization Prospects: The 2014 Revision (ST/ESA/SER.A/366)*, New York: United Nations. Available at: https://esa.un.org/unpd/wup/publications/files/wup2014-report.pdf (accessed 16 October 2017).

3 Sustainable intensification and maize value chain improvements in sub-Saharan Africa

Isaiah Nyagumbo, Mehreteab Tesfai, Udaya Sekhar Nagothu, Peter Setimela, James K. Karanja, Munyaradzi Mutenje and Connie Madembo

Introduction

Maize has the most potential to contribute to food and nutrition security (FNS) of the growing population and livelihood improvements in sub-Saharan Africa (SSA) where it is the staple diet for the majority of the population (AGRA, 2016). However, it will be a major challenge to produce enough maize due to several biophysical and socioeconomic constraints, including climate change induced weather extremes, land degradation, poor soil fertility, lack of quality seed and new pest and disease outbreaks (e.g., the recent fall armyworm (FAW) outbreak). In addition, lack of capacity to adopt improved technologies (inefficient extension services, poor input and output market incentives and overall lack of appropriate policy measures further make it difficult to improve maize value chains (VCs) for smallholders who constitute the majority in SSA.

In future, there is a need to focus on introducing crops and production systems that are more resilient to increased frequencies of erratic rainfall, dry spells, and late onset of rains, increasing temperatures and terminal heat stress among others. Maize farmers have resisted shifting to other resilient crops such as sorghum and millets, mainly due to lack of market demand of these crops and consumer preferences. Farmers in SSA continue to grow maize although the risk of crop failure is high since it is the staple diet (Kostandini *et al.*, 2013). We need improved maize varieties and technologies if maize monocultures have to continue in the extreme weather scenarios where drought and heat stress is likely to increase. A recent study by Steward *et al.* (2017), suggested the use of the conservation agriculture (CA) approach to address climate uncertainty, as it improves adaptive capacity of maize farmers to cope with increasing drought and heat stresses.

The complex and vulnerable nature of smallholder farming in the region needs targeted innovative technological, institutional and policy measures to improve and sustain maize production and productivity. However, there has been little attention given so far to integrating the various components to promote sustainable intensification (SI) of smallholder maize-based cropping systems in SSA. The main objective of this chapter, therefore, is to analyse the main drivers that constrain maize productivity and show the need for an integrated approach for maize VC enhancement. The chapter first provides an introduction describing the potential of maize towards addressing FNS, production trends and patterns. Following this, the main biophysical, institutional and policy drivers that influence maize productivity and measurers implemented and promoted by the International Centre for Maize and Wheat Improvements (CIMMYT) and other agencies are described. The chapter also presents some experiences of SI approaches such as CA and crop diversification to promote maize production in SSA. Towards the end, the chapter discusses the necessary improvements to be made for maize VC enhancement and finally presents some recommendations and future strategies to promote SI of maize.

Maize production and productivity trends

Globally, the average maize grain yield per hectare showed an increasing trend between 1961 and 2014. Yet in Africa, yield slightly increased only in 2014, but it is still three times lower than the world average (Figure 3.1). On the other hand, the area under cultivation of maize has increased dramatically both globally and in Africa (Figure 3.2). This shows that the recorded maize production increase in Africa was due to new area brought under cultivation (extensification) rather than productivity per unit of land (intensification). Expanding maize into new areas in the future is not a sustainable

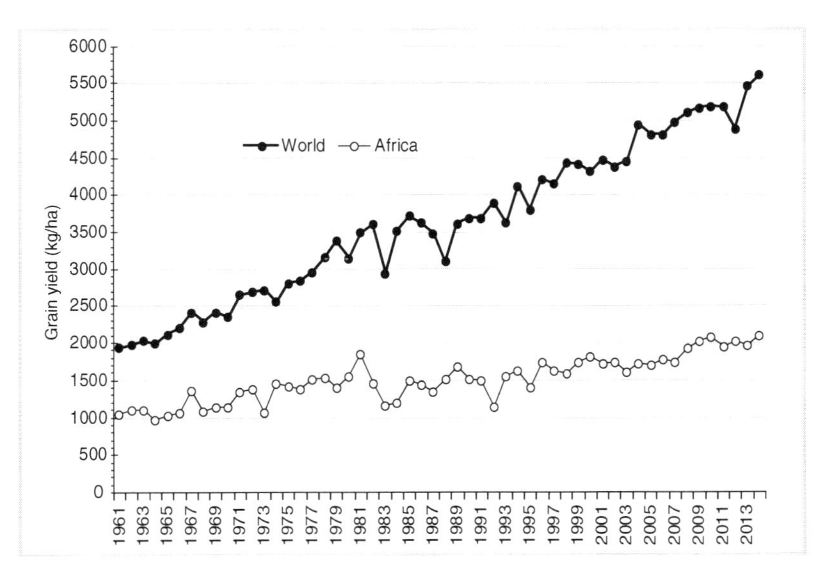

Figure 3.1 Average yield of maize grain in the world versus in Africa during 1961–2014

Source: Authors' own compilation based on FAOSTAT database: www.fao.org/faostat/en/#data/QC.

option as it often comes with an environmental cost in terms of increased land degradation (Shiferaw *et al.*, 2011).

SSA continues to experience food insecurity due to low yields from staple food crops (e.g., maize). Agricultural production in SSA in general continues to lag behind population growth estimated at 2.4 per cent, one of the highest growth rates for any sub-region. Therefore, the need and justification to move towards SI of maize-based production systems remains apparent. Maize is now the staple food in many of the countries in Eastern and Southern Africa (ESA) but it is a relatively a new world crop that was introduced after the Second World War and widely cultivated now in many countries in Africa (MacCann, 2005). In ESA, maize is mostly grown in the mid-altitude humid and hot zones while the highlands account for the smallest proportion (Setimela *et al.*, 2005). The mid-altitude zones have a low to medium risk of drought and are therefore considered ideal for growing maize. Due to its increasing popularity, maize is now widely grown also in the semi-arid environments where sorghum and millets were once the dominant crops. These agro-ecological zones are highly drought prone, not ideal for growing maize, and maize is harvested only once in every five seasons in some cases (Kostandini *et al.*, 2013).

Maize yields under smallholder farmer conditions in SSA remain among the lowest in the world, ranging from 2 t ha^{-1} to less than 1.5 t ha^{-1} (Figure 3.1). This is much below the potential, despite the advent of improved

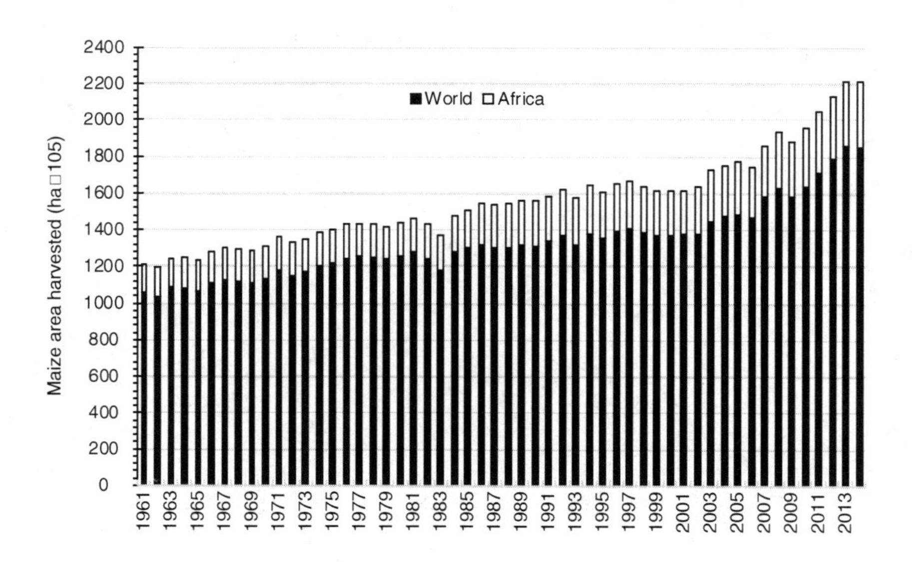

Figure 3.2 Total cultivation area of maize harvested during 1961–2014 in the world versus in Africa

Source: Authors' own compilation based on FAOSTAT database: www.fao.org/faostat/en/#data/QC.

maize technologies. A majority of the countries in SSA grow maize in more than 50 per cent of their farms meant for cereal production (FAO, 2010).

Furthermore, though the total cultivation area under maize has shown an increasing trend in all the regions of Africa, the productivity has not increased proportionately (Figure 3.3). Maize accounts for almost half of the calories and protein consumed in Eastern and Southern Africa (ESA) region and one-fifth of calories and protein consumed in Western Africa (Shiferaw *et al.*, 2011; Macauley, 2015). Yet, maize consumption constitutes only between 19 and 55 per cent of the human daily energy intake with Lesotho, Malawi and Zambia ranking as the region's top maize consumers (Nuss and Tanumihardjo, 2011). Between 2011 and 2014, Ethiopia was the only country in SSA other than South Africa that has attained an average of >3 t ha^{-1} yield; while Zambia and Uganda have >2.5 t ha^{-1}, followed by Malawi with >2 t ha^{-1} (Abate *et al.*, 2015). The SSA average is about 1.8 MT ha^{-1} (Abate *et al.*, 2015) based on FAOSTAT data analysis. In most of the Southern African countries, maize occupies more than 80 per cent of the cultivated land area, but the yields are low. For example, in Zambia average yields are 2.64 t ha^{-1}, 2.19 t ha^{-1}in Malawi, 0.97 t ha^{-1} in Mozambique and in Zimbabwe it is only 0.93 t ha^{-1} (FAO, 2007). This is attributed to poor management, lack of improved technologies, droughts and inaccessibility of inputs. As domestic production is unable to satisfy the demand, some countries are even importing maize.

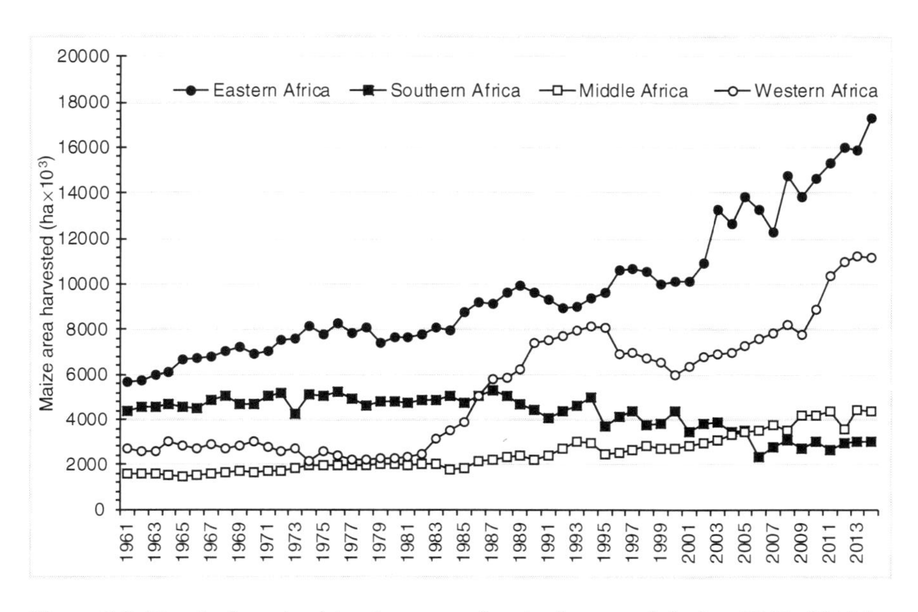

Figure 3.3 Trend of total cultivation area of maize harvested during 1961–2014 in Eastern, Southern, middle and Western Africa

Source: Authors' own compilation based on FAOSTAT database: www.fao.org/faostat/en/#data/QC.

Apart from the above reasons, the average farm size of more than 80 per cent of the farms in SSA is less than 2 ha (Harris and Orr, 2014), thereby limiting viability. Smallholder systems are characterized by low input and low output levels. In Malawi, a typical cultivated farmland holding size is even smaller, ranging between 0.2 and 0.3 ha (Ellis *et al.*, 2003). Whereas in Zimbabwe, smallholder arable land averages 2.5 ha (Gambiza and Nyama, 2006). In Mozambique, 97 per cent of production comes from about 3.2 million subsistence farms, with an average size of 1.2–1.4 hectares (Anderson *et al.*, 2016). Thus, farm size and the subsistence nature of smallholders limit their ability to increase productivity, besides lack of technology and policy options.

Biophysical, institutional and policy drivers and adaptation measures

A number of biophysical, institutional and policy drivers analysed in the following sections are responsible for the low productivity of maize. Possible adaptation measures to be implemented and promoted in the SSA are also discussed in the following sections.

Biophysical drivers to low productivity

Climate change

Extreme climate change has been highlighted as one of the major constraints of maize production in SSA, grown entirely under rainfed conditions (Lobell *et al.*, 2008). It is predicted that in future, the region will witness more extreme weather events such as the El Niño experienced in Southern Africa in the 2015–16 season that impacted negatively on the region's food security. According to some estimates, 25 per cent of the maize crop suffers frequent droughts, and the region incurs losses of up to 50 per cent of the potential harvest (CIMMYT, 2013).

Soil fertility

In addition, problems of declining soil fertility are widespread in SSA, largely as a consequence of continued cultivation of crops with low levels of nutrient inputs. Soil fertility is probably an even more serious constraint than climate change in most maize growing environments in Africa, due to very low and untimely use of fertilizer and the demise of bush-fallow systems of soil fertility replenishment (Morris *et al.*, 2007). The decline in soil fertility has been well recognized as a severe constraint to maize production in a number of studies (Okalebo *et al.*, 2006) coupled with increased and severe land degradation (Nkonya *et al.*, 2015) caused by an array of factors including climate change and poor soil/land management practices under a rapidly increasing population (Godfray *et al.*, 2010).

Chemical inputs alone for intensification of maize production cannot be in the best interest of the environment, and this raises questions about sustainability. Long-term field crop experiments in Kenya (Nandwa and Bekunda, 1998) and Zimbabwe (Waddington *et al.*, 2007) have clearly shown declining yield trends for mineral fertilizer-intensive treatments in a maize monocropping system. In addition to natural challenges, fertilizer use for most countries in SSA is around 10 kg ha^{-1} of NPK fertilizer that still falls far below the 50 kg ha^{-1} per year recommended dosage and reaffirmed by heads of government in the Malabo declaration (Malabo Declaration, 2014). One thing is about the right dosage, but the other concern is also about the timing and method of fertilizer application that may affect soil fertility. In general, smallholders in SSA lack access and cannot afford the recommended fertilizer doses, a major constraint besides seed in the production phase of the VC.

Pests and diseases

Maize also keeps facing new challenges from new diseases and pests. In 2011, maize lethal necrosis disease caused by interaction of *Maize chlorotic mottle virus* and *Sugarcane mosaic virus* emerged as a serious threat to maize production in Eastern Africa. The grain yield losses in 2012 were estimated to be 90 per cent in farmers' fields in Eastern Africa (Mahuku *et al.*, 2015). Losses due to abiotic stresses (recurrent drought, low levels of soil fertility) are often compounded by high incidence of diseases, insect pests and weeds that on average can reduce yields by more than 30 per cent (Shiferaw *et al.*, 2011). The frequency and severity of pests and disease epidemics is dynamic. Some pests and diseases that are problematic now may become less important as resistant cultivars are developed and deployed. Similarly, those pests and diseases presently considered unimportant may become more prevalent in the future with changes in climate, cropping practices and the introduction of new germplasm. For instance, the study by Ward *et al.* (1997) has shown that increased adoption of zero tillage and retention of residues cover results in increased incidences and severity of grey leaf spot. The recent epidemic by FAW across Africa is another typical example of how insect pests can cause serious damages that spread over large areas (Box 3.1).

Box 3.1 The fall armyworm (FAW)

A new lepidopteran invasive pest of American origin, the fall armyworm (*Spodoptera frugiperda* (J.E. Smith)) has been reported for the first time in January 2016 in Nigeria (Goergen *et al.*, 2016), and has since then assumed epidemic proportions in several southern African

(continued)

(continued)

countries, including Zambia, Zimbabwe, Malawi, Mozambique, Namibia and South Africa. FAW was already confirmed in 11 African countries, and was suspected to be present in at least 14 other countries, by April 2017. Across SSA, estimates indicated 13.5 million tonnes of maize, valued at US$ 3 billion, to be at risk during 2017–18 from FAW from a regional meeting organized by AGRA, CIMMYT and partners in Nairobi in April 2017. The FAW pest thus presents a new and serious challenge to maize production in the SSA region. Proposed control strategies include cultural, chemical and in the long term the breeding and selection of resistant maize varieties to FAW. However, combating this insect pest requires a cross border cooperation, integrated approach and investments supported by both private and public sector agencies in the region.

Source: Authors' own compilation.

Measures to address biophysical constraints

In response to *climate change and variability*, since the mid-90s scientists have been trying to develop high yielding drought resistant (DT) maize varieties for climate vulnerable zones (Fisher *et al.*, 2015; Setimela *et al.*, 2017; Lunduka *et al.*, 2017). A DT maize variety is defined as one that can produce approximately 30 per cent of its potential yield (1–3 t ha^{-1}) after suffering water stress for six weeks before and during flowering and grain filling stages (Magorokosho *et al.*, 2009). The Drought Tolerant Maize for Africa project released about 160 DT maize varieties between 2007 and 2013 (Setimela *et al.*, 2017).

The programme developed maize hybrids and Open Pollinated Varieties (OPVs) for drought, low nitrogen and important abiotic stresses. Elite hybrids and OPVs were tested and validated through the regional testing network under optimal conditions, managed and random drought stress, low N stress and MSV infection (Setimela *et al.*, 2017). Subsequently various studies have also shown that the adoption of DT maize varieties, for example, could go a long way towards overcoming dry spell induced crop failures (Lunduka *et al.*, 2017). Research thus has played a crucial role in improving maize VCs in SSA by developing DT and high yielding varieties.

Apart from the above measures, farmers in southern Africa are also implementing adaptation measures to climate change such as dry and early planting, substituting maize with other drought resistant crops, changing planting dates (Nyagumbo *et al.*, 2017), optimizing crop mixes during the rainy season and using water harvesting technologies and irrigation methods (Mano and Nhemachena, 2007).

To address *soil degradation and fertility decline*, CIMMYT and other national research institutions have been working continuously to improve soil fertility schedules for maize across different agro-ecological zones and socioeconomic conditions in Africa and other parts of the world. A range of climate smart soil improving technologies such as conservation agriculture (Vogel, 1995; Thierfelder *et al.*, 2016; Nyagumbo *et al.*, 2016), intercropping maize with legumes (Nyagumbo *et al.*, 2009), and Integrated Soil Fertility Management (Vanlauwe *et al.*, 2011) were implemented on research station sites and farmer fields that showed positive results.

The role of CA is described in detail in the later parts of this chapter. However, the problem is not the technology *per se* but rather the timely accessibility and affordability of inputs (seeds and fertilizers) when it comes to applying chemical fertilizers; shortage of land when practising intercropping or crop rotation of maize with legumes; the competing demands of the crop residues (especially in dry climates) when it comes to leaving the required residues in the field under a CA cropping system.

As maize became a staple food in Africa (McCann, 2005), there is a need to increase research to develop high yielding varieties that can perform well under increased biotic and abiotic stress factors (Morris, 2002). Exploitation of genetic diversity and using the available germplasm to develop good quality seed with better *tolerance to pest and diseases* seed was seen as an important strategy to support maize production in SSA. Significant cultivar improvement development efforts in Southern Africa only started in earnest between the 1950s and 1960s (Ricker-Gilbert *et al.*, 2013). To sustain varietal development and address new challenges in the late 1990s, CIMMYT together with some private seed companies introduced new maize germplasm, resistant to different biotic and abiotic stresses. The new germplasm developed was resistant to: a) southern corn leaf blight (*Bipolaris maydis*); b) southern rust (*Puccinia polysora*); c) northern corn leaf blight (*Exserohilum turcicum*); d) common rust *(Puccinia sorghi*); e) grey leaf spot (*Cercospora spp.*); f) stalk and ear rots caused by *Diplodia ssp.* and *Fusarium ssp.* and g) kernel rot caused by several *Fusarium* and *Aspergillus* species. These were some of the common diseases affecting maize crop during those years. From these efforts, CIMMYT successfully released a number of multi-stress tolerant germplasms that were further used by seed companies for commercial development and release of seed varieties or to extract genes for further improvement (Bänziger *et al.*, 2006).

With regard to *pests and diseases*, an integrated pest management (IPM) approach is necessary to combat pest outbreaks and minimize damages. If applied with a diverse range of methods they can be more suitable to smallholders. The push–pull pest control system, for example, is one of the success stories made in pest control measures in SSA (Box 3.2). The push–pull technology established as a reliable IPM, is applicable without needing conventional pesticides. Although the area given to the cereal crop itself is reduced under the push–pull system, overall, higher yields are produced per unit area (Hassanali *et al*., 2008). Though it is additional work, and needs training of farmers to improve adoption, the results looks promising.

Box 3.2 The push–pull pest control

Push–pull is a system that uses intercropping of cereals with a legume, in this case maize intercropped with Desmodium that repels (pushes) pests such as the stem borer from the maize crop towards the border crop, Napier grass, which lures (pulls) stem borers so that they lay their eggs in the grass instead of the maize (Biovision, 2015). The freshly hatched stem borers get trapped in the gummy substance produced by Napier grass allowing only a few larvae to survive to adulthood. Since 1997, the technique has benefited almost 40,000 farmers in East Africa. Consequently, the yields also increased up to 3.5 tonnes per hectare, and supported 300,000 people out of hunger and poverty, according to the International Centre of Insect Physiology and Ecology. The Centre intends to disseminate this technique further, and explore new push and pull plants with drought and temperature tolerance, since Desmodium and Napier grass cannot perform well in dry conditions.

Source: Biovision (2015).

Besides breeding for multi stress tolerant varieties, CIMMYT also started to develop new hybrids and OPVs with improved nutritional qualities through pro vitamin A (PVA) and quality protein maize (QPM). The PVA enriched maize varieties have vitamin A levels of 9–14µg/g, while the QPM varieties have higher levels of lysine and tryptophan compared to normal maize. Both amino acids (lysine and tryptophan) are essential for humans and mono-gastric animals who cannot synthesize them. In addition, the QPM varieties contain an equivalent of 90 per cent of the nutritive value in milk protein, as compared to 40 per cent in ordinary varieties. The nutritional impact of both QPM and PVA varieties were evident in malnourished children in other countries (Ganon *et al*., 2014). Ongoing efforts to breed maize for increased PVA carotenoids in Zambia have resulted in varieties with 9–17µg/g total PVA carotenoids, primarily as β-carotene. Recent studies analysing bioefficacy and food retention of the PVA varieties suggest that this target is adequate to meet the needs of the most vulnerable populations and address the serious malnutrition problems among the poor whose staple diet is maize. This was an important contribution and demonstrated how genetic diversity and plant breeding can address FNS issues with impacts on a larger scale, especially for the poor.

Institutional and policy drivers and adaptation measures

The agricultural sector in SSA overall remains fragmented and subsistence oriented, and most governments are unable to meet the variations in production, demands and price fluctuations despite efforts made at various levels (Barrett, 2007). The African maize market, in particular, is partially disconnected from global trends. Africa generally accounts for 1.5 to 3.5 per cent of global exports of maize; by comparison, the value of the continent's exports of maize flour represented 20 per cent of worldwide trade in 2013. Policy uncertainty in general is a major limiting factor for maize production and marketing. Given the importance of maize as a staple food, governments often justify intervening in the markets to ensure FNS. Sometimes, it can be in the form of input subsidies, export or import bans, purchases for national food reserves, or punitive import duties, policies favouring private sector that can provide them market advantages with large contracts to supply public entities with maize.

Despite the policy interventions, maize VCs in SSA are seriously impacted by market information asymmetry, lack of marketing skills and collective organization, poor enforcement of regulations that protect farmers and increased monopsony power on the part of agro-dealers and buyers, which in return lead to involuntary exclusion of smallholders from participation and low financial returns and increased marketing risks and financial loss to farmers. Limited access to information and knowledge about agro-inputs suppliers and micro-credit providers further increases the smallholder vulnerability to make profits (Box 3.3). Lack of grading and quality control checks on the maize input and output market segments due to poor capacity of the public institutions allows the private sector to exploit the smallholders.

Box 3.3 Maize input sector, accessibility and quality

The main maize markets are found in the district centres and provincial capitals, often far from the small farms. Most farmers have to meet their demands from village and roadside markets where they are exposed to limited choices, poor quality products and higher prices from the informal traders. There is poor enforcement of seed quality control regulations particularly in countries such as Malawi and Mozambique. Some vendors sell counterfeit seeds, thereby eroding farmers' trust of the seed. Seed companies tend to produce varieties based on market demands rather than suitability to agro-climatic needs, thereby limiting farmers' choices. For example, in the 2015/2016 season, the long duration maize varieties were the ones mostly available in the markets in Malawi, though the climate information available had indicated that the rainfall season was short. Furthermore, the supply of inputs is often quite erratic and there are periods when agro-dealers fail to get the inputs from the wholesalers, thus affecting availability of the inputs in the market. There is limited market information from both the supply and demand side. It seems agro-dealers, seed and fertilizer manufacturers are not responsive to market demands. Similarly, farmers also have

(continued)

(continued)

limited knowledge on the markets, different products available and prices. Because of these challenges, organizations such as CIMMYT have sought to capacitate seed systems through the training of seed producers, agro-dealers as well as ensuring that seed companies have the capacity to produce and multiply seed varieties in demand. This has been implemented mainly through the Drought Tolerant Maize for Africa Seed Scaling project (DTMASS) since 2014.

Source: Adapted from http://dtma.cimmyt.org

Seed is one of the most critical inputs for maize farmers in SSA. Smallholders depend both on the *formal* and *informal seed sector* for their needs. Therefore, during the 1990s, the private sector increased their activity throughout ESA as governments reformed agricultural production services through the World Bank led structural adjustments. To date, private seed companies now dominate and control more than 97 per cent of all commercial maize seed sales in southern Africa (Rashid *et al.*, 2001). The value of maize seed sales is estimated at around US$ 500 million followed by horticulture crops at US$ 250 million as reported by the African Centre for Biodiversity in 2015. Private sector dominance in the seed sector has affected the local informal maize seed sector in many of the countries in the SSA. On the production side, farmers also face challenges accessing other inputs, e.g. farm power sources such as draught power, labour (Houmy *et al.*, 2013), mechanization for land preparation and planting (Rurinda *et al.*, 2014). However, the major constraint for smallholders appears to be accessing seed from the formal sector (Box 3.4) due to high cost, inaccessible input and product markets, and lack of options and information with regard to suitable varieties (Bänziger *et al.*, 2006).

Box 3.4 Maize seed systems

Farmers access seed through the formal and informal seed systems (Sperling and McGuire, 2010). The formal seed system has an established seed value chain starting with the development, testing and registration of new varieties, maintenance of parental lines, seed production, and marketing and distribution (MacRobert *et al.*, 2014). On the other hand the informal seed system is less organised although similar activities may take place as an integrated part of crop production rather than discrete activities (Sperling and McGuire, 2014). The informal seed sector is an important source of seed for smallholders and hence it should be protected.

Source: (MacRobert *et al.*, 2014; Sperling and McGuire, 2014).

The planting window is short in the semi-arid regions. Hence ready accessibility to the inputs at the time of planting is crucial for smallholders. A study in Zimbabwe suggested that the time of planting is more important than the type of tillage system employed (Nyagumbo, 2008). Due to the costs involved in hiring or owning such inputs, smallholders tend to compromise on planting time that significantly affects yields. Farmer collectives and community sharing of farm implements, labour and other inputs including seeds has been tried and was successful in crops such as rice and potato, for e.g., the integrated seed delivery system in Ethiopia that could be suitable for maize. However, it requires some organization of farmers and cooperation of other relevant actors in the maize VC, including scientific agencies such as CIMMYT and national agriculture agencies.

With regard to quality seeds access, governments need to ensure that good quality seeds are made available to farmers on time. However, government-based input subsidy programmes could influence agro-dealer–farmer relationships and the number of transactions in the input market. Most small and medium range locally based agro-dealers look to expand their business within the current markets and expand into new potential markets only if there is adequate capital and high demand.

Issues such as input price policies and subsidies especially seed and fertilizers are highly politicized in many countries including Malawi (Chinsinga, 2011a, b), Tanzania (Wilson and Lewis, 2015), Kenya (Grow Africa, 2015) and Zambia and significantly impact maize production. In general, there is lack of adequate knowledge to use fertilizers resulting in weak fertilizer response. Inadequate capacity for local manufacturers in fertilizer blending and granulation is also an issue. Better quality blending could be achieved if the blending plants were stationed in major agricultural production regions to reduce transportation and transaction costs and problems of the fertilizer desegregation. Zambia and Malawi have made considerable progress in uptake of improved maize seed due to government seed and fertilizer subsidies programmes and a large number of seed companies when compared to Mozambique and Angola where there are only a few (Langyintuo *et al.*, 2008). Fisher *et al.* (2015) found that the major barriers to adoption were lack of availability of improved seed, inadequate information, lack of resources, high seed price, and perceived attributes of different varieties. Generally, the highest adoption rates of improved varieties have been by big commercial farmers (Langyintuo *et al.*, 2008).

Input subsidies on fertilizer, seeds, equipment, herbicides and other related inputs often dictate technology uptake by farmers as was the case in Zimbabwe since the mid-1990s (Nyagumbo and Rurinda, 2012). The study further showed that the macro, market-related policy constraints, for example, in the ESA region included:

- lack of harmonized seed policy;
- devaluation of the local currencies following market liberalization resulting in significant input price increases;

- limited availability of foreign currency particularly at times when currency is needed to import fertilizer;
- high interest rates and stringent collateral requirements;
- poor transport infrastructure including rural roads resulting in transport constituting more than 30 per cent of input costs;
- policy uncertainty relating to government supported programmes for input supplies, which discourage small scale private sector development;
- inadequate human capital development – mainly market and business skills – resulting in weak markets including rural credit;
- inadequate market and climate information; and
- a poorly functioning regulatory system. Such an unstable policy environment that does not protect the interests of smallholders thus makes it difficult to promote SI of maize in the region.

At the broader policy level, the Comprehensive Africa Agriculture Development Program (CAADP) agenda of the African Union aims at ensuring food security and economic growth through agricultural intensification, and places emphasis on protection of the natural resource base through the judicious application of sustainable land and water management practices (Bwalya *et al.*, 2009). Regional initiatives can play a key role in strengthening maize VCs and promoting cooperation among relevant actors at the SSA level. However, it is the national governments that need to make investments and implement the policies.

The other issue concerning maize is the maize input prices that are correlated to maize producer prices, which are often regulated by governments. For example, in Mozambique, favourable maize producer prices often result in increased area cropped to maize while maize yields also respond positively to increased producer prices (Figure 3.4). These figures emphasize the fact that favourable policies have been important food security prime drivers in this region.

Some countries in SSA are now in the process of developing their long-term strategies integrating the agriculture sector into overall economic growth.

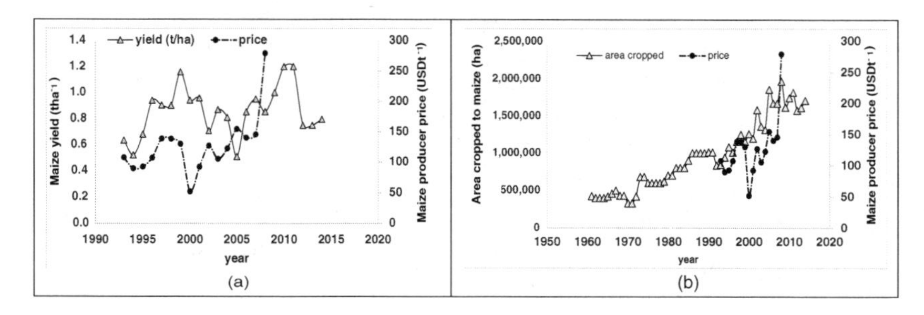

Figure 3.4 The relationship between (a) maize yield and (b) area cropped (ha) against producer price per tonne in Mozambique

Source: Adapted from FAO (2016).

For example, Kenya (Box 3.5) developed the Agricultural Sector Development Strategy (ASDS) in 2010, and the overall aim of this strategy is to strategically make the agricultural sector a key driver for achieving the 10 per cent annual economic growth rate expected under the economic pillar of the Kenya Vision 2030 launched in 2008. Through the ASDS, the government aims at transforming the agricultural sector into a profitable economic activity capable of attracting private investment and providing gainful employment for the people. Since maize is the major food crop, any new policy initiatives in the agriculture sector influence maize production and consumption.

Box 3.5 Price elasticity and consumption of maize in Kenya

A survey conducted in Kenya indicated that the wealthiest 20 per cent of households consumed about 37 per cent less maize per person compared to the poorest 20 per cent of households in 2009 (Kamau *et al.*, 2011). Kenya's per capita incomes have however not grown significantly enough to fully explain the 3 per cent average annual decline in consumption. For every 1 per cent increase in income, maize consumption is expected to decline by 0.4 per cent (Musyoka *et al.*, 2010). This translates to 0.8 per cent decline per year and a total of 5 per cent over six years, which is significantly less than the 16.8 per cent decline recorded between 2003 and 2009. The other possible reason for declining per capita consumption of maize is price inflation. This effect can manifest itself in average prices rising beyond the levels most consumers can afford (leading to reduced consumption) and/or relative maize prices rising to make its alternatives more attractive to consumers (enhance substitution). The former is explained by the effect of own price elasticity for maize products, while the latter functions through cross price elasticity and should lead to greater substitution into other foods. Musyoka *et al.* (2010) found substantial substitution effect for maize, in both the poor and non-poor households, with the latter the most likely to substitute maize when prices increase. If Kenya can increase its average maize yield by just 15 per cent it can meet expected national demand.

Source: Adapted from Kamau *et al.* (2011).

Investment in research and development in African agriculture is not adequate, except in a handful of countries. It has been almost 15 years since African heads of state and government pledged to allocate 10 per cent of their national budgets to the agricultural sector as part of the CAADP. The commitment, also known as the Maputo Declaration Target, rallied African governments to increase spending in the sector to stimulate agricultural growth, reduce poverty, and build food and nutrition security. Despite 20 per cent increase in agricultural research and development in the SSA during the last decade, national investment levels in the poorer countries have fallen quite low. New investments in the future need a long-term vision that is sustainable and tailored to not only meet the needs of the private sector but also address the limitations faced by economically disadvantaged farmers (Beintema and Stads, 2011).

Sustainable intensification of maize cropping system

SI is broadly defined as the investment of inputs and capital to increase crop productivity over the long-term, while protecting the underlying resource base (Pretty *et al.*, 2011). Several authors, including Giller *et al.* (2015) have emphasized the need to identify technologies that are appropriate for the SI of smallholder agriculture in SSA. Practices associated with SI often include integrated soil fertility management combining various forms of organic matter inputs and mineral fertilizers as recommended for Southern Africa by Snapp *et al.* (1998); in Eastern Africa by Okalebo *et al.* (2006); and for Western Africa by Schlecht *et al.* (2006) and resource conserving technologies such as conservation agriculture (CA) (e.g. Pretty *et al.*, 2011). In this section, exemplary research results achieved from the CA practices in southern Africa are discussed in relation to their contribution to promote SI of maize-based cropping system.

Conservation agriculture

Conservation agriculture (CA) is one of the climate smart agriculture (CSA) approaches which is aimed at improving productivity and adaptation to climate change (Khatri-Chhetri *et al.*, 2017). CSA is hinged on three key pillars: i) *Productivity*: Sustainable increase in agriculture productivity and income; ii) *Adaptation*: Enhancing resilience or adaptation of ecosystems and livelihoods to climate change and iii) *Mitigation*: Reducing and removing greenhouse gases emission (GHG) from the atmosphere and sequestering carbon (FAO, 2016).

Conservation agriculture practices such a mulching and minimum tillage support adaptation through increased infiltration, reduced run-off and soil erosion (Thierfelder and Wall, 2009). Reduced run-off from CA systems contributes to increased water storage in the soil profile and this stabilizes the yields in seasons of extreme weather (Ladha *et al.*, 2006). CA positively addresses climate variability impacts (Thierfelder *et al.*, 2017; Steward *et al.*, 2017). Yet, there is still limited knowledge on its mitigation effects. CA through adjustments in planting dates can help adapt to climate variability. In this case the delay in the onset of rains as described in Box 3.6.

Box 3.6 Effects of climate variability on CA maize productivity and planting date

The main objective here is to establish effects of climate variability on CA maize productivity and planting date. Effects of different CA systems on six location in three southern African countries were tested. Maize planting dates were evaluated using a 30-year historical meteorological data series.

Results showed that:

- Animal traction mechanised CA systems using the ox-drawn ripper tine (CA-Ripper) and the Fitarelli direct seeder (CA-Direct seeder) as well as the winter prepared basins (CA-Basins) improved timeliness of operations and enabled earlier planting across all locations ($P<0.05$) compared to the conventional mouldboard ploughing (CMP), late prepared basins (CA-Basin late) and CA established using dibble stick systems (CA-Dibble systems).
- Mechanized CA systems potentially offered farmers the earliest opportunity to plant (Figure 3.5) and better flexibility on when to plant. Timely planting of CA systems however only translated to higher maize yields when plating was carried out during periods of low rainfall variability with yield benefits of such early plantings being only apparent from Zimbabwe sites.
- Monthly rainfall coefficients of variation during the cropping season could vary from as high as 250 per cent down to about 20 per cent on some sites thereby showing the risky nature of rainfed cropping in this region.
- Maize yield benefits were observed when planting coincided with the optimum planting date for that location in all six sites.
- Analysis of planting dates suggested season onset was being progressively delayed over time by 0.29 in Chitedze and 0.40 days/yr in Chitala Research Stations, Malawi due to climate change. However, this delay in season onset was not apparent on the sites in Zimbabwe and Mozambique.

Source: Adapted from Nyagumbo *et al.*, (2017)

The results shown in Box 3.6 provide evidence-based research on the benefits of CA systems in terms of capacity to shift planting dates forward thereby enabling to adapted timely planting, particularly mechanized ones and those in which land preparation is carried out in winter. A more recent study evaluating CA technologies performance across the globe has also

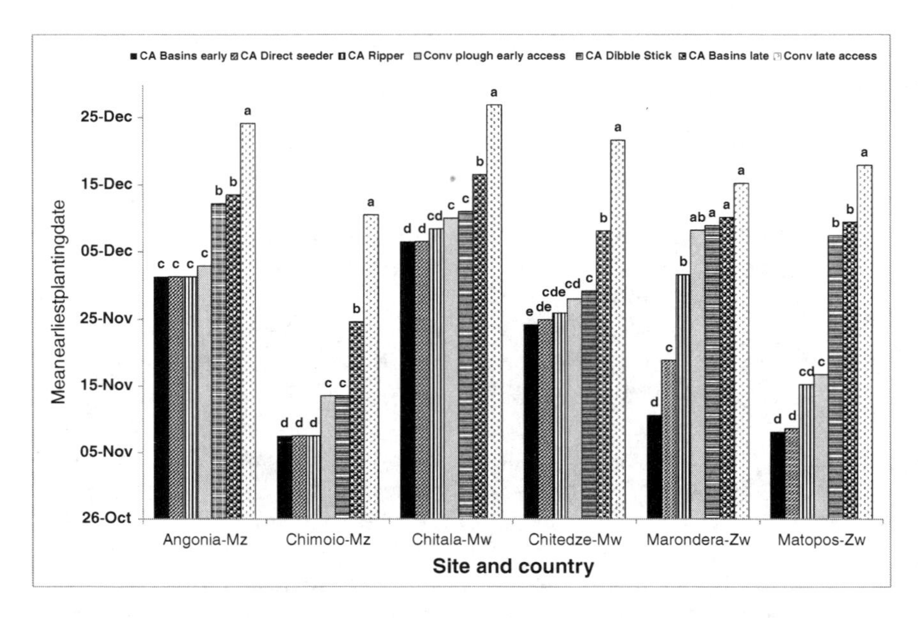

Figure 3.5 Mean earliest planting dates for different conventional and conservation agriculture cropping systems across agro-ecologies in Malawi, Mozambique and Zimbabwe. *Means with the same letter at each location are not significantly different at 5 per cent significance level.*

Note: *CMP-early* = Conventional mouldboard ploughing used by farmers with early access to draft power; *CMP-late* = Conventional mouldboard ploughing used by farmers with delayed access to draft animals; *CA-basins early* = CA basins prepared in winter; *CA-Basins late* = CA basins prepared at the onset of rains; *CA Direct seed* = CA planted with the Fitarelli animal traction direct seeder; *CA-ripper* = CA with animal traction ripper; *CA dibble* = CA seeding using the dibble stick system.

Source: Adapted from Nyagumbo *et al.* (2017).

provided further evidence that CA technologies contribute to improved adaptive capacities to drought and increased heat stresses associated with climate change (Steward *et al.*, 2017). In the latter study, CA practices out yielded conventional tillage systems in drier environments with seasonal rainfall deficits.

CA improves maize productivity

Research studies by the project Sustainable Intensification of Maize legume systems in Eastern and Southern Africa (SIMLESA) over the last seven years agree with findings from other related studies that yield increases from CA became statistically significant after 2 to 5 years of CA implementation. In some cases CA yields were more than twice higher than conventional yields (Thierfelder *et al.*, 2017; Nyagumbo *et al.*, 2016). Yield differences between CA and conventional practices tended to be more pronounced in maize-legume rotation systems and in seasons when serious water stresses were experienced in the season (Figure 3.6). Yet, the use of CA basins tended to depress yields in soils prone to waterlogging or during excessively wet seasons as was experienced in Mozambique's Angonia district over a four year period (2010/11 to 2013/14) where the use of CA basins depressed yields by some 2 per cent (Figure 3.6d). These results therefore give merits to CA systems in terms of improving productivity although there may be yield depressions, but contribute to climate smartness. Yet farmers implementing CA systems still face serious challenges associated with residue cover provision, weed control and lack of inputs (Thierfelder *et al.*, 2016). This is mostly due to competing use of the crop residues for fodder in some regions, or where lands are opened for grazing after harvest, and lack of proper management of crop residues after crop harvest.

Data points above the 1:1 line suggest a relative advantage of the respective CA cropping system while those above the 1:2 indicate CA yields that were more than double those of the respective conventional farmer practice were (Figure 3.6).

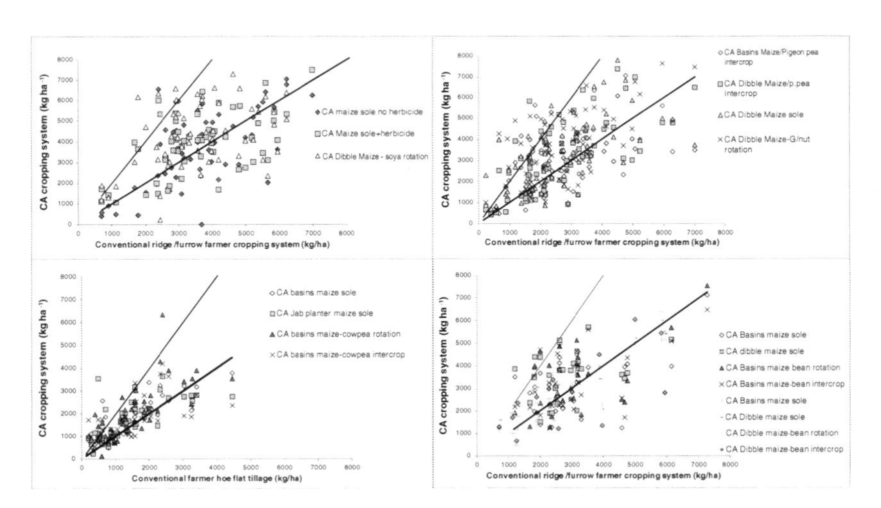

Figure 3.6 Comparisons of relative advantages of different conservation agriculture cropping systems against conventional farmer practices over four cropping seasons in four agro-ecologies of (a) Malawi mid-altitude agro-ecology (Lilongwe, Mchinji and Kasungu districts), (b) Malawi lowlands (Balaka, Ntcheu and Salima districts), (c) Mozambique central region (Sussundenga and Gorongosa districts) and (d) Mozambique north-western region (Angonia district).

Source: Adapted from Nyagumbo, *et al.* (2016).

Maize-legume crop rotations

Legumes not only improve soil fertility but also supplement proteins to the diets of poor households in rural Africa, as evident from Malawi. Introducing agro-ecological approaches to maize farmers, has led to maize-legume cropping systems, and improvement in the dietary patterns of rural households in Ekwendeni district (Bezner-Kerr and Patel, 2015). Research studies by the Sustainable Intensification of Maize legume Systems in Eastern and Southern Africa (SIMLESA) project in the lowlands of Malawi showed that the largest yield increases in some cases exceeding 50 per cent were often derived from maize-legume rotations involving cowpea or groundnuts. Intercrops involving pigeon pea and cowpea although failing to give maize yield increases in the short term tended to give better yield stability in both wet and dry seasons in Malawi and Mozambique (Nyagumbo, *et al.*, 2016). The shortage of nitrogen (N) is a key factor that limits productivity of smallholder maize-based systems across the SSA (Morris *et al.*, 2007), and legumes address this challenge in a sustainable manner through the biological fixation of N that builds soil organic matter (SOM) and soil N (Drinkwater *et al.*, 1998). The suppression of weeds, the breaking of pest cycles and biophysical rehabilitation of soil by deep taproot systems have also been demonstrated through the integration of legumes (Chikowo *et al.*, 2003).

Findings from the SIMLESA project and other studies thus suggest that intercropping tends to depress maize yields although the practice offers an opportunity to boost overall farm output, soil fertility, and adds needed diversity to farmers' fields.

Improved agronomic practises

Taking two examples from Kasungu district (Malawi) and Sussundenga district (Mozambique), it can be shown that maize yields from true farmer practices were often much lower than those from corresponding farmer practices under improved management regimes arising from improved agronomy involving better varieties, recommended fertilization rates and better planting configurations (Figures 3.7 and 3.8).

For Kasungu, mean yields computed over six years show that the relative yield increase compared with the local average, amounted to 88 and 109 per cent for the improved conventional ridge/furrow farmer practice and CA with herbicides plus rotations, respectively (Figure 3.7). Similarly, for Mozambique, corresponding yield increases relative to the local averages amounted to 100 and 162 per cent for the conventional flat tillage compared with the CA maize–cowpea rotation. It is apparent in both cases that the largest yield 'jump' was derived from the improved farmer practices compared to use of CA practices (Figure 3.8).

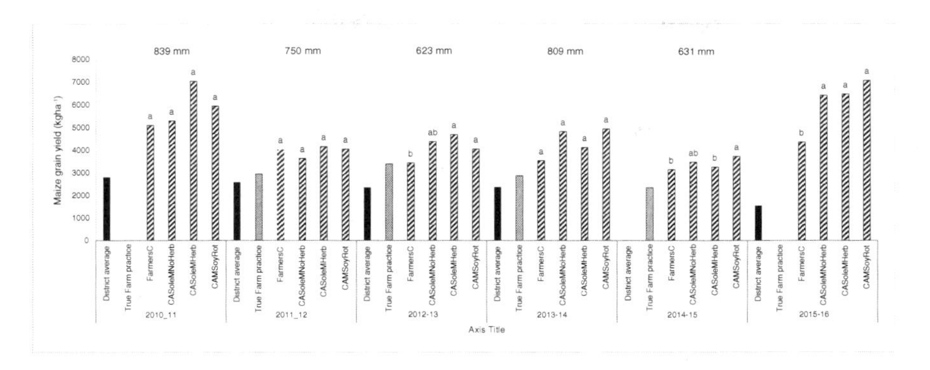

Figure 3.7 Mean maize yields by different cropping systems relative to local district averages in Kasungu district, Malawi (2010–11 to 2015/16)

Source: Authors' own compilation from SIMLESA project data, CIMMYT.

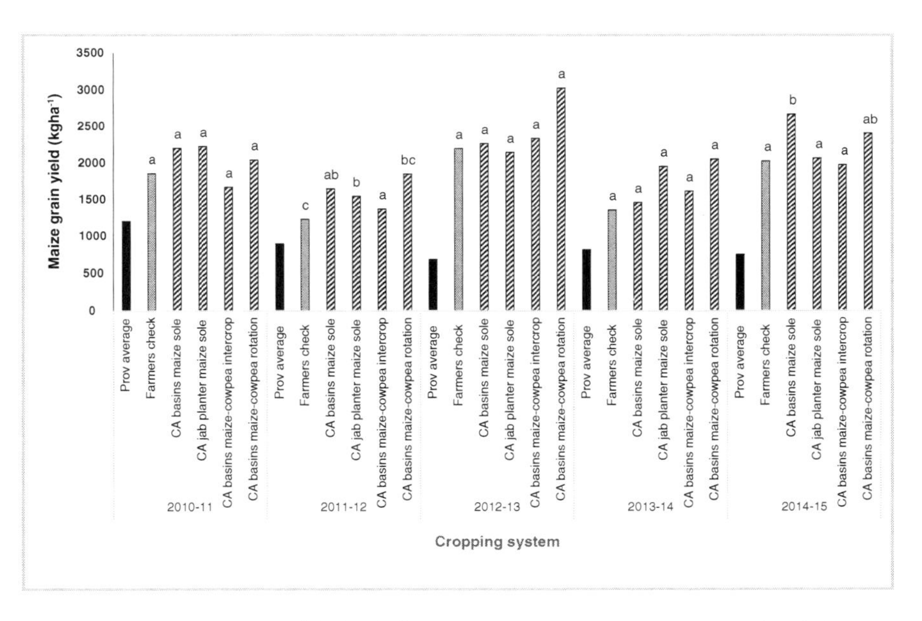

Figure 3.8 Mean maize yields by different cropping systems relative to local district averages in Sussundenga district, Mozambique (2010–11 to 2014/15)

Source: Authors' own compilation from SIMLESA project data, CIMMYT.

These results therefore suggest that the most significant yield improvements in these smallholder communities could come from investments in improved agronomy. For example in Malawi where corresponding actual farm yields were measured at each farm where experiments were conducted, more than 58 per cent yield increase between improved farm practice and the conventional 'true' farmer practices was observed over the six year period on average. Improved agronomy in this case constituted improved maize varieties, use of recommended fertilizers and better planting configurations but no CA. The findings here were significant in that they clearly suggested that investment in good agronomic practices potentially offers farmers the largest return on investments, although CA practices adoption gives them an extra increase. The use of good agronomic practices by farmers therefore could be the 'lowest hanging fruit' for guiding policy towards closing the yield gap (Van Ittersum *et al.*, 2013) in maize productivity in SSA.

Maize value chain and improvement measures

Maize value chains (VCs) in a majority of the maize growing countries in Africa is fragmented and complex with inadequate access to inputs, poor drying, fumigation, storage and quality grading facilities. It is dominated by small-scale unorganized agro-dealers, millers, periodical political influences on trade, price support measures and low capacity of smallholders to use fertilizers or purchase improved seeds (Meredian Institute, 2010; Wilson and Lewis, 2015). Thus, a VC approach should be the basis to improve maize productivity and income of smallholders in the future, as it enables analysis of the challenges and opportunities on both production and post-harvest phases simultaneously. It helps VC actors to identify where in the VC the investments are best in order to realize the most returns. In some situations, focus may be needed on improving access and affordability of inputs (seed, fertilizer, mechanization), whereas in others, it may be storage, marketing or ensuring price stability.

This argument is supported by a new VC project initiated in Kenya by the Grow Africa Secretariat aimed at increasing growth in sales and revenue in maize VCs in Africa (Grow Africa, 2015). The results showed that maize farmers in Kenya lose up to 53 per cent of potential net profit from their harvest, mainly due to lack of suitable inputs, mechanization and poor storage facilities leading to losses due to disease, e.g. aflatoxins, theft, pests and rodents. In situations where there are no proper storage facilities, and where farmers are in need of cash or credit, they cannot wait for a better price, and are forced to sell their produce immediately after harvest. Though VC actors are aware of the reasons for losses, prioritization becomes difficult where the investments are limited. A decision support tool developed in the project, the Grow Africa Gap Analysis tool for Agriculture (GATA) is supposed to guide VC actors to prioritize investments that would yield the highest returns and most impact on improving productivity and smallholder profitability (Grow Africa, 2015). The project also recommended the setting up of a VC actor platform that could bring together all relevant actors and foster collaboration, build trust, enable simple innovative interventions such as the warehouse receipt system that gives farmers access to credit when their crop is stored in the warehouses, and allow them to sell when the price is attractive. The Grow Africa case demonstrated how cost-effective and viable investments in Kenya were found to relate to storage and improved access to markets, reduced production losses by up to 95 per cent, while streamlining farmers' access to local markets (Grow Africa, 2015). Creating demand-driven partnerships across the maize VC is thus important, while simultaneously improving market information and communication through mobile phones – which allows farmers to take advantage of the input and output markets.

In another case, the Science and Innovation for African Agricultural Value Chains project in Africa found that significant value for smallholder farmers could be increased by reducing primarily postharvest inefficiencies (Meredian Institute, 2010). The key issues identified were gender and market dynamics, cultural and other contextual aspects that influence adoption of specific solutions. The focus thus was put on training women on use of proper technologies, improving access to good quality seed, targeted extension and micro-credit facilities for women. Whereas, a study in Uganda, by Kaganzi *et al.* (2009) showed that rapid urbanization provides new market opportunities for smallholders to supply higher value markets. In such situations, farmers need to work closely with service providers and other actors to meet the market quality. Thus collective action and collaboration among relevant actors helps in improving VCs in the growing urban markets. Wilson and Lewis (2015) recommend promoting Green Growth approaches for sustainable maize production and VC improvement – through better farming systems to improve soil fertility, on-farm water harvesting systems to improve available soil moisture, agroforestry and improved tillage systems such as conservation agriculture. In addition, they recommend an overall improvement of the capacity of farmers' groups, traders and millers.

As discussed in the previous sections, it is the policy environment regarding staple grains in most countries that largely influences and creates uncertainty with VC stakeholders, especially the commitment from the private sector to develop agricultural markets and provide smallholders the necessary services and markets (Grant *et al.*, 2012). Studies show that stabilizing food markets requires a stable policy environment (Jayne *et al.*, 2006).This may require regional cooperation, for e.g. the Southern African Development Community (SADC) that works across countries can increase access to technology, markets, finance and transparency in market information and trade. Mather *et al.* (2012) through their econometric analysis of nationally representative smallholder panel data sets from Kenya, Mozambique and Zambia suggested new design for the public policies and investments that will enable smallholders to increase their marketed surpluses of maize. The analysis also demonstrated the need for increased access to input and output markets, especially land, storage and road networks. In Malawi, studies suggest that smallholder farmers efficiency gains of as much as 40 per cent could be derived simply from investments in improving markets, extension and farmer organization (Tchale, 2009).

Conclusion and recommendations

Maize is the main staple diet for a majority of the population in SSA and thus the potential crop to contribute to FNS. Yet, maize productivity in the region still falls below the potential yield due to small sized farms that operate on a subsistence basis and several biophysical and socioeconomic constraints.

Hence, there is an immediate need to address these issues in order to meet the market demand for maize and future FNS. This would require an integrated and holistic VC analysis and interventions that address those bottlenecks in different countries. Suitable innovative technology, institutional and policy options need to be identified for each situation. Some broad recommendations for maize VC improvement include:

- enhanced access to land and regularizing land rights for smallholders;
- promoting SI approaches to improve soil fertility and water conservation at the farm level (e.g. CA, maize–legume crop rotations);
- capacity building of smallholder farmers to adopt new technologies;
- increasing access to quality inputs and mechanization through organized local agro-dealers and State Extensions Services;
- developing efficient agro-advisory services to farmers using mobile phones and other faster means;
- setting up functional VC actor platforms at national and district levels for active cooperation;
- investing in better storage, grading, drying, fumigation facilities for smallholders in the maize growing areas;
- introducing innovative financing schemes such as credit voucher systems to smallholders who store their grains;
- mainstreaming gender and youth through gender sensitive policy environment to actively take part in maize VCs;
- providing a stable and favourable policy environment for significantly improving the efficiency of maize VCs;
- advancing cross border or regional cooperation to promote research and development of stress-tolerant, high yielding maize varieties, trade and managing pest and disease outbreaks.

References

Abate, T., Shiferaw, B., Menkir, A., Wegary, D. and Kebede, Y. (2015) Factors that transformed maize productivity in Ethiopia. *Food Security*, 7 (5), pp. 965–981.

AGRA (Alliance for a Green Revolution in Africa) (2016) Africa agriculture status report 2016: Progress towards Agricultural Transformation in Africa. Available at: https://agra.org/aasr2016/public/assr.pdf (accessed 05 July 2017).

Anderson, J., Learch, C.E., Gardner, S.T., Bishop, C., Fitzgerald, M.A., Mukemi, L., Musiime, D. and O'Reilly, D. (2016) National Survey and Segmentation of Smallholder. Households in Uganda: Understanding their demand for financial, agricultural, and digital solutions. National Survey. CGAP, 94. Available at: www.cgap.org/sites/default/files/Working%20Paper_CGAP%20 Smallholder%20Household%20Survey_UGA_April%202016.compressed.pdf (accessed 06 January 2018).

Bänziger, M., Setimela, P. S., Hodson, D. and Vivek, B. (2006) Breeding for improved abiotic stress tolerance in maize adapted to southern Africa. *Agricultural Water Management*, 80 (1–3), pp. 212–224.

Barrett, C.B. (2007) Displaced distortions: Financial market failures and seemingly inefficient resource allocation in low-income rural communities. In: Bulte, E. and Ruben, R. (eds) *Development Economics Between Markets and Institutions: Incentives for growth, food security and sustainable use of the environment.* Wageningen, Netherlands: Wageningen Academic Publishers.

Beintema, N.M. and Stads, G.J. (2011) *African Agricultural R&D in the New Millennium: Progress for some, challenges for many.* International Food Policy Research Institute (IFPRI).

Bezner Kerr, R. and Patel, R. (2015) Food security in Malawi: Disputed diagnoses, different prescriptions. In: Sekar, N. (ed.) *Food Security and Development: Country cases.* London: Earthscan, pp. 205–229.

Biovision (2015) Push–pull East Africa. Available at: www.biovision.ch/en/projects/kenya/push-pull-east-africa/ (accessed 10 October 2017).

Bwalya, M., Diallo, A.A., Phiri, E. and Hamadoun, M. (2009) *Sustainable Land and Water Management:* The CAADP Pillar1 framework. Midrand: The Comprehensive Africa Agriculture Development Programme (CAADP).

Chikowo, R., Mapfumo, P., Nyamugafata, P., Nyamadzawo, G. and Giller, K.E. (2003) Nitrate-N dynamics following improved fallows and maize root development in a Zimbabwean sandy clay loam. *Agroforestry Systems*, 59, pp. 187–195.

Chinsinga, B. (2011a) Seeds and subsidies: The political economy of input subsidies in Malawi. *IDS Bulletin*, 42 (4), pp. 59–69.

Chinsinga, B. (2011b) Agro-dealers, subsidies and rural market development in Malawi: A political economy enquiry, FAC Working Paper 31, Future Agricultures Consortium, Brighton.

CIMMYT (International Centre for Maize and Wheat Improvements) (2013) The Drought Tolerant Maize for Africa project. DTMA Brief, September. Available at: http://dtma.cimmyt.org/index.php/about/background (accessed 10 October 2017).

Drinkwater, L.E., Wagoner, P. and Sarrantonio, M. (1998) Legume-based cropping systems have reduced carbon and nitrogen losses. *Nature*, 396, pp. 262–265.

Ellis, F., Kutengule, M. and Nyasulu, A. (2003) Livelihoods and rural poverty reduction in Malawi. *World Development*, 31, pp. 1495–1510.

FAO (Food and Agriculture Organization of the United Nations) (2007) Promoting integrated and diversified horticulture production in Maputo Green Zones towards a stable Food Security System. Rome: FAO.

FAO (Food and Agriculture Organization of the United Nations) (2010) FAO global information and early warning system on food and agriculture world food programme. Special report FAO/WFP crop and food security assessment mission to Zimbabwe.

FAO (Food and Agriculture Organization of the United Nations) (2016) Eastern Africa Climate-Smart Agriculture scoping study Ethiopia, Kenya and Uganda. FAO, Addis Ababa, Ethiopia.

Fisher, M., Abate, T., Lunduka, R., Asnake, W., Alemayehu, Y. and Madulu R. (2015) Drought tolerant maize for farmer adaptation to drought in sub-Saharan Africa: Determinants of adoption in eastern and southern Africa. *Climatic Change*, 133(2), pp. 283–299.

Gambiza, J. and Nyama, C. (2006) *Country Pasture/Forage Resource Profiles Zimbabwe*. Rome: FAO.

Ganon, B., Kaliwile, C., Arscott, S., Schmaelzle, S., Chileshe, J., Kalungwana, N., Mosonda, M., Pixley, K., Masi, C. and Tanumihardjo, S. (2014) Biofortified orange maize is as efficacious as a vitamin A supplement in Zambian children even in the presence of high liver reserves of vitamin A: A community-based, randomized placebo-controlled trial. *American Journal of Clinical Nutrition*, 6, pp. 1541–1550.

Giller, K.E., Andersson, J.A., Corbeels, M., Kirkegaard, J., Mortensen, D., Erenstein, O. and Vanlauwe, B. (2015) Beyond conservation agriculture. *Frontiers in Plant Science*, 6, pp. 870.

Godfray, H.C.J., Beddington, J.R., Crute, I.R., Haddad, L., Lawrence, D., Muir, J.F., Pretty, J., Robinson, S., Thomas, S.M. and Toulmin, C. (2010) Food security: The challenge of feeding 9 billion people. *Science*, 327, pp. 812–818.

Goergen G., Kumar, P.L. and Sankung, S.B. (2016) First report of outbreaks of the Fall Armyworm Spodoptera frugiperda (J.E. Smith) (Lepidoptera, Noctuidae), a new alien invasive pest in Western and Central Africa. *PLoS ONE*, 11 (10), pp. e0165632.

Grant, W., Wolfaardt, A. and Louw, A. (2012) Maize value chain in the SADC region. Available at: http://satradehub.org/images/stories/downloads/pdf/technical_reports/TECH20120229_Grant%20et%20al_Maize%20Value%20Chain%20in%20the%20SADC%20Region.pdf (accessed 12 October 2017).

Grow Africa (2015) Maize Value Chain Partnership launched. Available at: www.growafrica.com/value-chain/maize/feed (accessed 12 October, 2017).

Harris, D. and Orr, A. (2014) Is rainfed agriculture really a pathway from poverty? *Agricultural Systems*, 123, pp. 84–96.

Hassanali, A., Herren, H., Khan, Z.R, Pickett, J.A. and Woodcock, C.M. (2008) Integrated pest management: The push–pull approach for controlling insect pests and weeds of cereals, and its potential for other agricultural systems including animal husbandry. *Philosophical Transactions of the Royal Society B*, 363, pp. 611–621.

Houmy, K., Clarke, L., Ashburner, J. and Kienzle, J. (2013) Agricultural mechanization in sub-Saharan Africa: Guidelines for preparing a strategy. *Integrated Crop Management*, 22, pp. 105.

Jayne, T.S., Zulu, B. and Nijhoff, J.J. (2006) Stabilizing food markets in Eastern and Southern Africa. *Food Policy*, 31 (4), pp. 328–341.

Kaganzi, E., Ferris, S., Barham, J., Abenakyo, A., Sanginga, P. and Njuki, J. (2009) Sustaining linkages to high value markets through collective action in Uganda. *Food Policy*, 34, pp. 23–30.

Kamau, M., Olwande, J. and Githuku, J. (2011) Consumption and expenditures on key food commodities in urban households: The case of Nairobi. Working Paper WPS 41/2011a. Nairobi, Kenya: Tegemeo Institute. Available at: www.tegemeo.org/index.php/researchresources/publications/item/113 (accessed 12 October 2017).

Khatri-Chhetri, A., Aggarwal, P.K., Joshi, P.K. and Vyas, S. (2017) Farmers' prioritization of climate-smart agriculture (CSA) technologies. *Agricultural Systems*, 151, pp. 184–191.

Kostandini, G., Rovere, R.L. and Abdoulaye, T. (2013) Potential impacts of increasing average yields and reducing maize yield variability in Africa. *Food Policy*, 43, pp. 213–226.

Ladha, J.K., Rao, A.N., Raman, A. and Noor, S. (2006) Agronomic improvements can make future cereal systems in South Asia far more productive and results in a lower environmental footprint. *Global Change Biology*, 22, pp. 1054–1074.

Langyintuo, A.S., Mwangi, W., Diallo, A.O., MacRobert, J., Dixon, J. and Bänziger, M. (2008) Challenges of the maize seed industry in eastern and southern Africa: A compelling case for private–public intervention to promote growth. *Food Policy*, 35(4), pp. 323–331.

Lobell, D.B., Burke, M.B., Tebaldi, C., Mastrandrea, M.D., Falcon, W.P. and Naylor, R.L. (2008) Prioritizing climate change adaptation need for food security in 2030. *Science*, 80 (319), pp. 607–610.

Lunduka, R.W., Mateva, K.I., Magorokosho, C. and Manjeru, P. (2017) Impact of adoption of drought-tolerant maize varieties on total maize production in south Eastern Zimbabwe. *Climate Development*, 0, pp. 1–12.

Macauley, H. (2015) Background paper: Cereal crops: Rice, maize, millet, sorghum, wheat. African Development Bank. Available at: http://www.afdb.org/fileadmin/uploads/afdb/documents/Events/DakAgri2015/Cereal_Crops-Rice__Maize__Millet__Sorghum__Wheat.pdf (accessed 12 October 2017).

McCann, J.C. (2005) *Maize and Grace: Africa's encounter with a new world crop, 1500–2000*. Cambridge, MA: Harvard University Press.

MacRobert J., Setimela P., Gethi J. and Regasa M.W. (2014) *Maize Hybrid Seed Production Manual*, Mexico, D.F: CIMMYT.

Magorokosho, C., Vivek, B. and MacRobert, J. (2009) Characterization of maize germplasm grown in eastern and southern Africa: Results of the 2008 regional trials coordinated by CIMMYT, Harare, Zimbabwe. Available at: http://repository.cimmyt.org/xmlui/bitstream/handle/10883/3780/96060.pdf?sequence=1 (accessed 25 October 2017).

Mahuku, G., Lockhart, B.E., Wanjala, B., Jones, M.W., Kimunye, J.N., Stewart, L.R. and Redinbaugh, M.G. (2015) Maize lethal necrosis (MLN), an emerging threat to maize-based food security in sub-Saharan Africa. *Phytopathology*, 105(7), pp. 956–965.

Malabo Declaration on Accelerated Agricultural Growth and Transformation for Shared Prosperity and Improved Livelihoods (2014). Available at: https://au.int/en/documents/31247/malabo-declaration-201411-2.

Mano, R. and Nhemachena, C. (2007) Assessment of the Economic Impacts of Climate Change on Agriculture in Zimbabwe: A Ricardian Approach, The World Bank Development Research Group, Sustainable Rural and Urban Development Team. Available at: http://documents.worldbank.org/curated/en/584431468168844535/pdf/Wps4292.pdf (accessed 20 October 2017).

Mather, K. and Jinks, J.L. (2012) *Introduction to Biometrical Genetics*. Berlin: Springer Science & Business Media.

Meredian Institute (2010) Science and innovation for African Agricultural Value Chains. Available at: www.merid.org/~/media/Files/Projects/Value%20Chains%20Microsite/Maize_Value_Chain_Overview.pdf (accessed 12 October 2017).

Morris, M.L. (2002) *Impacts of International Maize Breeding Research in Developing Countries, 1966–98*. Mexico, D.F: CIMMYT.

Morris, M.L., Kelly, V.A., Kopicki, R.J. and Byerlee, D., (2007) *Fertilizer Use in African Agriculture: Lessons learned and good practice guidelines*. Washington, DC: World Bank Publications.

Musyoka, M.P., Lagat, J.K., Ouma, D.E., Wambua, T. and Gamba, P. (2010) Structure and properties of urban household food demand in Nairobi, Kenya: Implications for urban food security. *Food Security*, 2 (2), pp. 179–193. Available at: http://link.springer.com/article/10.1007/s12571-010-0063-6 (accessed 12 October 2017).

Nandwa, S.M. and Bekunda, M.A. (1998) Research on nutrient flows and balances in East and Southern Africa: State-of-the-art. *Agriculture, Ecosystems and Environment*, 71, pp. 5–18.

Nkonya, E. and Anderson, W. (2015) Exploiting provisions of land economic productivity without degrading its natural capital. *Journal of Arid Environment*, 112, pp. 33–43.

Nuss, E.T. and Tanumihardjo, S. A. (2011) Quality protein maize for Africa: Closing the protein inadequacy gap in vulnerable populations. *Advances in Nutrition: An International Review Journal*, 2, pp. 217–224.

Nyagumbo, I. (2008) A review of experiences and developments towards conservation agriculture and related systems in Zimbabwe. In: Goddard, T., Zoebisch, M.A., Gan, Y.T., Ellis, W., Watson, A. and Sombatpanit, S. (eds) *No-till Farming Systems*. Bangkok: World Association of Soil and Water Conservation, pp. 345–372.

Nyagumbo, I., Mvumi, B.M. and Mutsamba, E.F. (2009) Conservation agriculture in Zimbabwe: Socio-economic and biophysical studies. In: *Sustainable Land Management Conference*, Country Pilot Programme UNDP, GEF, unpublished, Windhoek, Namibia.

Nyagumbo, I. and Rurinda, J. (2012) An appraisal of policies and institutional frameworks impacting on smallholder agricultural water management in Zimbabwe. *Physics and Chemistry of the Earth*, 47–48, pp. 21–32.

Nyagumbo, I., Mkuhlani, S., Pisa, C., Kamalongo, D., Dias, D. and Mekuria, M. (2016) Maize yield effects of conservation agriculture based maize–legume cropping systems in contrasting agro-ecologies of Malawi and Mozambique. *Nutrient Cycling in Agroecosystems*, 105, pp. 275–290.

Nyagumbo, I., Mkuhlani, S., Mupangwa, W. and Rodriguez, D. (2017) Planting date and yield benefits from conservation agriculture practices across Southern Africa. *Agricultural Systems*, 150, pp. 21–33.

Okalebo, J.R., Othieno, C.O., Woomer, P.L., Karanja, N.K., Semoka, J.R.M., Bekunda, M.A., Mugendi, D.N., Muasya, R.M., Bationo, A. and Mukhwana, E.J., (2006) Available technologies to replenish soil fertility in East Africa. *Nutrient Cycling in Agroecosystems*, 76, pp. 153–170.

Pretty, J.N., Toulmin, C. and Williams, S. (2011) Sustainable intensification in African Agriculture. *International Journal of Agricultural Sustainability*, 9 (1), pp. 5–24.

Rashid, M.H., Mulugetta, M. and Wilfred, M. (2001) *Maize Breeding Research in Eastern and Southern Africa: Current status and impacts of past investments made by the public and private sectors 1966–97*. Mexico, D.F: CIMMYT.

Ricker-Gilbert, J., Mason, N.M., Darko, F.A. and Tembo, S.T. (2013) What are the effects of input subsidy programs on maize prices? Evidence from Malawi and Zambia. *Agricultural Economics*, 44 (6,) pp. 671–686.

Rurinda, J., Mapfumo, P., van Wijk, M.T., Mtambanengwe, F., Rufino, M.C., Chikowo, R. and Giller, K.E. (2014) Sources of vulnerability to a variable and changing climate among smallholder households in Zimbabwe: A participatory analysis. *Climate Risk Management*, 3, pp. 65–78.

Schlecht, E., Buerkert, A., Tielkes, E. and Bationo, A. (2006) A critical analysis of challenges and opportunities for soil fertility restoration in Sudano–Sahelian West Africa. *Nutrient Cycling in Agroecosystems*, 76, pp. 109–136.

Setimela, P.S., Chitalu, Z., Jonazi, J., Mambo, A., Hodson, D. and Bänziger, M. (2005) Environmental classification of maize testing sites in the SADC region and its implication for collaborative maize breeding strategies in the subcontinent. *Euphytica*, 145, pp. 123–132.

Setimela, P.S., Magorokosho, C., Lunduka,R., Gasura,E., Makumbi, D., Amsal Tarekegne, A.,. Cairns, J.E., Ndhlela,T., Erenstein, O. and Mwangi, W. (2017) On-farm yield gains with stress-tolerant maize in Eastern and Southern Africa. *Agronomy Journal*, 109 (2), pp. 406–417.

Shiferaw, B., Hellin, J. and Muricho, G. (2011) Improving market access and agricultural productivity growth in Africa: What role for producer organizations and collective action institutions? *Food Security*, 3(4), pp. 475–489.

Snapp, S.S., Mafongoya, P.L. and Waddington, S. (1998) Organic matter technologies for integrated nutrient management in smallholder cropping systems of southern Africa. *Agriculture, Ecosystems and the Environment*, 71, pp. 185–200.

Sperling, L. and McGuire, S.J. (2010) Persistent myths about emergency seed aid. *Food Policy*, 35, pp. 195–201.

Sperling, L. and McGuire, S.J. (2014) Persistent myths about emergency seed aid. *Food Policy*, 35, pp. 195–201.

Steward, P.R., Dougill, A.J., Thierfelder, C., Pittelkow, C.M., Stringer, L.C., Kudzala, M. and Shackelford, G.E. (2017) Agriculture, ecosystems and environment: The adaptive capacity of maize-based conservation agriculture systems to climate stress in tropical and subtropical environments : A meta-regression of yields. *Agriculture, Ecosystems and the Environment*, 251, pp. 194–202.

Tchale, H. (2009) The efficiency of smallholder agriculture in Malawi. *African Journal of Agriculture and Resource Economics*, 3 (2), pp. 101–121.

Thierfelder., C. and Wall, P.C. (2009) Effects of conservation agriculture techniques on infiltration and soil water content in Zambia and Zimbabwe. *Soil and Tillage Research*, 105 (2), pp. 217–227.

Thierfelder, C., Matemba-Mutasa, R., Bunderson, W.T., Mutenje, M., Nyagumbo, I. and Mupangwa, W. (2016) Evaluating manual conservation agriculture systems in southern Africa. *Agriculture, Ecosystems and the Environment*, 222, pp. 112–124.

Thierfelder, C., Chivenge, P., Mupangwa, W., Rosenstock, T.S., Lamanna, C. and Eyre, J.X. (2017) How climate-smart is conservation agriculture (CA)? – its potential to deliver on adaptation, mitigation and productivity on smallholder farms in southern Africa. *Food Security*, 9, pp. 537–560.

Van Ittersum, M.K., Cassman, K.G., Grassini, P., Wolf, J., Tittonell, P. and Hochman, Z. (2013) Yield gap analysis with local to global relevance – A review. *Field Crops Research*, 143, pp. 4–17.

Vanlauwe, B, Kihara, J. Chivenge, Pypers, P., Coe, R. and Six, J. (2011) Agronomic use efficiency of N fertilizer in maize-based systems in sub-Saharan Africa within the context of integrated soil fertility management. *Plant Soil*, 339, pp. 35–50.

Vogel, H. (1995) The need for integrated weed management systems in smallholder conservation farming in Zimbabwe. *Journal of Agriculture in the Tropics and Subtropics*, 96 (1), pp. 35–56.

Waddington, S.R., Mekuria, M., Siziba, S. and Karigwindi, J. (2007) Long-term yield sustainability and financial returns from grain legume-maize intercrops on a sandy soil in subhumid north Central Zimbabwe. *Experimental Agriculture*, 43 (4), pp. 489–503.

Ward, J.M.J., Laing, M.D. and Nowell, D.C. (1997) Chemical control of maize grey leaf spot. *Crop Protection*, 16, pp. 265–271.

Wilson, R.T and Lewis, J. (2015) The maize value chain in Tanzania: A report from Southern Highlands food systems programme. Available at: www.saiia.org.za/value-chains-in-southern-africa/1055-008-tanzania-maize/file (accessed 12 October 2017).

4 The role of wheat in global food security

Maricelis Acevedo, Jason D. Zurn, Gemma Molero, Pawan Singh, Xinyao He, Meriem Aoun, Philomin Juliana, Harold Bockleman, Mike Bonman, Mahmoud El-Sohl, Ahmed Amri, Ronnie Coffman and Linda McCandless

Introduction

Wheat is the most widely cultivated cereal in the world, a staple food for 40 percent of the world's population that contributes 20 percent of total dietary calories and proteins worldwide (Braun *et al.*, 2010). Wheat demand has doubled since the 1980s, with most of the demand coming from developing countries, which harvest 50 percent of global wheat production annually. Consequently, wheat is at the epicenter of global food security. Tremendous gains in wheat productivity occurred in the 1960s and '70s as a direct result of the high-yielding, short-stature, disease-resistant and fertilizer-responsive wheat lines that were developed and distributed during the Green Revolution, led by Dr. Norman Borlaug that saved millions of people from starvation. Global wheat production during the first decade following the Green Revolution increased by over 4 percent with increases of over 8 percent in South Asia, East Asia, Mexico and Central America. Grain production needs to double to feed a world population that is estimated to reach approximately 9 billion by 2050. Challenges imposed by climate change and the related abiotic (heat, drought, salinity) and biotic (pathogens, insects, weeds) stresses means that substantial efforts are needed to incorporate resistance or tolerance to abiotic and biotic stresses as well as improve gains in productivity and quality to ensure the food security of the fast-growing world population (Rosegrant *et al.*, 1995; Solh, 2016). To be able to meet the predicted demands for wheat in the near future, farmers across the globe—and especially smallholder farmers in the developing world—are in need of a second Green Revolution. Scientists at national and international research institutions are exploring new technologies to increase the rate of genetic gain in wheat by reducing yield losses caused by biotic and abiotic stresses and mining genetic variability and diversity that can lead to better adaptation under current and future climatic conditions.

Evolution of wheat and early breeding efforts

The creation and spread of cultivated wheat

The modern cultivated wheat species durum, *Triticum turgidum* subsp *durum* (2n = 4x = 28), and common wheat, *T. aestivum* (2n = 6x = 42), formed through a series of amphiploidization and domestication events over the course of 2 million years (Dubcovsky and Dvorak, 2007; Dvorak *et al.* 2012; Salamini *et al.*, 2002). The diploid A and B genome donors were the first to hybridize creating the allotetraploid, wild emmer wheat (Dubcovsky and Dvorak, 2007; Dvorak *et al.*, 2012; Salamini *et al.*, 2002). Wild emmer was domesticated by selecting individuals with higher yields, uniform maturation, and a non-brittle rachis (Dubcovsky and Dvorak, 2007; Salamini *et al.*, 2002). Free-threshing wheat, where the glume was easily removed, was created as domestication of emmer continued, resulting in the creation of durum (Dubcovsky and Dvorak, 2007; Salamini *et al.*, 2002). Cultivated tetraploids were eventually spread northward where a hulled allohexaploid *T. aestivum* progenitor is thought to have arisen through an amphiploidization event between a durum progenitor and the D genome donor, *A. tauschii* (Figure 4.1; Dubcovsky and Dvorak, 2007; Dvorak *et al.*, 2012; Salamini *et al.*, 2002). Further domestication would occur to remove the hulled phenotype inherited from *A. tauschii*, resulting in common wheat (Dvorak *et al.*, 2012). Over time, cultivated wheat spread from the Middle East to Europe, Africa, and Eastern Asia (Figure 4.1;

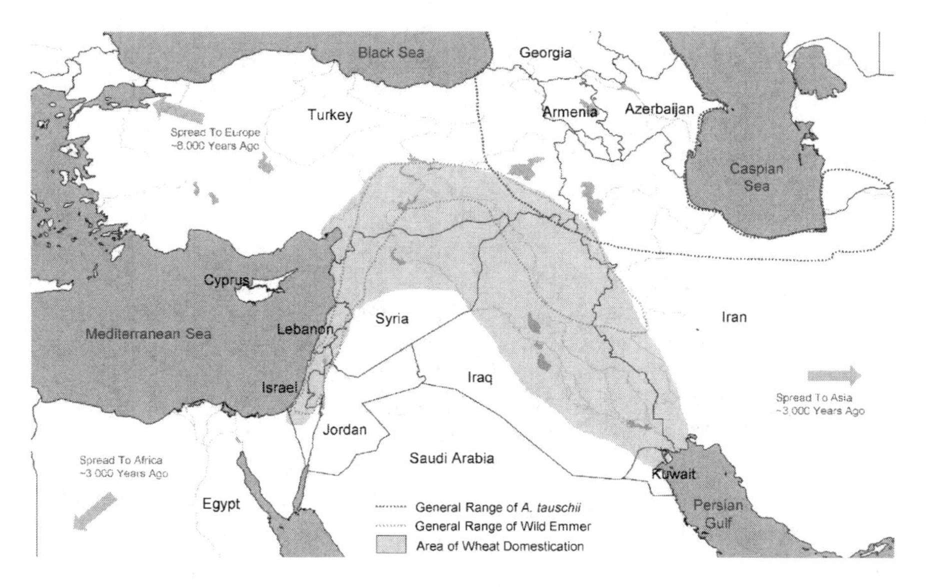

Figure 4.1 Area of wheat domestication, approximate range of *A. tauschii*, approximate range of wild emmer, and routes of dissemination

Source: Adapted from Dubcovsky and Dvorak, 2007; and Salamini *et al.*, 2002; Shewry, 2009.

Shewry, 2009). European colonists would later introduce cultivated wheat into North America and Australia, making wheat a globally produced crop (Olmstead and Rhode, 2011; Shewry, 2009).

Early wheat breeding efforts

Prior to the twentieth century, little was done to improve wheat through breeding. The wheat of the time consisted of landraces uniquely adapted to a region through selection imposed by the environment and the occasional astute farmer. A concerted effort toward creating improved varieties would not occur until the rediscovery of Mendel's laws near the beginning of the twentieth century (Lupton, 1987; Percival, 1921).

Early breeding efforts near the beginning of the twentieth century in Europe consisted of crossing wheat variety "Squarehead" with locally adapted, tall landraces (Lupton, 1987). "Squarehead" was known for having both a high yield and straw strength, which was important to prevent lodging (Lupton, 1987). By the mid-1920s, many European countries had established public breeding programs focused on the needs of farmers in the region. Private breeding programs became more prominent in many European countries in the mid-1930s (Lupton, 1987).

In North America, numerous challenges faced breeders as settlers expanded into the Great Plains. The varieties cultivated during the 1860s were adapted to Western Europe and the eastern coast of North America and were ill-suited for the Great Plains (Olmstead and Rhode, 2011; Paulsen and Shroyer, 2008). Eastern European germplasm was vital to creating wheat varieties that could survive in the Great Plains (Olmstead and Rhode, 2011; Paulsen and Shroyer, 2008). Within two decades, wheat production in North America shifted from the East Coast to the Great Plains (Olmstead and Rhode, 2011; Paulsen and Shroyer, 2008). Many of the new varieties were tolerant of the environment in the Great Plains, matured early, and were of good quality. As such, improved yield and disease resistance would become the driving forces for breeding efforts in North America (Lupton, 1987; Paulsen and Shroyer, 2008).

Wheat breeding efforts did not begin until the 1930s in eastern Asia (He *et al.*, 2001; Hoshino and Seko, 1996). In China, most wheat grown prior to the 1950s consisted of early-maturing, low-yielding landraces that were susceptible to diseases and prone to shattering and lodging (He *et al.*, 2001). These landraces were crossed with germplasm from the U.S. and Italy to improve many qualities that were lacking. Due to China's rising population, there was a strong focus on improving yield (He *et al.*, 2001). In Japan, wheat breeding was conducted at numerous research stations that focused on different traits for improvement depending on environmental needs (Hoshino and Seko, 1996; Nonaka 1983). Since the founding of the breeding program in Japan, a strong focus has been placed on breeding for cultivars that mature early, have high yields, and strong short straw (Hoshino and Seko, 1996; Nonaka, 1983). This focus on short straw has greatly influenced modern wheat production worldwide.

The short-strawed Japanese varieties were one-half to two-thirds the height of ordinary varieties, resistant to lodging, and produced more tillers resulting in a higher yield potential (Reitz and Salmon, 1968). In early breeding efforts, two short-strawed Japanese cultivars, "Aka Komughi" and "Norin 10," were of great importance (Lupton, 1987; Nonaka, 1983; Reitz and Salmon, 1968). "Aka Komughi" was exported to Italy in the 1930s and used to create numerous short-strawed European varieties that were grown throughout the Mediterranean, Eastern Europe, and South America (Lupton, 1987). "Norin 10" is an improved variety derived from the short-statured Japanese variety "Shiro Daruma" and the U.S. varieties "Glassy Fultz" and "Turkey Red" (Reitz and Salmon, 1968). "Norin 10" was noticed by the U.S. agricultural advisor, Samuel C. Salmon, following World War II and exported to the U.S. in 1946 (Reitz and Salmon, 1968). Orville Vogel, a USDA wheat breeder, used "Norin 10" in his breeding program and distributed semi-dwarf varieties throughout North America. Eventually, Norman Borlaug would use progeny from Vogel's breeding program to produce semi-dwarf varieties that was later be exported to India and Pakistan during the Green Revolution (Olmstead and Rhode, 2011; Reitz and Salmon, 1968).

Norman Borlaug and the Green Revolution

Norman Borlaug is one of the most celebrated agriculturalists of the twentieth century for his work as a scientist and humanitarian. Borlaug was hired as a wheat breeder in 1944 by the Rockefeller Foundation to alleviate Mexico's annual wheat production shortage (Olmstead and Rhode, 2011; Ortiz *et al.*, 2007; Vietmeyer, 2011). During his tenure in Mexico, Borlaug pioneered the shuttle breeding technique. Using this technique, scientists grow successive generations of plants in two locations—one location during the summer and the other during the winter. This process was repeated, selecting individuals over time, resulting in new varieties (Olmstead and Rhode, 2011). The shuttle breeding technique reduced breeding time by half while creating varieties adapted to multiple environments. Borlaug's early cultivars matured early and were disease resistant, however, they were tall and prone to lodging (Ortiz *et al.*, 2007; Vietmeyer, 2011). To solve the lodging problem, progeny from a "Norin 10" by "Brevor" cross was used in Borlaug's breeding program to produce semi-dwarf lines with strong straw and high yield potential (Olmstead and Rhode, 2011; Ortiz *et al.*, 2007; Reitz and Salmon, 1968; Vietmeyer, 2011). The resulting cultivars were catalysts for the Green Revolution and transformed Mexico into a net wheat exporter by the 1960s (Olmstead and Rhode, 2011).

The Green Revolution was an amazing feat of research capacity and wheat improvement in Southwestern Asia. The Green Revolution began when Borlaug visited wheat-breeding programs in North Africa and Western Asia. Most of these programs were ill-equipped and ill-staffed to develop varieties that met the needs of the farmers in their regions (Ortiz *et al.*, 2007; Vietmeyer, 2011). To improve breeding programs in these countries, Borlaug proposed the formation of an international training center in Mexico. The program started in 1961 and hundreds of researchers were trained by 1970 (Ortiz *et al.*, 2007; Vietmeyer, 2011). These researchers brought their newly gained experience and seeds from the Mexican breeding program back to their home countries. By the late 1960s, wheat yields had doubled in Pakistan and India, and new varieties spread to nearby countries (Ortiz *et al.*, 2007; Vietmeyer, 2011).

Borlaug was awarded the Nobel Peace Prize in 1970 for his wheat research and efforts to bring improved varieties to the developing world. During his Nobel lecture, Borlaug warned that many areas of the world remained untouched by the Green Revolution and new solutions would be needed to improve food security (Ortiz *et al.*, 2007; Vietmeyer, 2011). In the mid-1980s, Borlaug turned his attention to sub-Saharan Africa in an attempt at a second Green Revolution (Ortiz *et al.*, 2007). To Borlaug's dismay, the lack of infrastructure in Africa at the time prevented a second revolution from occurring. In his later years, Borlaug helped facilitate countless international research projects, one of the most notable being the Borlaug Global Rust Initiative, formed to address the emergence of a new race of stem rust, TTKSK, also known as Ug99, that came out of East Africa. Borlaug was a strong advocate for using new agricultural technology to improve food security (Ortiz *et al.*, 2007; Vietmeyer, 2011). The incorporation of new technologies in conjunction with systems Borlaug pioneered may inspire a second Green Revolution to combat future food security challenges.

Wheat genetic resources: status and prospects for efficient conservation and use of wheat genetic resources

Genetic resources are fundamental building blocks to global wheat production, past, present, and future. These resources represent a unique source of genetic diversity vital to increasing wheat yield potential and yield stability, providing sources of resistance to new pathogens and pathogen races, and tolerance to abiotic stresses (Skovmand *et al.*, 2002). Looking to the future, wheat accessions preserved in germplasm banks will be even more important as plant breeders seek new sources of diversity to keep pace with changes in the world's climate. Modern high-yielding wheat cultivars were produced by plant breeders using systematic methods to pyramid genes for adaptation to specific production environments, generally using cultivars and other advanced breeding lines from their regions. Introgression of additional variation found in genetic resources is necessary to increase yield and yield stability and to further improve other important traits of modern wheat cultivars.

Domestication and subsequent crop improvement resulted in genetic bottlenecks that have reduced the levels of genetic diversity in modern crop cultivars (Prada, 2009). In view of such threats of constant genetic erosion, gene banks have been established and maintained to preserve the wide array of wheat genetic resources (Figure 4.2). Worldwide, there are 1750 gene banks with total holdings of more than 7.5 million accessions. Of those, 470 gene banks contribute to conservation of wheat genetic resources with more than 900,000 accessions including duplicates (FAO-WIEWS, 2007; FAO, 2012). These resources include wild plant species related to the wild ancestors from which modern crops were derived, the wild ancestor species, landraces developed by farmers, and modern cultivars and breeding lines produced through scientific breeding. A better understanding of the genetic origin of cultivated wheat is important for understanding the genetic diversity currently available in wheat's primary and secondary gene pools and for

Figure 4.2 Diversity of domesticated wheat head morphology: (Clockwise from upper left) Examples of domesticated wheat from diploid to hexaploid: Einkorn *(T. monococcum ssp. monococcum*); Emmer (*T. turgidum ssp. dicoccon*); Durum (*T. turgidum ssp. durum*); Common Bread Wheat (*T. aestivum ssp. aestivum*); Spelt (*T. aestivum ssp. spelta*); and Macha (*T. aestivum ssp. macha*)

Source: Photos provided by Harold Bockleman.

assessing the potential to introgress new sources of disease and insect resistance and abiotic stress tolerance into new cultivars (Smale, 1997).

Although wheat collections are among the most comprehensive in the world, there are still gaps that need to be filled. The major geographical areas rich in wheat wild relatives diversity and species richness remains the Fertile Crescent, Central Asia and Caucasus, and around the Mediterranean basin. Gap analysis using DIVA-GIS (DIVA 2006, www.diva-gis.org) showed the need for further collecting of landraces in Oasis and mountainous areas. Even though landraces are still being used within some traditional farming systems, their number and acreage is decreasing and on-farm conservation should be implemented. For *in situ* conservation of wild relatives, there are only four protected areas that target wheat wild relatives. This type of conservation should be promoted more to complement the *ex situ* conservation efforts.

Use of wheat genetic resources

The value of genetic resources resides in using them to supply valuable genes for breeding efforts, but they can also be used to rehabilitate degraded farming systems. Distribution of accessions is a key gene bank activity to ensure the continuum between conservation and use of genetic resources. Wheat is among the few crops where high numbers of genetic resources are extensively used. Crosses between winter and spring varieties adopted by CIMMYT and ICARDA breeding programs allowed national programs to release high yielding varieties in several countries (Tadesse *et al.*, 2015).

Future work within germplasm banks must go beyond the baseline activities of collection, maintenance, distribution, and evaluation of accessions. These conventional activities will remain vital, but will no longer be sufficient to keep collections relevant to plant breeding in the twenty-first century. New approaches are needed to add value to collections through innovative research and the application of technology. With continued support, germplasm collections will continue to contribute to efforts to improve wheat to meet the daunting challenge of feeding a world of 9 billion humans whose population continues to grow.

Abiotic constraints to wheat production

Plants in agricultural systems frequently encounter abiotic stresses caused by temperature, water availability, high salt levels, as well as mineral deficiency and toxicity diminishing productivity of cereal crops worldwide. It is estimated that approximately 70 percent of yield reduction worldwide is the direct result of abiotic environmental stresses (Acquaah, 2012). Climate change exacerbates the frequency and severity of many abiotic stresses, particularly drought and high temperatures. The Intergovernmental Panel on Climate Change projects that global mean temperatures may rise between

1.5° to 4.5°C or more by 2050 with changes in precipitation patterns—that will increase water shortage—together with higher frequency of extreme weather events (IPCC, 2013). Impacts from recent climate-related extremes, such as heat waves, droughts, floods, among others, have increased over the last few years (Christidis *et al.*, 2015; Donat *et al.*, 2013; Gourdji *et al.*, 2013; Hanley and Caballero, 2012) with major negative impacts on agricultural production in broad regions of the world (Lobell and Gourdji, 2012; Zampieri *et al.*, 2017). Experts predict a reduction in wheat yields by 6 percent for each 1°C rise in temperature (Asseng *et al.*, 2014), though this figure could be much higher if accompanied by water shortages. Salinity is expected to increase as a consequence of drought as well as land degradation multiplying the negative impacts on wheat production. In addition, unexpected climatic events (i.e. heat waves) can reduce wheat productivity by ~0.2 t/ha for each day with maximum temperature above 34°C (Asseng *et al.*, 2011).

Future climatic and environmental changes emphasize the need for breeding strategies that deliver both substantial increases in yield potential and resilience to abiotic constraints including extreme weather events such as heat waves, late frost, and drought. Therefore, adaptation of wheat germplasm to current and future climate scenarios is vital to ensuring food security worldwide (Hubert *et al.*, 2010; Tester and Langridge, 2010). On the positive side, dramatic increases in wheat yields can be expected from innovative research that builds on more than five decades of investments in plant biotechnology, physiology, and genetics.

Effects of abiotic stresses on wheat production

Abiotic stresses reduce wheat growth by upsetting various physiological, biochemical, and molecular processes, thereby adversely affecting plant growth, productivity, and, ultimately, reducing yield. Heat, drought, and salinity are the main abiotic stresses that induce severe cellular damage in wheat plants. These environmental constraints require plants to continuously adapt and exercise specific tolerance mechanisms to cope with these events to maintain growth and productivity. Plant resistance to abiotic stresses can be grouped into (i) escape (e.g., short life cycle), (ii) avoidance (i.e., maintenance of good water status through stomatal closure and senescence of older leaves), or (iii) tolerance mechanisms (i.e., ability of plants to function under stress). Escape and avoidance mechanisms occur at the whole plant level and are associated with physiological whole-plant responses while tolerance mechanisms are generally those that occur at the cellular and metabolic level (Levitt, 1972). Stress episodes can be temporally variable, ranging from transient stress (i.e., midday high temperatures) to chronic stress (i.e., high Na^+ in sodic soils). Timing of stress relative to the time of the day and the developmental stage determines the impact on viability and yield (Mickelbart *et al.*, 2015). Transient stress during early vegetative stages slows growth (cell division and expansion) but may not greatly reduce the yield, whereas stress during reproductive development can considerably diminish productivity (Talukder *et al.*, 2014).

Drought

Water scarcity causes imbalances in osmotic and ionic homeostasis in the plant, loss of cell turgidity, and damage to functional and structural cellular proteins and membranes. As a consequence, water-stressed plants lose photosynthetic capacity and nutrient uptake reducing nutrient supply into the appropriate plant organs, thus resulting in yield reductions. Breeding for mechanisms related with drought resistance depends strongly on the drought duration, severity, and time of occurrence (van Ginkel *et al.*, 1998). Usually, drought escaping mechanisms (by shortened crop cycle) or early vigor can be effective mechanisms in environments with terminal drought, while avoidance (i.e., deeper root systems, leaf area reduction, stomatal closure, cuticular wax, pubescence and glaucousness) can help in environments with decreasing rainfall over the vegetation period (Blum, 2009). Tolerance mechanisms (i.e., accumulation of compatible solutes or stem carbohydrates pre-anthesis) are relevant for continuous and high intensity stress conditions that are found in residual moisture environments or soils with low water storage capacity (van Ginkel *et al.*, 1998). Different selection strategies may be required for different drought regimes.

Heat

Heat stress is an agricultural problem in many areas in the world, and heat stress around sensitive stages of wheat development has been identified as a possible threat to wheat production in Europe (Stratonovitch and Semenov, 2015). Upper threshold temperature for wheat during grain filling period is 26°C where even a short period (5–6 days) of exposure to temperatures of 28–32°C can result in significant decreases in yield of 20 percent or more (Stone and Nicolas, 1994). However, in many wheat production areas in developing countries, the temperature during the grain filling period tends to be much higher. Temperature accelerates developmental and growth rates in plants affecting final grain yield and quality of wheat. Wheat is particularly sensitive to hot temperatures during the reproductive stage (Prasad and Djanaguiraman, 2014) where warmer temperatures typically are associated with a reduction in leaf area index and green area duration.

Many of the traits associated with tolerance to high temperature conditions are associated with reproductive stages, and include variation in the time of day of flowering, maintenance of pollen viability, and variation in the ability of ovaries and grains to continue to grow under high temperature conditions (Chapman *et al.*, 2012). Lines with improved yield under terminal heat stress (Lopes and Reynolds, 2012) and with specific heat adaptive traits (chlorophyll retention, early ground cover and high biomass) have been successfully developed (Pask *et al.*, 2014).

Salinity

Increased salinization of arable land is expected to have considerable global effects, resulting in more than 50 percent land loss by 2050 (Wang *et al.*, 2003). In wheat, salt tolerance inhibits plant growth associated with two main factors (Munns *et al.*, 2006). As a first instance, the presence of salt in the soil reduces the ability of the plant to take up water, producing the osmotic or water-deficit effect of salinity. In addition, salt could enter in the transpiration stream, injuring transpirative tissues due to the excess of ions. Effects of salinity in the plant depend on the timing and severity of the salt treatment (Francois *et al.*, 1994). Low concentrations of salt in the soil may not reduce grain yield even though the leaf area and shoot biomass are reduced, which is reflected in a harvest index that could even increase with salinity (Husain *et al.*, 2004). As described for other abiotic stresses, early vigour, transpiration efficiency (useful in soils with transient salinity but not in dryland or irrigation salinity), early flowering, and rapid grain filling are useful traits for selection of salt tolerant plants (Colmer *et al.*, 2005). Despite the extensive work invested in screening materials for identifying salt-tolerant wheat cultivars, the development of new lines through breeding as a consequence of the screening work is scarce, especially in the case of salt tolerance (Munns *et al.*, 2006).

Biotic constraints to wheat production

Wheat diseases are a constant threat to wheat production. Wheat is subject to a number of diseases and pests, of which about 50 are economically important (Weiss 1987). Moreover, potential yield losses due to wheat diseases have been estimated at 18 percent and actual losses with current disease control strategies at 13 percent (Oerke, 2006). Improved disease control can mitigate many of the yield losses caused by the most destructive wheat diseases, most of which are caused by fungi. Many of the most destructive wheat diseases can be found globally and are easily disseminated to different production regions via wind currents (Bolton *et al.*, 2008; Chen, 2005; Dubin and Duveiller, 2011; Kohli *et al.*, 2011; Singh *et al.*, 2011, 2016). As such, if we hope to have an abundant and secure wheat crop by 2050, it is imperative to understand the numerous biological aspects of the host and pathogen as well as the effects that the environment can have on disease progress.

Rusts

Leaf rust is caused by *Puccinia triticina* Erikss. (*Pt*), stripe rust by *Puccinia striiformis* Westend. f.sp. *tritici* Erikss. (*Pst*), and stem rust by *Puccinia graminis* f.sp. *tritici* Erikss. and E. Henn (*Pgt*; Kolmer, 2005). These three pathogens can have large economic impacts due to their wide distribution, fast mutation rate, potential to cause epidemics, and rapid spread. Of the three rusts, leaf (brown) rust is the most commonly distributed and continues to threaten wheat production in many countries (Bolton *et al.*, 2008; Knott, 2012). Estimates of 5–15 percent or even higher yield losses of susceptible cultivars have been reported (Roelfs *et al.*, 1992). This proportion of losses depends mainly on time of disease onset, the genotype of the host, and the environmental conditions (Chu *et al.*, 2009). Over 50 virulent phenotypes of *Pt* are detected in North America annually (Hughes and Kolmer, 2016; Kolmer, 2013; McCallum *et al.*, 2007). Since the alternate hosts in many wheat-producing regions are absent or resistant to leaf rust, *Pt* reproduces mostly asexually. *Pt* populations on hexaploid wheat are more diverse for virulence and at the molecular level than those collected from durum wheat (Goyeau *et al.*, 2011; Kolmer and Acevedo, 2016; Ordoñez and Kolmer, 2007).

Stem (black) rust, is the most destructive of the wheat rusts and has been the culprit behind many crop failures and famines in many regions of the world throughout history. *Pgt* attacks common wheat, durum wheat, barley, Triticale, and wild wheat relatives (Singh *et al.*, 2011). Severe stem rust infections can result in up to 100 percent yield losses (Roelfs, 1985). Historically, stem rust has been found in almost all wheat producing zones (Saari and Prescott, 1985). In the U.S., the stem rust epidemics of 1916 resulted in reduction of wheat production by approximately 60 percent, mainly in the states of Minnesota and the Dakotas (Peterson, 2013). From the 1950s, this disease was effectively managed through deployment of resistant cultivars. The eradication of the alternate host barberry in the U.S. from 1918 to 1977 was another key factor that helped to reduce the virulence diversity in *Pgt* population (Roelfs, 1982). However, the emergence of race TTKSK (Ug99) in Uganda in 1998 with virulence to many stem rust resistance genes, including the widely deployed *Sr31*, has raised alarm around the world (Jin *et al.*, 2007; Pretorius *et al.*, 2000). Since 1998, numerous, highly virulent *Pgt* races have been identified in East and North Africa, the Middle East, Central Asia, and Europe.

Stripe rust, also called yellow rust, causes significant yield losses to wheat production around the world. In susceptible cultivars, stripe rust can cause up to 100 percent yield loss, especially when disease onset starts at seedling stage (Chen, 2005). Even though *Pst* generally thrives at low temperatures, recent devastating epidemics have occurred in warmer areas (Hovmøller *et al.*, 2010; Mboup *et al.*, 2009). For instance, in the U.S., stripe rust used to be a problem only in the Pacific Northwest and California where the environmental conditions are conducive for disease development. However, stripe rust has now become a serious problem throughout the U.S., due to the appearance of new virulent races that are better adapted to warm temperatures (Chen, 2005; Chen *et al.*, 2010). Similarly, in recent years, the number of regional outbreaks of yellow rust has increased in East and North Africa (Ezzahri *et al.*, 2009; Singh *et al.*, 2016) as well as in Central and West Asia (El Amil, 2015; Rahmatov *et al.*, 2016).

Fusarium head blight

Fusarium head blight (FHB) or head scab, caused mainly by *Fusarium graminearum*, has been a major threat for wheat production in China, Japan, Brazil, Argentina, as well as many European countries (Buerstmayr *et al.*, 2012). In the U.S. and Canada, however, it had not been regarded as a major disease until the 1990s when FHB re-emerged with heavy epidemics (McMullen *et al.*, 2012). In the U.S., FHB caused great economic losses that were estimated to be $7.67 billion in wheat and barley between 1993 and 2001 (McMullen *et al.*, 2012). In the great FHB epidemics in China in 2012, approximately 1,640,000 hectares or 68.3 percent of wheat acreage was severely affected in the Yellow and Huai River valley region, the No. 1 wheat production zone in China. Yield reductions in Europe and South America can be as high as 60 percent and 70 percent, respectively (Singh *et al.*, 2016). The disease usually occurs under warm and humid weather conditions, with the most susceptible stage being at anthesis (Singh *et al.*, 2016). In addition to yield loss and quality deterioration, FHB causes mycotoxin contamination of the grains. Deoxynivalenol (DON) is the most frequently encountered mycotoxin, and is toxic to both humans and animals (Buerstmayr *et al.*, 2012).

Wheat blast

Wheat blast (WB), or brusone, is a devastating disease in the tropical parts of South America, including Brazil, Bolivia, Paraguay, and Argentina (Kohli *et al.*, 2011). In 2016, a WB outbreak was first recorded in Bangladesh (Malaker *et al.*, 2016), causing global concern on its further spread to neighboring countries. The disease is caused by the *Triticum* pathotype of *Magnaporthe oryzae* B. C. Couch and is thought to have originated through the loss of functionality of genes that control host specificity of *M. oryzae* isolates that infect *Avena* and *Lolium* species (Inoue *et al.*, 2017). Highest yield losses occur when the fungus infects the rachis at the base of the spike affecting total or partial grain filling depending upon the time of infection (Duveiller *et al.*, 2016). Yield losses of 10–100 percent have been reported, depending on years, genotypes, and planting date (Duveiller *et al.*, 2016).

Tan spot

Tan spot (TS), caused by *Pyrenophora tritici-repentis*, occurs in all major growing areas of the world including North America, South America, Australia, Asia, and Northern Africa (Faris *et al.*, 2013). In Australia, TS is the most important disease of wheat, causing average annual losses of $212 million (Murray and Brennan, 2009). Both quantitative and qualitative resistance to TS of wheat have been reported. The fungus produces at least three HSTs—Ptr ToxA, Ptr ToxB, and Ptr ToxC—that interact directly or indirectly with the products of the dominant host genes *Tsn1, Tsc2*, and *Tsc1*, respectively (Faris *et al.*, 2013).

Septoria nodorum blotch

Septoria nodorum blotch (SNB) causes substantial yield losses in wheat-growing areas in Europe, North America, and Australia. In Australia, SNB causes average annual yield losses of $108 million (Murray and Brennan, 2009). The pathogen produces several HSTs that increase disease severity, and, so far, seven HSTs have been identified and characterized (Gao *et al.*, 2015). The pathogen population shows great diversity differing in ability to produce the HSTs but race specialization still remains inconclusive.

Septoria tritici blotch

Epidemics of Septoria tritici blotch (STB), caused by *Zymoseptoria tritici*, are most severe in areas with extended periods of cool, wet weather, particularly North America (U.S., Canada, Mexico), East Africa (Ethiopia, Kenya), South America (Brazil, Chile, Uruguay, Argentina), with the most damage occurring in Europe, Central and West Asia, and North Africa (Goodwin, 2012). Yield losses of 30–50 percent can be observed during severe epidemics (Eyal *et al.*, 1987). Both qualitative and quantitative resistance to STB have been observed.

Spot blotch

Spot blotch (SB), caused by *Cochliobolus sativus*, is most severe in warmer wheat growing regions including Bangladesh, Nepal, Eastern India, Bolivia, Brazil, Southeast China, Southeast Australia, Northeast Argentina, Paraguay, Zambia, Northern Kazakhstan, and the Great Plains of the U.S. and Canada (Villareal *et al.*, 1995). Heat, drought stresses, and low inputs of nutrients and water lead to increased disease severity; yield losses between 40 and 85 percent have been reported during severe epidemics (Singh *et al.*, 2016).

New technologies to increase the rate of genetic gain in wheat

Integration of innovative cutting-edge technologies in breeding programs to accelerate genetic yield gains and boost the efficiency is critical for sustaining global wheat production. Crop breeding has undergone major changes over the last 30 years thanks to the utilization of molecular biology techniques in crop improvement. Challenges such as food insecurity, climate change, biotic and abiotic stresses, have demanded the incorporation of technologies such as marker assisted selection (MAS), high-throughput phenotyping, and genomics-assisted breeding.

Marker assisted selection (MAS)

MAS uses DNA sequence variation in the genome to track specific genetic regions of interest; these regions can be then selected or bred for or against it. MAS is most useful when the traits of interest are of low heritability, are recessively inherited, require costly and complex phenotyping, and where pyramiding multiple genomic locations is necessary to achieve an acceptable phenotype (Tester and Langridge, 2010). For MAS to be most useful, the genetic marker needs to be tightly linked to the gene(s) controlling the phenotype and should be consistent in predicting phenotype. Phenotyping prediction is a challenge when dealing with complex and environmentally variable traits. In wheat, MAS has been effectively used to identify, characterize, clone, and pyramid genes for disease resistance, insect resistance, quality and yield, abiotic stresses, and agronomic traits (http://maswheat.ucdavis.edu/protocols/index.htm).

Genomic selection

Over the last few years, tremendous increases in genetic gains for economically important traits in dairy cattle have been achieved by implementing genomic selection (GS), which uses dense genome-wide marker information to predict the breeding value of individuals prior to phenotyping (Hayes *et al.*, 2013; Meuwissen *et al.*, 2001). With GS, a "training population" (TP) comprising individuals that have been genotyped and phenotyped is used to calibrate a prediction model, which is then used to estimate the breeding value of individuals that have only been genotyped, referred to as "selection candidates" (SC) or "validation population". These genomic-estimated breeding values (GEBVs) enable the breeder to make selections simultaneously on several desired superior alleles. This increases genetic gains, selection intensity, and the proportion of top performers in the breeding population, while eliminating poor-performing progenies before the next generation of costly field testing, thereby reducing cost, time, labor, and resources (Heffner *et al.*, 2009; Jannink *et al.*, 2010; Lorenz *et al.*, 2011).

Although GS is a promising strategy to increase the rate of genetic gain in wheat, several factors such as the choice of prediction models, genetic architecture. and heritability of the trait, marker type, and density, linkage disequilibrium between the causal loci and the markers, size of the training population, relatedness between the individuals in the training and validation populations, genotype x environment interactions, have to be carefully considered for successful implementation. (Daetwyler *et al.*, 2010a, 2010b; Howard *et al.*, 2014; Jarqu'n *et al.*, 2014; Lorenzana and Bernardo, 2009).

Genomic prediction in wheat

Several studies explored different models and marker platforms for genomic prediction of various traits in wheat (Table 4.1). Although accuracies from these studies are promising, practical implementation and comparison of genetic gains with phenotypic selection are still lacking (except for Rutkoski *et al.*, 2015).

High throughput phenotyping (HTP)

Recently, advances in high throughput genotyping complemented by high throughput phenotyping strategies have become an exciting area of wheat research. While current approaches for field-based phenotyping of a large number of breeding lines require a lot of time, cost, and labor, HTP using unmanned aerial systems (UAS) has the potential to provide rapid, precise and high-resolution phenotype measurements, thereby boosting genetic gains. Several imaging systems such as Red Green Blue, hyper-spectral, multi-spectral, and thermal cameras can be used on these UAS platforms (Haghighattalab *et al.*, 2016). Images processed from these systems can be used to derive vegetation indices (VI) like canopy temperature (CT) and normalized difference vegetation index (NDVI), that are known to be correlated with components of grain yield (Reynolds and Langridge, 2016). Using correlated or secondary traits to improve genomic prediction accuracies for low-heritability traits such as grain yield has been suggested; HTP platforms can be an excellent resource providing measurements of such traits (Jia and Jannink, 2012; Mackay *et al.*, 2015). It has been reported that incorporating secondary traits within environments increased accuracies for grain yield by 70 percent in genomic prediction models (Rutkoski *et al.*, 2016; Sun *et al.*, 2017).

Despite promising results obtained from MAS, GS, and HTP, an ongoing challenge is to effectively implement these technologies in wheat breeding programs and also improve the accuracy of predictions, which is directly proportional to the rate of genetic gain. A cost–benefit approach is required to compare conventional breeding strategies with these advanced technologies; greater impact can be achieved only by successful deployment and delivery of these technologies. We hope that wheat varieties developed through these advanced technologies incorporating superior allelic combinations can substantially increase wheat productivity, bridge the demand gap, improve livelihoods, and increase the economic gain for smallholders in developing countries.

Policy interventions to support sustainable intensification of wheat production for food security

In evaluating the role of wheat in food security, it is important to consider the policy, societal, and regulatory challenges that must be met to

Table 4.1 Genomic prediction accuracies for different traits in wheat

Traits	*Genomic prediction accuracies*	*Comments*	*Reference*
Flour yield	0.76	Accuracies for multi-family prediction using ridge-regression and DArT markers.	Heffner *et al.* (2011)
Water solvent retention capacity	0.58		
Heading date	0.75		
Height	0.74		
Lactic acid solvent retention capacity	0.75		
Lodging	0.28		
Sodium carbonate solvent retention capacity	0.66		
Pre-harvest sprouting	0.55		
Protein	0.45		
Softness	0.66		
Sucrose solvent retention capacity	0.74		
Test weight	0.56		
Grain yield	0.20		
Yield (irrigated)	0.32	Accuracies using BLUP and GBS markers imputed with the multivariate-normal expectation maximization method.	Poland *et al.* (2012)
Yield (drought)	0.42		
Heading days	0.35		
Thousand kernel weight	0.33		
Days to heading	0.65	Mean accuracies using RKHS model and DArT markers.	Pérez-Rodríguez *et al.* (2012)
Grain yield	0.67		
Days to heading	0.4	Accuracies within a year using RKHS model and DArT markers.	Rutkoski *et al.* (2012)
Fusarium damaged kernels	0.43		
Incidence, severity and kernel quality index	0.54		
Deoxynivalenol concentration	0.27		
Incidence	0.53		
Severity	0.59		
Stem rust	0.62	Mean accuracies from five bi-parental populations using GBLUP and DArT markers.	Crossa *et al.* (2014)
Stem rust (adult plant resistance)	0.57	Accuracies using GBLUP model and GBS markers.	Rutkoski *et al.*, (2014)

Leaf rust (Adult plant resistance)	0.35	Mean accuracies across years using GBLUP and SNPs from Illumina iSelect 9K bead chip.	Daetwyler *et al.* (2014)
Stem rust (Adult plant resistance)	0.27		
Stripe rust (Adult plant resistance)	0.44		
Heading time	0.4	Accuracies using RR-BLUP and SNPs from Illumina 9K array.	Zhao *et al.* (2014)
Height	0.4		
Fusarium Head Blight – index	0.52	Accuracies using RR-BLUP and the highest GBS marker density.	Arruda *et al.* (2015)
Fusarium damaged kernels	0.82		
Incidence, severity and kernel quality index	0.73		
Deoxynivalenol concentration	0.64		
Thousand kernel weight	0.83	Accuracies using a Gaussian kernel model and GBS markers.	Battenfield *et al.* (2016)
Test weight	0.89		
Grain hardness	0.82		
Flour yield from milling	0.82		
Grain protein	0.91		
Flour protein	0.90		
Flour sodium dodecyl sulfate	0.87		
sedimentation volume	0.86		
Optimum mix time	0.87		
Torque at the integral of midline peak	0.85		
Work value from Alveograph curve	0.77		
Alveograph P (tenacity)/L (extensibility)	0.85		
Loaf volume			
Leaf rust (Seedling)	0.64	Mean accuracies using GBLUP and GBS markers.	Juliana *et al.* (2017a)
Leaf rust (Adult plant resistance)	0.47		
Stem rust (Adult plant resistance)	0.55		
Stripe rust (Seedling)	0.73		
Stripe rust (Adult plant resistance)	0.55		
Septoria tritici blotch (Adult plant	0.45	Mean accuracies using GBLUP and GBS markers.	Juliana *et al.*, (2017b)
resistance)	0.55		
Stagonospora nodorum blotch (Seedling)	0.66		
Tan spot (seedling)	0.48		
Tan spot (APR)			

Source: Adapted from various sources.

increase the benefits of sustainable agriculture intensification in wheat, to maximize scientific inputs and use efficiencies, and to best manage abiotic and biotic stresses in improving yield. Three areas to be addressed in wheat include gender, germplasm exchange, and investment in wheat research and technology. Ignoring these critical control points will increase the constraints faced by smallholders and the challenges of meeting the rising global demand for food.

Gender equality

It is well understood that gender inequalities and a lack of attention to gender in agricultural development contributes to lower productivity, higher levels of poverty and under-nutrition (FAO, 2011; ILRI, 2010; Quisumbing and Maluccio, 2003). Policy initiatives to strengthen the position of women in agriculture should include gender equality policies giving women farmers access to innovations such as improved varieties, including women farmers in participatory variety selection, and improving educational opportunities so that the gender ratio of men/women scientists in wheat might approach 1/1.

Uptake of new technologies and improved crop varieties by men and women is a pertinent concern of global funding organizations. In general, although both men and women benefit from their adoption, according to the research, it is men who tend to benefit more (ILRI, 2010). ILRI, through its Improving Productivity and Market Success (IPMS) of Ethiopian Farmers' Project, maintains that activities must address fundamental imbalances in women's access to inputs and services as well as increase women's opportunities to strengthen their skills and knowledge base. It also emphasizes increasing women's participation in market-oriented agricultural production, strengthening women's decision-making roles, improving well-being and easing workloads, and strengthening opportunities to mainstream gender considerations into operational procedures. Complementary services such as access to credit, literacy classes, and farmer organizations need to be offered—and accepted—by stakeholders (ILRI, 2010).

Creating an enabling environment for women scientists in wheat is a fundamental element in developing the next generation of wheat breeders. To address this challenge, the Borlaug Global Rust Initiative developed the Jeanie Borlaug Laube Women in Triticum Early Career Awards, which, to date, have provided professional development opportunities for 40 women from 22 countries working in wheat during the early stages of their careers. Recipients receive support to attend the BGRI Technical Workshop and are eligible to attend a wheat breeding and pathology training at CIMMYT, in Obregon, Mexico (BGRI, 2016). An ancillary award, the Jeanie Borlaug Laube Mentor Award, recognizes men and women who have proved to be excellent mentors of women working in Triticum and its nearest relatives. Efforts such as the Jeanie Borlaug Laube awards and opportunities for training contribute to building a stronger, better-integrated community of global wheat scientists who collaborate across gender and international boundaries.

Investment in research and technology

In the future, transnational wheat research will require expertise in attracting and leveraging research funding. One strategy to leverage research dollars, suggested by researchers at the International Maize and Wheat Improvement Center (CIMMYT), is to develop a Global Crop Improvement Network (GCIN) that provides access to well-controlled and coordinated "field laboratories," where technologies and data are shared. Testing wheat genotypes at hundreds of field sites in dozens of wheat-producing countries and various agro-ecological environments will allow scientists to test thousands of high-yielding, disease-resistant lines against various temperature, moisture, disease and pest profiles (Reynolds *et al.*, 2017). An example of such a facility is the international stem rust screening platform at the Kenya Agriculture and Livestock Research Organization (KALRO) in Njoro, Kenya, established in 2005. A similar screening platform was established later at the Ethiopian Institute for Agricultural Research (EIAR) research center in Debre Zeit, Ethiopia. These hubs, established with support from the Durable Rust Resistance in Wheat (DRRW) project (www.globalrust.org/page/durable-rust-resistance-wheat), play a vital role in evaluating global wheat germplasm; facilitating pre-breeding, testing, and releasing stem rust resistant varieties; pathogen surveillance; and gene discovery and characterization studies (Singh *et al.*, 2015).

The International Wheat Improvement Network (IWIN) that emerged after the Green Revolution is a baseline from which to evolve and expand transnational field testing and shared research platforms, or GCIN. Well-coordinated multilateral partnerships involving field-based research projects include national agricultural research services as well as other downstream arenas such as radio, video, mobile, and social media communications; private seed multiplication and distribution networks; and extension programs (Reynolds *et al.*, 2017). This would thereby create or repair linkages that currently disrupt the adoption process of improved wheat varieties and new technologies.

It has been the experience of International Programs in the College of Agriculture and Life Sciences at Cornell University in their Durable Rust Resistance in Wheat, and Delivering Genetic Gain in Wheat projects over the last 10 years that the major funding for wheat research comes from private funding rather than core university funding. Philanthropic organizations such as The Bill & Melinda Gates Foundation, the Department for International Development of the UK, Carlos Slim, Warren Buffet (to name a few), and private industries such as Syngenta, Maharashtra Hybrid Seeds Company, Sathguru Ltd., etc., are increasingly willing to fund large multi-institution projects with clear work plans and outputs that are focused on particular problems that affect food security—such as developing new varieties of wheat that are resistant to wheat rust, or generating genetic gains in wheat in the face of multiple biotic and abiotic stresses.

Germplasm and biodiversity exchange

Global food security depends on the free movement and open sharing of plant genetic resources. Norman Borlaug depended on germplasm from Mexico, Japan, China, India and elsewhere to develop the dwarf varieties of wheat and rice that contributed to the Green Revolution. Borlaug's success protected wheat farmers from rust diseases for over 40 years (Coffman *et al.*, 2016).

The Borlaug Global Rust Initiative that Dr. Borlaug helped found in 2005 depends on free germplasm exchange. Increasingly, such exchanges are unduly burdened by the forms and legal documents associated with the multilateral Convention on Biological Diversity (www.cbd.int) and related treaties and conventions, including the International Treaty on Plant Genetic Resources for Food and Agriculture (ITPGRFA), the Nagoya Protocol (www.cbd.int/abs/), and the Access and Benefit-sharing Clearing House (ABS) (www.cbd.int/abs/theabsch.shtml). Increasingly, it is up to the researchers to determine that the germplasm they are deploying has been acquired legally (Coffman *et al.*, 2016).

The Convention on Biological Diversity protects bio-resources from around the world but can impair the breeding and pathogen testing associated with developing new varieties of wheat that will be resistant to biotic and abiotic stresses. Stringent country-specific regulations on the movement of country-specific pathogens such as Ug99 among collection sites and biosafety testing labs in the U.S., Denmark, and elsewhere unnecessarily delay testing, identification, and the resulting response—whether for pathogen identification or the resultant breeding of wheat varieties that include the genes resistant to the particular pathogen (Coffman *et al.*, 2016).

Policies and regulations related to germplasm and intellectual property are uneasy bedfellows when it comes to plant breeding. Related to arguments regarding policies that regulate germplasm exchange are arguments about intellectual property rights (IPRs) that impact the activities of plant breeders involved in wheat. In 1961, plant breeders' rights (PBRs) were established through the Union Internationale pour la Protection des Obtentions Végétales. Similar plant variety protections were established in the U.S. in 1970. In both cases, issues related to plant breeders' rights have been playing out in the courts ever since (Bjørnstad, 2016). Both conventions give the breeder the exclusive right to market a certain variety while keeping its genes available to competing breeders as germplasm for crossing. In this case, germplasm is frequently "leased," not "bought."

Patent and licensing laws in the U.S., EU, India, and elsewhere can be contradictory. Seeds are further regulated by international biodiversity laws that regulate biodiversity rights on a nation-by-nation basis. Signatories to the Convention on Biological Diversity (CBD) instituted "facilitated access" to 35 crops including wheat (Bjørnstad, 2016). Patent holders are in a competitive and often costly business because it can take decades to develop and test new varieties. Legal transaction costs deter small breeders and favor the "Big Six"—Monsanto, Syngenta, DuPont, Bayer, Dow, and BASF. Not surprisingly, universities are increasingly in the game of patenting as a means of funding expensive breeding programs and, in some cases, generating millions of dollars of income.

Gene banks have a dual mission of conservation and use of genetic resources. International germplasm exchange is involved in those transactions and is subject to various conventions. Many of these genetic resources include wheat wild relatives that contain desirable traits related to yield potential and tolerance to abiotic and biotic stresses. These traits are important in generating new wheat varieties adapted to climate change and other as yet unforeseen threats. For global food security in wheat and other crops, the thorny thicket of policies and regulations related to germplasm exchange and intellectual property will have to be negotiated.

Conclusion

In summary, an increase in sustainable wheat production is essential for global food security and reduction of poverty today and, especially, in the future. Estimates of poverty reduction have shown that in low-income countries the impact on poverty reduction as a result of agricultural development is 2.3 to 4.25 times greater than for the equivalent investment in non-agricultural sectors (Christiaensen *et al.*, 2011). Moreover, from 1994 to 2014 the return for each $1 invested into wheat research is $73 to $103 (https://blogs.worldbank.org/voices/global-wheat-breeding-returns-billions-benefits-stable-financing-remains-elusive).

The success of the Green Revolution was in great part the result of large investments in crop research, infrastructure, and market development (Pingali, 2012). To be able to develop wheat varieties for the future, a second green revolution is needed. This second green revolution should focus not only on increases in production and yield, but also take into consideration crop input efficiency, carbon footprint, and malnutrition. Most importantly, it should seek higher rates of variety replacement and aggressive adoption of newly developed, more nutritious, higher yielding, and disease-resistant wheat germplasm by national wheat breeding programs and farmers. Increases in variety adoption and replacement at the country level requires the existence of an honest data-and-fact-driven variety release process where farmers are aware of the benefits and risks of the newly developed germplasm, that seed of new germplasm is available in a timely manner, and industry works in collaboration with the national programs to multiply only the most appropriate wheat varieties. Beyond the technical constraints of production, the new green revolution should also emphasize policy development, capacity building, and gender equity.

Governments, national and international institutions must work together to address wheat production challenges on and off the farm. In addition to lack of access to machinery or proper inputs in developing countries, current wheat production barriers involve market access, food safety, and post-harvest losses. Providing better infrastructure such as roads, storage facilities, crop management information, and training opportunities will increase productivity and income, which can then be then utilized to access other goods and service reducing poverty. Gender awareness and equality are paramount when considering agricultural technologies for on-farm adoption. Women farmers consistently encounter barriers to accessing agricultural resources and technologies (FAO, 2011). By removing these barriers, we can increase agricultural productivity and provide better opportunities for women and their families to get out of poverty.

To withstand current and future challenges to wheat production globally, we must harness the best scientific knowledge and technological breakthroughs while ensuring that the achievements of these advances reach farmers' fields. Future scientific innovations in wheat will require long-term investments in research, human capacity building, and infrastructure. New technologies such as genomic selection, high throughput phenotyping, and genome editing are promising and may play a larger role in wheat breeding in the future as scientists decode the complexity of the wheat genome. With the right policies in place and political and scientific commitment, sustainable intensification of wheat production can alleviate food insecurity and provide a better future for generations to come.

References

Acquaah, G. (ed.) (2012) *Principles of Plant Genetics and Breeding*, 2nd edn. Oxford: Wiley-Blackwell.

Arruda, M.P., Brown, P.J., Lipka, A.E., Krill, A.M., Thurber, C., Kolb, F.L. (2015) Genomic selection for predicting head blight resistance in a wheat breeding program. *Plant Genome*, 8, pp. 1–12.

Asseng, S., Ewert, F., Martre, P., Rötter, R., Lobell, D., Cammarano, D., Kimball, B.A., Ottman, M.J., Wall, G.W., White, J.W., Reynolds, M.P., Alderman, P.D., Prasad, P.V.V., Aggarwal, P.K., Anothai, J., Basso, B., and Biernath, C. (2014) Rising temperatures reduce global wheat production. *Nature Climate Change*, 5, pp. 143–147.

Asseng, S., Foster, I., and Turner, N. (2011) The impact of temperature variability on wheat yields. *Global Change Biology*, 17, pp. 997–1012.

Battenfield, S.D., Guzman, C., Gaynor, R.C., Singh, R.P., Pe–a R.J., Dreisigacker, S., Fritz A.K., Poland, J.A. (2016) Genomic selection for processing and end-use quality traits in the CIMMYT spring bread wheat breeding program. *Plant Genome*, 9, p. 2.

Blum, A. (2009) Effective use of water (EUW) and not water-use efficiency (WUE) is the target of crop yield improvement under drought stress. *Field Crops Research*, 112, pp. 119–123.

BGRI, 2016. About BGRI. www.globalrust.org/about-bgri (accessed September 2017).

Bjørnstad, A. (2016) "Do not privatize the giant's shoulders": Rethinking patents and plant breeding. *Cell Press*, 34, pp. 609–617.

Bolton, M.D., Kolmer, J.A., and Garvin, D.F. (2008) Wheat leaf rust caused by *Puccinia triticina. Molecular Plant Pathology*, 9, pp. 563–575.

Braun, H.J., Atlin, G., and Payne, T. (2010) Multi-location testing as a tool to identify plant response to global climate change. In: Reynolds, M.P. (ed.) *Climate Change and Crop Production*. Wallingford, UK: CABI, pp. 115–138.

Buerstmayr, H., Adam, G., and Lemmens, M. (2012) Resistance to head blight caused by *Fusarium spp.* in wheat. In: Sharma, I. (ed.) *Disease Resistance in Wheat*. Wallingford, UK; Cambridge, MA: CABI, pp. 236–276.

Chapman, S.C., Chakraborty, S., Dreccer, M.F., and Howden, S.M. (2012) Plant adaptation to climate change: Opportunities and priorities in breeding, crop and pasture. *Science*, 63, pp. 251–268.

Chen, X.M. (2005) Epidemiology and control of stripe rust *Puccinia striiformis f. sp. tritici* on wheat. *Canadian Journal Plant Pathology*, 27, pp. 314–337.

Chen, X.M., Penman, L., Wan, A.M., and Cheng, P. (2010) Virulence races of *Puccinia striiformis f. sp. tritici* in 2006 and 2007 and development of wheat stripe rust and distributions, dynamics, and evolutionary relationships of races from 2000 to 2007 in the United States. *Canadian Journal of Plant Pathology*, 32, pp. 315–333.

Christiaensen, L., Demery, L., and Kuhl, J. (2011) The (evolving) role of agriculture in poverty reduction: An empirical perspective. *Journal of Development Economics*, 96, pp. 239–254.

Christidis, N., Jones, G., and Stott, P. (2015) Dramatically increasing chance of extremely hot summers since the 2003 European heatwave. *Nature Climate Change*, 5, pp. 56–50.

Chu, C.G., Friesen, T.L., Xu, S.S., Faris, J.D., and Kolmer, J.A. (2009) Identification of novel QTLs for seedling and adult plant leaf rust resistance in a wheat doubled haploid population. *Theoretical Applied Genetics*, 119, pp. 263–269.

Coffman R., Acevedo, M., and McCandless, L. (2016) Rust, risk and germplasm exchange: The Borlaug Global Rust Initiative. *Indian Journal of Plant Genetic Resources*, 29, pp. 417–419.

Colmer, T.D., Munns, R., and Flowers, T.J. (2005) Improving salt tolerance of wheat and barley: Future prospects. *Australian Journal of Experimental Agriculture*, 45, pp. 1425–1443.

Crossa, J., Pérez, P., Hickey, J., Burgueño, J., Ornella, L., Cer-n-Rojas, J., Zhang, X., Dreisigacker, S., Babu, R., Li, Y., Bonnett, D., and Mathews, K. (2014) Genomic prediction in CIMMYT maize and wheat breeding programs. *Heredity*, 112, pp. 48–60.

Daetwyler, H.D., Bansal, U.K., Bariana, H.S., Hayden, M.J., and Hayes, B.J. (2014) Genomic prediction for rust resistance in diverse wheat landraces. *Theoretical and Applied Genetics*, 127, pp. 1795–1803.

Daetwyler, H.D., Hickey, J.M., Henshall, J.M., Dominik, S., Gredler, B., Van Der Werf, J.H.J., Hayes, B.J. (2010a) Accuracy of estimated genomic breeding values for wool and meat traits in a multi-breed sheep population. *Animal Production Science*, 50, pp.1004–1010.

Daetwyler, H.D., Pong-Wong, R., Villanueva, B., Woolliams, J.A. (2010b) The impact of genetic architecture on genome-wide evaluation methods. *Genetics*, 185, pp. 1021–1031.

DIVA. (2006) Homepage of DIVA-GIS mapping program providing a downloadable version of the DIVA GIS program, additional materials and data. Accessed throughout the project. www.diva-gis.org.

Donat, M.G., Alexander, L.V., Yang, H., Durre, I., Vose, R., Dunn, R.J.H., Willett, K.M., Aguilar, E., Brunet, M., Caesar, J., Hewitson, B., Jack, C., Klein Tank, A.M.G., Kruger, A.C., Marengo, J., Peterson, T.C., Renom, M., Oria Rojas, C., Rusticucci, M., Salinger, J., Sanhouri Elrayah, A., Sekele, S.S., Srivastava, A.K., Trewin, B., Villarroel, C., Vincent, L.A., and Zhai, P., (2013) Updated analyses of temperature and precipitation extreme indices since the beginning of the twentieth century: The HadEX2 dataset. *Journal of Geophysical Research Atmospheres*, 118 (5), pp. 2098–2118.

Dubcovsky, J., and Dvorak, J. (2007) Genome plasticity a key factor in the success of polyploid wheat under domestication. *Science*, 316, pp. 1862–1866.

Dubin, H.I. and Duveiller, E. (2011) Fungal, bacterial and nematode diseases of wheat: Breeding for resistance and other control measures. In: Bonjean, A.P., Angus, W.J., and Van Ginkel, M. (eds) *The World Wheat Book, Vol. 2*. Paris: Lavoisier, pp. 1131–1181.

Duveiller, E., He, X., and Singh, P.K. (2016) Wheat blast: An emerging disease in South America potentially threatening wheat production. In: Bonjean, A. and van Ginkel, M. (eds) *The World Wheat Book, Vol. 3*. Paris: Lavoisier.

Dvorak, J., Deal, K.R., Luo, M., You, F.M., von Borstel, K., and Dehghani, H. (2012) The origin of spelt and free-threshing hexaploid wheat. *Journal of Heredity*, 103, pp. 426–441.

El Amil, R. (2015) *Effet de l'hôte et de la Température sur la Structure de la Population de Puccinia striiformis f. sp. Tritici, Agent de la Rouille Jaune du blé au Moyen Orient*. Ph.D., Université Paris-Sud.

Ezzahiri, B., Yahyaoui, A., and Hovmøller, M.S. (2009). An analysis of the 2009 epidemic of yellow rust on wheat in Morocco, in *The 4th Regional Yellow Rust Conference for Central and West Asia and North Africa*. Antalya: Turkish Ministry of Agriculture and Rural Affairs, International Center for Agricultural Research in the Dry Areas, International Maize and Wheat Improvement Center, Food and Agriculture Organization of the United Nations.

Eyal, Z., Scharen, A.L., Prescott, J.M., and van Ginkel, M. (1987) The Septoria diseases of wheat: Concepts and methods of disease management, vi 52 pages, CIMMYT, Mexico, D.F. http://repository.cimmyt.org/xmlui/handle/10 883/1113.

FAO-WIEWS. (2007) World information and early warning system. http://apps3.fao.org/wiews/institute_query.htm

FAO. (2011) The State of Food and Agriculture 2011: Women in agriculture: Closing the gender gap for development. Rome: Food and Agriculture Organization of the United Nations.

FAO. (2012) The state of the world's plant genetic resources for food and agriculture. Second report, Rome, Italy: Food and Agriculture Organization of the United Nations. www.fao.org/AG/AGp/agps/Pgrfa/pdf/swrshr_e.pdf

Faris, J.D., Liu, Z., and Xu, S.S. (2013) Genetics of tan spot resistance in wheat. *Theoretical and Applied Genetics*, 126 (9), pp. 2197–2217.

Francois, L.E., Grieve, C.M., Maas, E.V, and Lesch, S.M. (1994) Time of salt stress affects growth and yield components of irrigated wheat. *Agronomy Journal*, 86, pp. 100–107.

Gao, Y., Faris, J.D., Liu, Z., Kim, Y.M., Syme, R.A., Oliver, R.P., Xu, S.S., and Friesen, T.L. (2015) Identification and characterization of the SnTox6-Snn6 interaction in the Parastagonospora nodorum–wheat pathosystem. *Molecular Plant Microbe Interaction*, 28, pp. 615–625.

Goodwin, S.B. (2012) Resistance in wheat to septoria diseases caused by *Mycosphaerella graminicola (Septoria tritici)* and *Phaeosphaeria (Stagonospora) nodorum*. In: Sharma, I. (ed.) *Disease Resistance in Wheat*. Wallingford, UK; Cambridge, MA: CABI, pp. 151–159.

Gourdji, S.M., Sibley, A.M., and Lobell, D.B. (2013) Global crop exposure to critical high temperatures in the reproductive period: historical trends and future projections. *Environmental Research Letters*, 8, pp. 24041.

Goyeau, H., Berder, J., Czerepak, C., Gautier, A., Lanen, C., and Lannou, C. (2011) Low diversity and fast evolution in the population of *Puccinia triticina* causing durum wheat leaf rust in France from 1999 to 2009, as revealed by an adapted differential set. *Plant Pathology*, 61, pp. 761–72.

Haghighattalab, A., González Pérez, L., Mondal, S., Singh, D., Schinstock, D., Rutkoski, J., Ortiz-Monasterio, I., Singh, R.P., Goodin, D., and Poland, J. (2016) Application of unmanned aerial systems for high throughput phenotyping of large wheat breeding nurseries. *Plant Methods*, 12, pp. 35.

Hanley, J., and Caballero, R. (2012) The role of large-scale atmospheric flow and Rossby wave breaking in the evolution of extreme windstorms over Europe. *Geophysical Research Letters*, 39, pp. 2–7.

Hayes, B.J, Lewin, H.A., Goddard, M.E. (2013) The future of livestock breeding: genomic selection for efficiency, reduced emissions intensity, and adaptation. *Trends in Genetics*, 29, pp. 206–214.

He, Z.H., Rajaram, S., Xin, Z.Y., and Huang, G.Z. (2001) *A History of Wheat Breeding in China*. Mexico, D.F.: CIMMYT. http://libcatalog.cimmyt.org/download/cim/74633.pdf (accessed July 2017).

Heffner, E.L., Sorrells, M.E., Jannink, J-L. (2009) Genomic selection for crop improvement. *Crop Science*, 49, pp. 1–12.

Heffner, E.L., Jannink, J., and Sorrells, M.E. (2011) Genomic selection accuracy using multifamily prediction models in a wheat breeding program. *Plant Genome*, 4, pp. 65–75.

Hoshino, T., and Seko, H. (1996) History of wheat breeding for a half century in Japan. *Euphytica*, 89, pp. 215–221.

Hovmøller, M.S., Walter, S., and Justesen, A.F. (2010) Escalating threat of wheat rusts. *Science*, 329, p. 369.

Howard, R., Carriquiry, A.L., and Beavis, W.D. (2014) Parametric and nonparametric statistical methods for genomic selection of traits with additive and epistatic genetic architectures. *G3*, 4, pp. 1027–1046.

Hubert, B., Rosegrant, M., van Boekel, M. a.J.S., and Ortiz, R. (2010) The future of food: Scenarios for 2050. *Crop Science*, 50, pp. 33–50.

Hughes, M., and Kolmer, J. (2016) Physiologic specialization of *Puccinia triticina* on wheat in the United States in 2014. *Plant Disease*, 100, pp. 1768–1773.

Husain, S., Von Caemmerer, S., and Munns, R. (2004) Control of salt transport from roots to shoots of wheat in saline soil. *Functional Plant Biology*, 31(11), pp. 1115–1126.

Inoue, Y., Vy, T.T., Yoshida, K., Asano, H., Mitsuoka, C., Asuke, S., Anh, V.L., Cumagun, C.J.R., Chuma, I., Terauchi, R., Kato, K., Mitchell, T., Valent, B., Farman, M., and Tosa, Y. (2017) Evolution of the wheat blast fungus through functional losses in a host specificity determinant. *Science*, 357, pp. 80–83.

ILRI, 2010. Opportunities for promoting gender equality in rural Ethiopia through the commercialization of agriculture, IPMS Working Paper 18, pp. xi–xiii.

IPCC (2013). Intergovernmental panel for climate change. Summary for policymakers. Available at: www.climatechange2013.org/images/report/WG1AR5_SPM_FINAL.pdf (accessed September 2017).

Jannink, J-L. (2010) Dynamics of long-term genomic selection. *Genetics Selection Evolution*, 42, pp. 35.

Jannink, J-L., Lorenz, A.J., and Iwata, H. (2010) Genomic selection in plant breeding: from theory to practice. *Briefings in Functional Genomics*, 9, pp. 166–177.

Jarqu'n, D., Kocak, K., Posadas, L., Hyma, K., Jedlicka, J., Graef, G., and Lorenz, A. (2014) Genotyping by sequencing for genomic prediction in a soybean breeding population. *BMC Genomics*, 15, p. 740.

Jia, Y., and Jannink, J-L. (2012) Multiple-trait genomic selection methods increase genetic value prediction accuracy. *Genetics*, 192, pp. 1513–1522.

Jin, Y., Singh, R.P., Ward, R.W., Wanyera, R., Kinyua, M.G., Njau, P., Fetch, T., Pretorius, Z.A., and Yahyaoui, A. (2007) Characterization of seedling infection types and adult plant infection responses of monogenic Sr gene lines to race TTKS of *Puccinia graminis f. sp. tritici. Plant Disease*, 91, pp. 1096–1099.

Juliana, P., Singh, R.P., Singh, P.K., Crossa, J., Huerta-Espino, J., Lan, C., Bhavani, S., Rutkoski, J.E., Poland, J.A., Bergstrom, G.C., and Sorrells, M.E. (2017a) Genomic and pedigree-based prediction for leaf, stem, and stripe rust resistance in wheat. *Theoretical Applied Genetics*, 130, pp. 1415–1430.

Juliana, P., Singh R.P., Singh, P.K., Crossa, J., Rutkoski, J.E., Poland, J.A., Bergstrom, G.C., and Sorrells, M.E. (2017b) Comparison of models and whole-genome profiling approaches for genomic-enabled prediction of Septoria tritici blotch, Stagonospora nodorum blotch, and tan spot resistance in wheat. *Plant Genome*, 10 (2), p. 0.

Knott, D.R. (2012) *The Wheat Rusts: Breeding for resistance* (Vol. 12). New York, NY: Springer Science & Business Media.

Kohli, M.M., Mehta, Y.R., Guzman, E., De Viedma, L., and Cubilla, L.E. (2011) Pyricularia blast - a threat to wheat cultivation. *Czech Journal of Genetics and Plant Breeding*, 47, pp. S130–S134.

Kolmer, J.A. (2013) Leaf rust of wheat: Pathogen biology, variation and host resistance. *Forests*, 4, pp. 70–84.

Kolmer, J.A. (2005) Tracking wheat rust on a continental scale. *Current Opinions in Plant Biology*, 8, pp. 1–9.

Kolmer, J.A., and Acevedo, M.A. (2016) Genetically divergent types of the wheat leaf fungus *Puccinia triticina* in Ethiopia, a center of tetraploid wheat diversity. *Phytopathology*, 106, pp. 380–385.

Levitt, J. (1972) *Responses of Plants to Environmental Stresses*. New York, NY: Academic Press.

Lobell, D.B., and Gourdji, S.M. (2012) The influence of climate change on global crop productivity. *Plant Physiology*, 160 (4), pp. 1686–1697.

Lopes, M.S., and Reynolds, M. P. (2012) Stay-green in spring wheat can be determined by spectral reflectance measurements (normalized difference vegetation index) independently from phenology. *Journal of Experimental Botany*, 63 (10), pp. 3789–3798.

Lopes, M.S., Reynolds, M.P., Manes, Y., Singh, R.P., Crossa, J., and Braun, H.J. (2012) Genetic yield gains and changes in associated traits of CIMMYT spring bread wheat in a "historic" set representing 30 years of breeding. *Crop Science*, 52, pp. 1123.

Lorenz, A.J., Chao, S., Asoro, F.G., Heffner, E.L., Hayashi, T., Iwata, H., Smith, K.P., Sorrells, M.E., and Jannink, J-L. (2011) Genomic selection in plant breeding: Knowledge and prospects. *Advances in Agronomy*, 110, pp. 77–123.

Lorenzana, R.E., and Bernardo, R. (2009) Accuracy of genotypic value predictions for marker-based selection in biparental plant populations. *Theoretical and Applied Genetics*, 120, pp. 151–161.

Lupton, F.G.H. (1987) History of wheat breeding. In: Lupton, F.G.H. (ed.) *Wheat Breeding: Its scientific basis*. London, UK: Chapman & Hall, pp. 51–70.

McCallum, B.D., Fetch, T., and Chong, J. (2007) Cereal rust control in Canada. *Crop and Pasture Science*, 58, pp. 639–647.

Mackay, I., Ober, E., and Hickey, J. (2015) GplusE: Beyond genomic selection. *Food Energy Security*, 4, pp. 25–35.

McMullen, M., Bergstrom, G., De Wolf, E., Dill-Macky, R., Hershman, D., Shaner, G. and Van Sanford, D. (2012) A unified effort to fight an enemy of wheat and barley: Fusarium head blight. *Plant Disease*, 96, pp. 1712–1728.

Malaker, P.K., Barma, N.C., Tewari, T.P., Collis, W.J., Duveiller, E., Singh, P.K., Joshi, A.K., Singh, R.P., Braun, H.-J., and Peterson, G.L. (2016) First report of wheat blast caused by Magnaporthe oryzae pathotype Triticum in Bangladesh. *Plant Disease*, 100, pp. 23–30.

Mboup, M., Leconte, M., Gautier, A., Wan, A.M., Chen, W.Q., de Vallavielle-Pope, C., and Enjalbert, J. (2009) Evidence of genetic recombination in wheat yellow rust population of a Chinese over-summering area. *Fungal Genetics and Biology*, 46, pp. 299–307.

Meuwissen, T.H.E., Hayes, B.J., and Goddard, M.E. (2001) Prediction of total genetic value using genome-wide dense marker maps. *Genetics*, 157, pp. 1819–1829.

Mickelbart, M.V., Hasegawa, P.M., and Bailey-Serres, J. (2015) Genetic mechanisms of abiotic stress tolerance that translate to crop yield stability. *Nature Reviews Genetics*, 16, pp. 237–251.

Munns, R., James, R.A., and Läuchli, A. (2006) Approaches to increasing the salt tolerance of wheat and other cereals. *Journal of Experimental Botany*, 57 (5), pp. 1025–1043.

Murray, G., and Brennan, J. (2009) Estimating disease losses to the Australian wheat industry. *Australasian Plant Pathology*, 38, pp. 558–570.

Nonaka, S. (1983) History of wheat breeding in Japan. In: Sakamoto, S. (ed.) *Proceedings of the 6th International Wheat Genetics Symposium, Maruzen Co., Ltd., Kyoto*, pp. 593–599.

Oerke, E.C. (2006) Crop losses to pests. *Journal of Agricultural Science*, 144, pp. 31–43.

Olmstead, A.L., and Rhode, P.W. (2011) Adapting North American wheat production to climatic challenges, 1839–2009. *Proceedings of the National Academy of Sciences of the United States of America*, 108, pp. 480–485.

Ordoñez, M.E., and Kolmer, J.A. (2007) Simple sequence repeat diversity of a worldwide collection of *Puccinia triticina* from durum wheat. *Phytopathology*, 97, pp. 574–583.

Ortiz, R., Mowbray, D., Dowswell C., and Rajaram S. (2007) Dedication: Norman E. Borlaug the humanitarian plant scientist who changed the world. *Plant Breeding Reviews*, 28, pp. 1–37.

Pask, A.J.D., Joshi, a.K.K., Manès, Y., Sharma, I., Chatrath, R., Singh, G. P.P., *et al.* (2014) A wheat phenotyping network to incorporate physiological traits for climate change in South Asia. *Field Crops Research*, 168, pp. 156–167.

Paulsen, G.M., and Shroyer, J.P. (2008) The early history of wheat improvement in the Great Plains. *Agronomy Journal*, 100, pp. S-70–S-78.

Percival, J. (1921) *The Wheat Plant*. London: Duckworth.

Pérez-Rodríguez, P., Gianola, D., González-Camacho, J.M., Crossa, J., Manès, Y., Dreisigacker, S. (2012) Comparison between linear and non-parametric regression models for genome-enabled prediction in wheat, *G3*, 2, pp. 1595–1605.

Peterson, P.D. (2013) *"The barberry or bread": The public campaign to eradicate common barberry in the United States in the early 20th century*. APS Features.

Pingali, P.L. (2012) Green Revolution: Impacts, limits, and the path ahead. *Proceeding of the Natural Academy of Science*, 109, pp. 12302–12308.

Poland, J., Endelman, J., Dawson J., Rutkoski, J., Wu, S.Y., Manes, Y., Dreisigacker, S., Crossa J., Sánchez-Villeda, H., Sorrells, M., and Jannink, J-L. (2012) Genomic selection in wheat breeding using genotyping-by-sequencing. *Plant Genome*, 5, pp. 103–113.

Prada, D. (2009) Molecular population genetics and agronomic alleles in seed banks: Searching for a needle in a haystack?. *Journal of Experimental Botany*, 60, pp. 2541–2552.

Prasad, P.V.V., and Djanaguiraman, M. (2014) Response of floret fertility and individual grain weight of wheat to high temperature stress: Sensitive stages and thresholds for temperature and duration. *Functional Plant Biology*, 41, pp. 1261–1269.

Pretorius, Z.A., Singh, R.P., Wagoire, W.W., and Payne,T.S. (2000) Detection of virulence to wheat stem rust resistance gene Sr31 in *Puccinia graminis* f. sp. *tritici* in Uganda. *Plant Disease*, 84, pp. 203.

Quisumbing, A.R., and Maluccio J.A. (2003) Resources at marriage and intrahousehold allocation: Evidence from Bangladesh, Ethiopia, Indonesia, and South Africa. *Oxford Bulletin of Economics and Statistics*, 65, pp. 283–327.

Rahmatov, M., Rouse, M.N., Nirmala, J., Danilova, T., Friebe, B., Steffenson, B.J., and Johansson, E. (2016) A new 2DS-2RL Robertsonian translocation transfers stem rust resistance gene Sr59 into wheat. *Theoretical and Applied Genetics*, 129, pp. 1383–1392.

Reitz, L.P., and Salmon, S.C. (1968) Origin, history, and use of Norin 10 wheat. *Crop Science*, 8, pp. 686–689.

Reynolds, M.P., and Langridge, P. (2016) Physiological breeding. *Current Opinion in Plant Biology*, 31, pp. 162–171.

Reynolds, M.P., Braun, H.J., Cavalieri, A., Chapotin, J.S., Davies, W.J., Ellul, P., Feuillet, C., Govaerts, B., Kropff, M.J., Lucas, H., Nelson, J., Powell, W.,

Quilligan, E., Rosegrant, M.W., Singh, R.P., Sonder, K., Tang, H., Visscher, S., and Wang, R. (2017) Improving global integration of crop research, *Science*, 357, pp. 359–360.

Roelfs, A.P. (1982) Effects of barberry eradication on stem rust in the United States. *Plant Disease*, 66, pp. 177–181.

Roelfs, A.P. (1985) Wheat and rye stem rust. In: Roelfs, A.P., and Buschnell, W.R. (eds) *The Cereal Rusts, Vol. II: Diseases, distribution, epidemiology and control*. Orlando, FL: Academic Press, pp. 3–37.

Roelfs, A.P., Singh R.P., and Saari, E.E. (1992) *Rust Diseases of Wheat: Concepts and methods of disease management*. Mexico, DF: CIMMYT.

Rosegrant, M.W., Agcaoili-Sombilla, M.C., and Perez, N.D. (1995) *Global Food Projections to 2020: Implications for investment*. Darby, PA: Diane Publishing.

Rutkoski, J., Benson, J., Jia Y., Brown-Guedira G., Jannink, J-L., and Sorrells, M. (2012) Evaluation of genomic prediction methods for Fusarium head blight resistance in wheat. *Plant Genome*, 5, pp. 51.

Rutkoski, J., Poland, J., Mondal, S., Autrique, E., González Pérez, L., Crossa, J., Reynolds, M., and Singh, R. (2016) Canopy temperature and vegetation indices from high-throughput phenotyping improve accuracy of pedigree and genomic selection for grain yield in wheat. *G3*, 6, pp. 1–36.

Rutkoski, J., Singh, R.P., Huerta-Espino, J., Bhavani, S., Poland, J., Jannink, J-L., and Sorrells, M.E. (2015) Genetic gain from phenotypic and genomic selection for quantitative resistance to stem rust of wheat. *Plant Genome*, 8, p. 0.

Rutkoski, J.E., Poland, J.A., Singh, R.P., Huerta-Espino, J., Bhavani, S., Barbier, H., Rouse, M.N., Jannink, J-L., and Sorrells, M.E. (2014) Genomic selection for quantitative adult plant stem rust resistance in wheat. *Plant Genome*, 7, p. 3.

Saari, E.E., and Prescott, J.M. (1985) World distribution in relation to economic losses. In: Roelfs, A.P., and Bushnell, W.R. (eds) *The Cereal Rusts. Vol. II: Diseases, distribution, epidemiology and control*. Orlando, FL: Academic Press, pp. 259–298.

Salamini, F., Ozkan, H., Brandolini, A. Schäfer-Pregl, R., and Martin, W. (2002) Genetics and geography of wild cereal domestication in the near east. *Nature Reviews Genetics*, 3, pp. 429–441.

Shewry, P.R. (2009) Wheat. *Journal of Experimental Botany*, 60: 1537–1553.

Singh, A., Pallavi, J.K., Gupta, P., and Prabhu, K.V. (2011) Identification of microsatellite markers linked to leaf rust adult plant resistance (APR) gene *Lr48* in wheat. *Plant Breeding*, 130, pp. 31–34.

Singh, R.P., Hodson, D.P., Jin, Y., Lagudah, E.S., Ayliffe, M.A., Bhavani, S., Rouse, M.N., Pretorius, Z.A., Szabo, L.J., Huerta-Espino, J., and Basnet, B.R. (2015) Emergence and spread of new races of wheat stem rust fungus: Continued threat to food security and prospects of genetic control. *Phytopathology*, 105, pp. 872–884.

Singh, R.P., Singh, P.K., Rutkoski, J., Hodson, D.P., He, X., Jørgensen, L.N., Hovmøller, M.S., and Huerta-Espino, J. (2016) Disease impact on wheat yield potential and prospects of genetic control. *Annual Review of Phytopathology*, 54, pp. 303–322.

Skovmand, B., Rajaram, S., Ribaut, J.M., and Hede, A.R.. (2002) Wheat genetic resources. FAO Document Repository, available at: www.fao.org/docrep/006/Y4011E/y4011e08.htm (accessed July 2017).

Smale, M. (1997) The Green Revolution and wheat genetic diversity: Some unfounded assumptions. *World Development*, 25, pp. 1257–1269.

Solh, M. (2016) The contribution of innovative agricultural research to sustainable development goals in dry areas. *Proceedings of the Tenth International Conference on Dryland Development: Sustainable development of drylands in the post 2015 world, Alexandria, Egypt, 21–24 August 2016*. International Dryland Development Commission (IDDC), Cairo, Egypt.

Stone, P.J., and Nicolas, M.E. (1994) Wheat cultivars vary widely in their responses of grain yield and quality to short periods of post-anthesis heat stress. *Functional Plant Biology*, 21 (6), pp. 887–900.

Stratonovitch, P., and Semenov, M.A. (2015) Heat tolerance around flowering in wheat identified as a key trait for increased yield potential in Europe under climate change. *Journal of Experimental Botany*, 66 (12), pp. 3599–3609.

Sun, J., Rutkoski, J.E., Poland, J.A., Crossa, J., Jannink, J-L., and Sorrells, M.E. (2017) Multitrait, random regression, or simple repeatability model in high-throughput phenotyping data improve genomic prediction for wheat grain yield, *Plant Genome*, 10 (2). doi: 10.3835/plantgenome2016.11.0111.

Tadesse, W., Amri, A., Ogbonnaya, F.C., Sanchez-Garcia, M., Sohail, Q., and Baum, M. (2015) *Wheat Genetic Resources and their Role in Breeding: Current status and future prospects*. Oxford: Oxford Academic Press, pp. 81–124.

Talukder, S.K., Babar, M.A., Vijayalakshmi, K., Poland, J., Prasad, P.V.V., Bowden, R., and Fritz, A. (2014) Mapping QTL for the traits associated with heat tolerance in wheat (Triticum aestivumL.). *BMC Genetics*, 15, pp. 97.

Tester, M., and Langridge, P. (2010) Breeding technologies to increase crop production in a changing world. *Science*, 327 (5967), pp. 818–822.

Van Ginkel, M., Calhoun, D.S., Gebeyehu, G., Miranda, A., Tian-you, C., Pargas Lara, R., Trethowan, R. M., Sayre, K., Crossa, J., and Rajaram, S. (1998) Plant traits related to yield of wheat in early, late, or continuous drought conditions. *Euphytica*, 100, pp. 109–121.

Vietmeyer, N. (2011) *Our Daily Bread: The essential Norman Borlaug*. Lorton, VA: Bracing Books.

Villareal, R.L., Mujeeb-Kazi, A., Gilchrist, L.I., and Del Toro, E. (1995) Yield loss to spot blotch in spring bread wheat in warm nontraditional wheat production areas. *Plant Disease*, 79, pp. 893–897.

Wang, W., Vinocur, B., and Altman, A. (2003) Plant responses to drought, salinity and extreme temperatures: Towards genetic engineering for stress tolerance. *Planta*, 218, pp. 1–14.

Weiss, M. (1987) *Compendium of Wheat Diseases*. St. Paul, MN: APS Press.

Zampieri, M., Ceglar, A., Dentener, F., and Toreti, A. (2017) Wheat yield loss attributable to heat waves, drought and water excess at the global, national and subnational scales. *Environmental Research Letters*, 12 (6). doi:10.1088/1748-9326/aa723b.

Zhao, Y., Mette, M.F., Gowda, M., Longin, C.F.H., and Reif, J.C. (2014) Bridging the gap between marker-assisted and genomic selection of heading time and plant height in hybrid wheat. *Heredity* (Edinb) 112, pp. 638–645.

5 Innovative practices in potato production for food and nutrition security

Ngo Tien Dung, Udaya Sekhar Nagothu, Alma Linda Morales-Abubakar, Jan Willem Ketelaar and Mehreteab Tesfai

Introduction

Potato (*Solanum spp.*) is the third most important root and tuber crop in the world, domesticated more than 7000 years ago in Peru and Bolivia (Maranzani, 2013; Devaux *et al.*, 2014). The Andes is the main source of origin for more than 4,000 types of potatoes showing the wide range of genetic diversity and the nutritional value of the crop. Potato is grown in more than 125 countries with annual production of about 300 million tons and consumed almost daily by more than a billion people. Millions of people in developing countries depend on potatoes for their survival. Developing countries are now the world's biggest producers as well as importers of potatoes and potato products (FAO, 2009; CIP, 2017). At the global level, production and trade in potatoes has increased dramatically, as consumers realize the nutritional benefits of the crop. Devaux *et al.* (2014) in their study show the untapped potential of potato to help improve food security and livelihoods in developing countries.

Potato is an important source of carbohydrates. One medium-sized potato provides about half of the daily adult requirement of vitamin C, whereas other staples, such as rice and wheat, have none. However, potato is very low in fat compared with just about 5 percent of the fat content of wheat. It has about one-fourth of the calories of bread. When boiled, potato provides more protein than maize and about twice as much calcium (The Crop Trust, 2017). Potato has the dual role of contributing to subsistence farming and reducing hunger on one hand and as a high value crop providing income and employment to farmers on the other (Thiele *et al.*, 2010). Potato production in Africa has started to increase in recent years and will further increase significantly (Okello *et al.*, 2017). With the fluctuating – but in general trends rising – prices of rice and other staples in the world, potatoes have the potential to deliver an alternative source of carbohydrates (*Independent*, 2016).

Potato production has become an input-intensive economic activity with excessive use of fertilizer and pesticides in regions of South and Southeast Asia which has led to environmental problems. To reduce environmental pollution in agricultural soils and water bodies, use is made of organic residues for, e.g. rice straw as soil fertilizers and for promoting soil health. There are several challenges related to the production and post-harvest management of potatoes that need to be addressed to realize the untapped potential of the crop, particularly in Africa, South America and Asia. More specifically, this is relevant for countries such as Peru, Columbia, Vietnam, India, Pakistan, Kenya, Uganda, Malawi and South Africa among others where potato consumption – particularly in processed format – is rapidly increasing in the urban areas and among youth.

Smallholder challenges in potato production

The main challenges of smallholders involved in potato production are related to procurement of and access to healthy seed material, soil fertility, disease and pest control, unorganized post-harvest handling operations, including cold storage, and low market prices of tubers at harvesting (Finckh *et al.*, 2006; Uddin *et al.*, 2010; Karanja *et al.*, 2014; Gebru *et al.*, 2017a). Some of these are well studied and known challenges to potato scientists and development workers. However, the research and development sector is still far from providing effective, efficient and sustainable solutions to smallholder farmers involved in potato production due to the following factors:

Seed material: Potato yield is largely influenced by the health status of seed tubers, including the physiological age of the tuber at the time of planting (FAO, 2008). The biological and physical quality of the tubers influence the yield and quality of the produce. A majority of smallholders does not have access to certified seed potato material and hence use part of their produce for seed purpose and at the same time supply to fellow farmers (Gebru *et al.*, 2017a, 2017b). Thus, in developing countries, farmers obtain seed mostly through informal systems or non-regulated channels (Lutaladio *et al.*, 2009). This may be a risk in case the seed material is infected by crop pests and disease pathogens, especially viruses but also late blight, nematodes or insect pests. Lack of clean seed material was found to be one of the major challenges facing potato production in the Oljoro–Orok Division of Kenya (Karanja *et al.*, 2014).There is a large scope for improvement in the seed potato value chain given the market potential for potato that is replacing the traditional diets in Africa and other regions. Recent studies show that potato yields continue to remain very low in parts of Africa and Asia mainly due to use of poor quality seed material (FAO, 2008; Okello *et al.*, 2017; Uddin *et al.*, 2010; Karanja *et al.*, 2014). By mere supply of healthy seed material, the potato productivity is expected to increase significantly.

Soil fertility management: Soil fertility management is important to maintain optimum tuber quality and yield since potato has a shallow root system (Rosen, 2016). Problems such as soil compaction, poor drainage and low soil pH can adversely affect potato production (McKenzie, 2013). Farmers do not pay much attention to the soil management due to lack of knowledge and resources. In the absence of systematic soil testing and analyses of the various nutrients, need based and efficient use of fertilizer is rarely seen on smallholder farms. If the fertilizer application is not done at the right time and with the right dosage, yield responses can be poor. Split applications of Nitrogen fertilizer is recommended at pre-planting, emergence and hilling phase of the plant for best results. The other major nutrients such as Phosphorus and Potassium applications should be done prior to planting. A need-based soil nutrient management is necessary to address the smallholder problems and enable sustainable intensification of potato production.

Disease and pest management: A major constraint limiting potato production is the late blight disease caused by *Phytophthora infestans*. The Irish Potato Famine of 1845 caused due to late blight resulted in severe shortages of potato, the staple diet, rendering millions of Irish food insecure leading to mass migrations from Europe to the United States and death of more than a million people (Maranzani, 2013). The report further reveals that one of the strains – the US-1 – was the main culprit and even today responsible for serious potato crop damages each year. We cannot afford the repetition of such extreme events that can be devastating. Predominantly, potato growers protect their crops with frequent preventive application of fungicides, which are costly and reduce profit margins. In organic potato farming, in fact, one of the main challenges is disease management, especially the late blight that is the most limiting factor. The study by Uddin *et al.* (2010) in Bangladesh showed that the two factors, late blight in the northern part of the country and poor seed quality in other regions, significantly affected potato yield. According to Finckh *et al.* (2006), resistance appears to be the most important strategy against late blight in organic potato production. According to Kromann *et al.* (2014), besides the potato late blight, arthropod pests, primarily the potato tuber moths (*Phthorimaea operculella*) cause serious problems negatively affecting tuber yield and quality during production and post-harvest stages at a global level. As a result, farmers resort to heavy use of chemical pesticides, which not only increases the costs of cultivation but also contributes to soil and environmental pollution and applicator health hazards. A location-specific integrated pest management is necessary to address the smallholder problems and enable sustainable intensification of potato production with optimal inputs of agro-chemicals.

Organic potato farming: Organic potato farming is another option initiated in some regions. Though yield from organic farming is comparatively lower than inorganic production, there is a growing market for organic potato farming. The early development and bulking behavior and the ability of a cultivar to make use of organic nutrients efficiently is another important factor to consider in potato production. However, farmers undertaking organic farming lack the proper training and capacity building and linkage to markets. It will be useful to address the constraints of organic potato production that have a potential market in the urban areas at premium farm gate prices. To promote organic farming, or other types of cultivation such as no tillage production, it is a prerequisite to address the current challenges related to technical as well as socio-economic and institutional factors including extension and market services to support smallholders in the potato production and consumption continuum.

Socio-economic and institutional constraints: A number of socio-economic characteristics and institutional factors influence potato production (Nyagaka *et al.*, 2009). These include, education, access to extension services, credit and membership in farmer associations as observed in their study in Kenya. A similar study by Kuyiah (2007) showed that limitation of cash and small land size were the two most important factors that inhibit farmers from realizing higher incomes and optimal production. In Malawi, potato farmers do not operate at full profit potential mainly due to socio-economic constraints. These farmers could increase their profits by 26 percent if they improved their input use efficiency (Assam *et al.*, 2012). Smallholder potato farmers as observed in Uganda face a number of production and marketing challenges including limited access to markets and low surpluses for sale according to a study by Sebatta *et al.* (2014). The study also showed that transaction costs play a big role in farmers entering into markets. Similar trends are seen in other countries in Africa where smallholders do not get adequate extension and market services to enable sustainable intensification of potato cultivation. In Asia, the services are comparatively well established compared to Africa, but the price fluctuations discourage farmers from increasing potato production. Dumping of potatoes due to low farm gate prices is not uncommon in some of the Asian countries.

Gender mainstreaming: Despite the increasing role of women in cultivation of high value crops, factors such as land ownership and other assets are gender-biased and limits their opportunity (Quisumbing *et al.*, 2015). Though the efforts to reduce gender differences in the smallholder farming sector are increasing, it is still not adequate enough. It is important that governments target the future strategies of potato cultivation to support the increasing involvement of women farmers. In developing countries, women have little access to modern technology, despite the fact that most farm activities such as weeding, planting and harvesting are done by them. In Uganda, women contribute labor significantly to production and harvesting of potato and other tuber crops, but it is men who make the key decisions for marketing (CGIAR, 2017). The Government extension programs, even in Asia and South America, do not adequately target women in these situations, putting them at a disadvantage to benefit from the new technologies and innovations that often cater to the needs of men. In reality, gender dynamics shape and influence the nature of participation in, as well as the ability to benefit from, seed and potato markets as seen in Malawi (Mudege *et al.*, 2015). Lack of gender mainstreaming and low membership in farmers' cooperatives significantly influence women in market participation. Technology, policy and market interventions in the agriculture sector that do not address underlying social structures will only benefit men and not women who are taking more responsibility in farming.

However, information on the above-mentioned factors that influence smallholders' potato production and productivity in developing countries is scattered in the literature. In this chapter, attempts were made to summarize the most important challenges hindering smallholder potato production. The objectives of this chapter are threefold: i) to review the technical, socio-economic, policy and institutional factors that constrain smallholder potato production and productivity, ii) to describe innovative practices that enhance sustainable potato production, taking the experiences of Zero tillage in Vietnam, and iii) to provide recommendations on potential areas of improvement for potato development.

The chapter started with a general introduction, followed by a review of the main challenges that smallholders encounter in potato cultivation. In the next part of this chapter, we present the case of the innovative practice of no-tillage potato production from Vietnam and show how it is widely promoted and applied and has brought positive results that address contemporary issues related to demographic and other social changes, food security, poverty alleviation and environmental health in rural areas of Vietnam. The lessons learned from the Vietnam experience are relevant to smallholders cultivating potatoes in other countries impacted by similar challenges including climate change and market constraints, especially in parts of Africa. Towards the end, the chapter provides some recommendations from other studies that have focused on different parts of the potato value chain (VC) and the need for *sustainable* intensification of potato cultivation in the future.

The innovative practice of no-tillage potato production: the Vietnam experience

In Vietnam, potato is an important winter-rotation and food crop, providing a stable source of income for smallholder farmers. Favorable soil and climatic conditions, especially in the northern plains (Red River Delta), mountainous north, northern-central and central highlands make it possible to grow potatoes every year on at least about 200,000 ha of land. However, potato productivity has been low and areas planted to the crop have declined due to lack of quality seeds and high labor costs – especially as many members of the main labor force in rural areas migrate to sell services in the labor market elsewhere. Local migration has now resulted in women – especially the elderly – being left behind to carry out farming activities in addition to looking after their families and attending to community and social obligations. Since conventional potato production is labor-intensive, many farm families shifted to planting other crops in recent decades. That is, until the concept of no-tillage potato Integrated Pest Management (IPM) in rice production systems was introduced through Farmer Field Schools (FFS) in Vietnam in 2008. This innovation was field-tested and promoted by the National IPM Programme, managed by the Plant Protection Department (PPD), Ministry of Agriculture and Rural Development (MARD) with assistance of the FAO

Asia Regional IPM Programme (www.vegetableipmasia.org/uploads/files/document/potato_one_file.pdf).

In northern Vietnam, potatoes are planted annually during the four month long winter season (i.e., between 20 October and 15 November) and spring season (i.e., between 20 December and 5 January). Conventional potato production is cumbersome and includes labor-intensive practices such as ploughing, raking and making beds and covering the seed tubers planted on the bed with a thick layer of soil.

The innovative process of no-tillage potato production entailed that the summer–autumn rice straw is cut close to the ground and the residues piled up in a corner of the field. Rice straw collected from about 3–4 ha is used for mulching 1 ha of potato field. The field is drained 7–10 days before the rice harvest but the field is not ploughed after harvesting. Instead, furrows are created at 100–120 cm intervals measuring 25–30 cm wide and 20–25 cm deep. The furrows serve as drainage for excess water and create elevated ridges that become the beds. The beds are ideal for growing potatoes without the usual need for labor-intensive ploughing, or tilling that is difficult and tedious, especially for elderly women.

No-tillage potato production requires about 1,200–1,600 kg/ha of seed tubers, depending on the size and planting density. Farmers buy non-certified but uniform-sized healthy tubers from the market, selecting those that are about 30–45 mm in diameter (i.e., 25–35 tubers/kg), each tuber with 2–3 sprouts measuring about 2–20 mm in length. If the tubers are bigger than 50 mm in diameter with many sprouts, these are sliced with clean (usually sterilized) cutting tools or knives at 5–7 days before planting. Based on expert recommendations, each slice – with at least two sprouts – is split further but not totally severed or remains attached by 2–3 mm to reduce entry points of disease pathogens. Farmers, however, have recently been applying a new method of fully severing the slices and sprinkling powdered cement on the surface to prevent rotting or disease infection (Hao, 2017).

Fully composted manure made of poultry or animal manure, peat and other organic materials, either purchased or coming from farmers' own resources, is used as basal fertilizer and applied at the point where the tuber is placed. If the organic fertilizer is not fully composted, it is applied as basal fertilizer between rows of seed tubers. If the soil is wet, fertilizers are applied around newly planted seed tubers. Nitrogenous fertilizers are not applied as basal fertilizer if tuber seed slices are planted because farmers have observed that it causes rotting. Seed tubers – sprouts upwards – are placed 30–35 cm from both edges of the ridge forming two horizontal rows with about 35–40 cm between them. The distance between seed tubers in a row is 30 cm. Direct-contact with chemical fertilizers and not fully composted manure is avoided by covering the seed tubers with a thin layer of powdered soil, humus or peat, rice husk or mature compost. A layer of straw about 7–10 cm thick is added to cover the surface of the whole bed.

No watering is needed if the soil humidity is high but the rows are watered, if needed, and the straw sprinkled with soil to prevent it from being blown away by strong winds. Furrow irrigation is applied 2–3 days after planting, if needed. However, usually the remaining water in the top soil from the recent rice cropping season provides sufficient moisture for crop establishment and growth at the start of the potato cropping season. Additional layers of straw about 10–12 cm thick are added (i.e., around the plants to avoid breaking sprouts and stems) after the first and second additional application of fertilizers at 15–20 days and 35–40 days after planting, respectively. During the same time, where facilities are available, furrow irrigation is applied with the water reaching up to one-third or two-thirds the height of the bed. However, irrigating the surface of the bed using a watering can is still practiced by most farmers. Soil from the furrows is dredged to improve drainage and added on top to keep the straw from being displaced (MARD, 2013).

Figure 5.1 The process of no-tillage potato production from no till to harvesting
Source: Photos by Ngo Tien Dung and Phi Ngoc Hung.

Harvesting the potatoes does not require any digging implements and is done simply by pulling the straw away from the bed to expose the clean tubers, an operation that women of all ages and children can easily perform (Figure 5.1). The straw residues are incorporated into the soil for ecological recycling to improve soil fertility and general soil health.

Farmer Field School approach to no-tillage potato production in Vietnam

An FAO-organized regional workshop for IPM trainers was organized in Dalat, Vietnam in October 2008, held in observance of the United Nations declared International Year of the Potato. Based on a presentation with prior field experience in Guangxi, China PR, this workshop provided initial ideas for Vietnam to explore the innovative no-tillage potato production system. An immediate priority for the Vietnamese was to engage interested IPM farmers to field test the innovative practice and embark on the development of a curriculum for Farmers Field Schools (FFS) so that farmers become aware and can learn about this innovative practice. The national and local government were keen to promote the no-tillage potato production using IPM for diversification of rice production systems. In the winter season 2008–2009, 25 farmers (8 percent male) from Thai Giang village, Thai Thuy district, Thai Binh province participated in the first pilot FFS (Figure 5.2). The curriculum covered the technical areas of: i) designing and implementing field studies; ii) crop planting ii) crop growth development; iii) soil and nutrient management; iv) insect pest; disease and weed management; v) water management; and vi) book keeping and analysis of economic benefits. Because the FFS is not an-end-in-itself but rather a means to develop critical thinking skills and empower the rural community. The FFS design also covered other topics such as: i) group formation; organization and management skills; ii) leadership development; iii) decision making; iv) folk media and mastering other communication tools for development; v) advocacy and knowledge sharing through field days; and v) skills for evaluation and planning.

Field studies were established to compare the innovative practices under the no-tillage potato production and the farmers' conventional practices (Table 5.1). At the end of the season, selected graduates from the pilot FFS were to comprise the core group of farmer trainers who would educate other farmers in the community on no-tillage potato production.

From the initial pilot FFS in Thai Giang village, the total number of FFS in no-tillage potato production in rice-based systems had reached 160 by 2014, training a total of over 4,000 farmers (70 percent women) in 22 provinces. The international civil society organization (CSO) Oxfam America had joined FAO to support FFS training for scaling out of the innovative potato growing method through Vietnam's National IPM Programme that is managed by PPD-MARD.

Figure 5.2 A female farmer harvesting potatoes by pulling away the rice
Source: Photo by Ngo Tien Dung.

Results and discussion

Potato productivity and profitability

The practices employed under no-tillage potato production addressed concerns about low potato productivity and decreasing areas planted to the crop from the use of conventional growing methods that required intensive labor – including weeding that was reduced by mulching. The labor involved in growing potatoes using the no-tillage practices was about 104 man days/ha/season less compared with conventional farmer's practices, a significant difference of about 31 percent (Table 5.2). More importantly, because the method did not require ploughing during land preparation and digging at harvest, it made growing potatoes possible again for women – especially the elderly – who are left to carry out farming activities due to urbanization and the migration of rural youth in search of better-paid employment opportunities.

Data on yield components from two fields collected for the duration of three years (i.e., 2009, 2010 and 2011) in Thai Giang village, Thai Thuy district and Vu An village, Kien Xuong district, Thai Binh province shows that the application of no-tillage practices resulted in more clusters of

Table 5.1 Similarities and dissimilarities between no-tillage potato production practices and conventional farmers' practices, Bac Giang, Vietnam 2016

No-tillage potato production practices	*Conventional farmers' practices*
Land preparation – no ploughing, furrows established	Farmers' local practice – ploughing, raking and making beds
Seed tubers from the markets or agents	Seed tubers from markets or agents
Mainly basal fertilizer: 13 ton manure, 555 kg phosphate, 167 kg Urea, 200 kg KCl, 350 kg lime	Mainly additional fertilizer: 11 ton manure, 700 kg NPK, 55 kg Urea, 90 kg KCl
Seed tubers measuring about 1.5–2 cm with 2-3 sprouts	Seed tubers measuring about 1.5–2 cm with 2–3 sprouts
Planting distance row to row: 35–40cm and seed to seed: 25–30 cm	Planting distance row to row: 30–35cm and seed to seed: 25–30 cm
Seed tubers planted on raised beds and covered by thin compost, fine soil and straw	Seed tubers planted on beds and covered by thick soil only
Furrow irrigation 2–3 times	Furrow irrigation 5–6 times
Rice straw added 2 times	Soil added 1–2 times depending on the weather
Integrated pest management including use of natural biological control such as predators and parasitoids, 1–2 pesticide for mildew	Pesticides applied 3–4 times for mildew, thrips, spider mites, etc.

Source: Adapted from Bac Giang PPSD (2016)

Table 5.2 Total labor (man day/ha/season) in FFS plots comparing no-tillage and conventional farmer's production, n = 7 (Nam Dinh, Yen Bai, Lao Cai and Hoa Binh provinces, Winter–Spring season 2013/14)

Village (1)	*No tillage (2)*	*Farmer's practice (3)*	*Difference (4) = (2) – (3)*	*% difference (5) = (4) / (3) × 100*
Giao Yen	224	350	–126	–36
Giao Phong	224	350	–126	–36
Huu Khanh	280	364	–84	–23
Gioi Phien	195	255	–60	–24
Thanh Nien 2	195	255	–60	–24
Ban Qua	274	430	–156	–36
Vinh Dong	295	415	–120	–29
Kim Bac	180	280	–100	–36
Mean n=7	233.3	337.4	–104.1	–31
Paired t-test			–8.57**	

Source: Adapted from Plant Protection Department (2015) ** (p<0.01).

potato tubers ranging from 37–38 compared with 35–36 using conventional farmer's practices. The weight of tubers/cluster was higher, 0.71kg/cluster compared with 0.67kg/cluster from conventional farmer's practices. This translated into higher yields with differences ranging from 2.20–2.42 tons/ha or yield differences of 10–11 percent. (Table 5.3). The results could possibly be attributed to environmental conditions (Wurr, 2001) and more specifically the amount of incident radiation during tuber initiation (Firman, 2008). The conventional practice of covering the tubers with thick soil could have hampered incident radiation that was available to the developing tubers grown under straw mulching.

Over the period 2012–2014, data on economic benefits from seven communes in four provinces (i.e., Nam Dinh, Yen Bai, Lao Cai and Hoa Binh) significantly showed differences in yields resulting from the application of no-tillage practices ranging from 5.5–7.78 tons per ha (Figure 5.3). The difference in yields between the innovative and conventional practices ranged from 6–57 percent (mean = 21 percent) higher yields across these provinces or an average significant difference of 2.69 tons/ha higher yields using paired t-test analysis (Table 5.4).

The difference in economic benefits resulting from the application of no-tillage practices ranged from $545–$2384 (Table 5.4). Incomes increased by 20–123 percent largely from costs of savings from reduced labor, especially for women (Figure 5.2). Not surprisingly, this enabled women in the IPM Farmers' Group in Thai Giang village, Thai Binh province to buy television sets with the extra money they made during just one season of growing potatoes in 2009. From their potato-farming income in 2010, they bought gas stoves while others reported saving the money to send their children to university. Using the data from the seven villages as a data set, a highly significant difference was obtained comparing profits from no tillage and farmer's practice (Table 5.5).

Table 5.3 Yield components of no-tillage and conventional farmer's practice plots, Thai Binh province, 2009–11

Parameter/ Unit	*2009*		*2010*		*2011*	
	No tillage	*Conventional tillage*	*No tillage*	*Conventional tillage*	*No tillage*	*Conventional tillage*
Cluster/ha	38.08	35.80	37.94	35.99	37.20	34.76
Weight of bulbs/ cluster (kg)	0.70	0.66	0.72	0.68	0.71	0.68
Yield (ton/ha)	22.50	20.30	23.21	20.79	22.45	20.08
Difference (%)	+9.77		+10.42		+10.55	

Source: Plant Protection Department (2012).

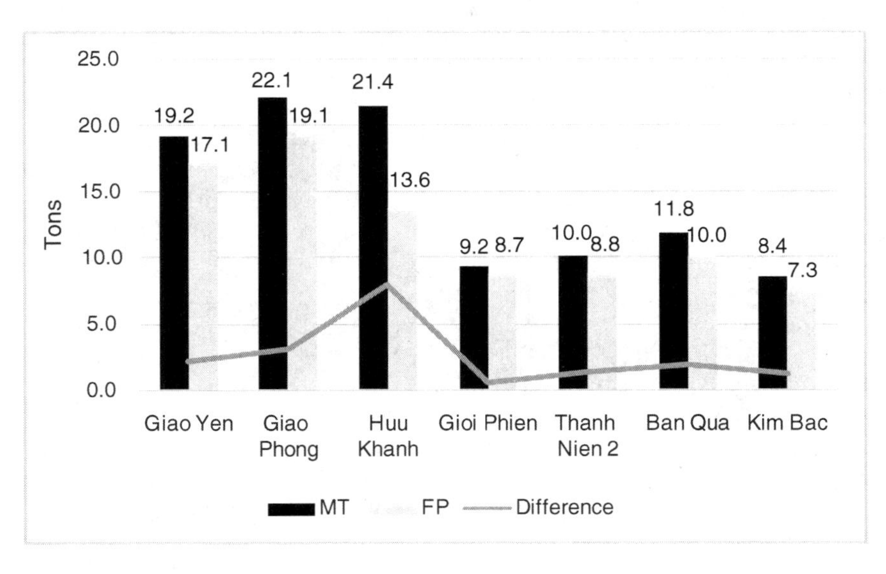

Figure 5.3 Yields, production costs and economic benefits from no-tillage and conventional farmer's practices, n = 7 (Nam Dinh, Yen Bai, Lao Cai and Hoa Binh provinces, Winter-Spring season 2013/14)

Source: Plant Protection Department (2015).

Table 5.4 Yields (tons/ha) from FFS no-tillage and conventional farmer's practice plots (Nam Dinh, Yen Bai, Lao Cai and Hoa Binh provinces, Winter–Spring season 2013/14)

Village (1)	*No tillage (2)*	*Farmer's practice (3)*	*Difference (4) = (2) – (3)*	*% difference (5) = (4) / (3) × 100*
Giao Yen	19.2	17.1	2.10	12
Giao Phong	22.1	19.1	3.02	16
Huu Khanh	21.4	13.6	7.78	57
Gioi Phien	9.2	8.7	0.50	6
Thanh Nien 2	10.0	8.8	1.25	14
Ban Qua	11.8	10.0	1.79	18
Kim Bac	8.4	7.3	1.10	15
Mean n=7	14.6	12.1	2.5	–21
Paired t-test			–2.69*	

Source: Plant Protection Department (2015).

In 2016, a visit to Nha village close to the first pilot village where the no-tillage potato production was first introduced in 2008 confirmed reports that farmers had been carrying out improvements in their houses – and providing

Table 5.5 Profits from FFS no-tillage and conventional farmer's practice plots, n = 7 (Nam Dinh, Yen Bai, Lao Cai and Hoa Binh provinces, Winter-Spring season 2013/14)

Village (1)	*No tillage (US$/ha) (2)*	*Farmer's practice (US$/ha) (3)*	*Difference (US$/ha) (4) = (2) – (3)*	*% difference (5) = (4) / (3) × 100*
Giao Yen	6,320	4,731	1,589	34
Giao Phong	7,521	5,573	1,947	35
Huu Khanh	4,317	1,933	2,384	123
Gioi Phien	3,339	2,793	545	20
Thanh Nien 2	3,768	2,807	961	34
Ban Qua	3,034	1,486	1,548	104
Kim Bac	2,217	1,263	955	76
Mean n=7	4,359	2,941	–1,419	** ($p<0.01$)

Source: Adapted from Plant Protection Department (2015).

assistance to their children to build their own homes – from profits and savings each year gained from no-tillage potato production. Economic benefits from application of the innovative practice and IPM had spread to farmers in adjacent villages. Core groups of farmers from the pilot FFS had educated farmers from close by villages about no-tillage potato production.

Climate change mitigation and enhancement of ecosystem services

Climate change mitigation

Burning rice straw residues in open fields after harvest is a serious health and environmental problem in rice-growing areas in Vietnam and other countries in Southeast Asia. For example, more than 80 percent of the rice straw produced in Vietnam is burnt in the field (Truc *et al.*, 2012). Consequently, organic C content in the straw is lost and a considerable amount of CO_2 and CH_4 is emitted into the atmosphere (Miura and Kanno, 1997). Such practices likely reduce organic carbon inputs into the soil as well as depleting soil organic matter levels, eventually leading to low yields. Moreover, the practice causes air pollution and contributes to global warming through emissions of greenhouse gases (GHGs) (Romasanta *et al.*, 2017). By 2020, Vietnam aims to reduce CH_4 and N_2O emissions from rice production systems by 20 percent (UN-Vietnam 2013). Interventions by using leftover rice straw for mulch (i.e., straw from 3–4 ha of rice) to 1 ha of potato field under no-tillage system, has provided additional benefits from reducing N_2O and CH_4 emissions (Bruun *et al.*, 2011; Wang *et al.*, 2012) caused by the traditional burning of rice straw. This practice will contribute to mitigating climate gases and to the achievement of Vietnams' emissions reduction targets.

Enhancement of ecosystem services

No-till potato practices provides several ecosystems services (ESS) and enhances biodiversity in the farming system. Under no-till practices, the rice straw mulch creates an important habitat for many of the potato's natural enemies, also known as "friends of farmers." These are insects, spiders and micro-organisms essential if the plant pest population is to be successfully regulated in a natural organic way. An equally significant result is that mulching with rice straw reduces the need for irrigation from 5 000 to just 900 cubic meters of water per hectare. The incorporation of straw into the soil under no-till potato practices sequesters carbon in the soils (i.e., regulating ESS), improves soil structure and puts nutrients back into the soil, thereby enhancing soil fertility and increasing the water holding capacity of soils and reducing soil erosion (i.e., supporting ESS). Eventually, no-till potato increases tuber yield and carbohydrate content of potatoes (i.e., provision of ESS).

Socio-economic benefits and enabling policy environment

Farmers have reported other unintended positive results from their FFS experience such as improving family relations, income and policy environment (Box 5.1).

Improving family life and income: No-tillage potato production has not only helped smallholders to increase their income but also to improve quality of family life as perceived by one of the farmers (Box 5.1). In the Vietnam case, the results also show that no tillage practice has in fact improved farmer organization and institution building.

Box 5.1 Contribution of FFS on no-tillage potato production to improving family life

Mr. Bui Dinh Hau, the Chief of the hamlet and husband of Mrs. Thuy, who joined the first pilot FFS, was one of two male farmers in his FFS. The rest were women. Each time he came home after attending an FFS session, he would inform his wife about what he had learned. That same season, together they applied what Mr. Hau learned from the FFS in their own fields. He claims "*Our relationship as husband and wife has become closer as a result of no-tillage potato IPM because it has made production easy and gives us more income. We have been practicing it since the time we learned it. We taught our sons what we have learned. It is only the members of our family – my husband and sons and myself – who work together in farming. Before, it would take one person a day to harvest one sào but now one person can harvest three sàos in a day . . . and not feel tired. We do not dig but simply pull the rice straw away."*

Source: Towards a Non-toxic Asia (2016), 1 sào = 360m^2 in northern Vietnam.

No-tillage potato production demonstrated how the core group of farmers, who are FFS alumni, play a key role in institution building at community level. Assisted by staff of the provincial Plant Protection Sub Department (PPSD), the core groups became local extension agents and resource persons. For example, led by members of the core groups, more farmers are now working with national research institutions in trying out no-tillage potato production practices to grow new varieties of potatoes as well as other crops. Government staff regularly share new information with core groups of farmers to assist them in developing sound and location-specific innovative strategies that could help them better cope with production challenges, changing environments and ecosystems. From an initial FFS on no-tillage potato production, core groups of farmers have emerged empowered with ecosystem knowledge, critical-thinking skills and stronger capabilities for employing community problem solving processes. This allows farmers to design strategies and assist other farmers in their community to diversify their rice-based farming systems. Such capacity in the hands of smallholder farmers is vital for addressing rapidly changing production environments for development of more profitable and sustainable livelihoods.

Enabling policy environment

The positive response of MARD to scale out the no-tillage potato IPM to other potato growing regions in the country clearly shows how the policy support can fast spread the innovation (Box 5.2). FAO has formally recognized the no-tillage potato production innovation as a Save and Grow Sustainable Intensification of Crop Production system of global relevance worth further promotion, adaptation and investments for scaling out (FAO, 2016). This will not only help to reduce the environmental impacts from burning of rice straw, but also support reduced costs of cultivation and improved income for farmers.

Price fluctuations discourage farmers from increasing potato production, and even resulted in dumping of potatoes. Hence, supportive policies are needed to strengthen the linkages among value chain actors in potato farming and to improve the pre-production as well as post production phases of potato farming by investing in research and development, e.g., in seed systems, cold-storage capacity and providing up-to-date market information to VC actors using smart phones and other online communication tools, and improving other related infrastructures.

Box 5.2 Development of an enabling policy environment

On 24 August 2011, Vietnam's Ministry of Agriculture and Rural Development (MARD) recognized no-tillage potato IPM as a promising model and issued Directive 1380/BVTV-TV for all potato producing provinces in the country to apply the practice in Spring-Season 2012. On 28 May 2013, MARD issued Decision Number 204/QD-TT-CLT recognizing *"no tillage potato IPM in combination with rice straw mulch"* as an *"agricultural technical advancement"* instructing all potato growing provinces to apply the practice. On 31 May 2013, the Plant Protection Department, MARD received the Vietnam Environment Award under Decision No. 832/QD-BTNMT issued by the Ministry of Natural Resources and Environment for its widely applied, economically and environmentally effective work on no-tillage potato. On 13 November, 2015 MARD presented the Vietnam Golden Rice Award to the National IPM Programme Coordinator and Deputy Director of Plant Protection Department, Mr. Ngo Tien Dung, the lead author of this chapter, for the pioneering and innovative work on no-tillage potato production in rice-based farming systems. The award recognizes groups and individuals who have contributed actively to serve the cause of industrialization and modernization of agriculture and rural areas in Vietnam.

Source: www.vegetableipmasia.org/news/view/125.

Lessons from the Vietnam case

The initiative on no-tillage potato has demonstrated how intensified crop production can be diversified and become more profitable and sustainable while addressing issues such as lack of labor and water, contemporary realities confronting farming communities in Vietnam and elsewhere in the world. The Vietnam case study also clearly demonstrates the key role national and local governments can play in promoting and investing in farmer education about crop production innovations and providing an enabling policy environment in tandem with strategic investments in quality farmer education through Farmers Field Schools for such innovations to be scaled out. Such support will be vital for smallholder farming communities to successfully adapt to rapidly changing production environments and a changing climate.

Improving potato value chains and markets

Before suggesting any interventions to improve VCs, analysis should be carried out to characterize, describe and understand the various actors along the VC (FAO and CFC, 2010). The chain actors in potato include growers/producers, processors, traders and consumers. The focus should not only be on the production side introducing good management practices, but also on the post-harvest handling and marketing. The main activities that can improve the potato VCs in the pre-production and post-production phases are listed in Box 5.3.

Box 5.3 Activities that can strengthen potato value chain

Pre Production side

- Good quality seed material accessible to smallholders,
- Promoting varieties that are resistant to climate extremes, and
- Training farmers on soil fertility management and late blight control.

Post production side

- Value addition through quality control at the farm level and beyond,
- Low cost storage devices, and
- Organizing farmers into collectives and linking to reliable markets.

Cross-cutting issues

- Private sector involvement,
- Gender empowerment,
- Capacity building, and
- Better services provision including research and extension.

Source: Authors' own interpretation.

In this case, we grouped the aforementioned activities into three broad drivers: technologies, enabling institutions and policies. The extent to which these activities contribute to the VC improvements are described below:

Technologies

The private sector can play an important role in developing a more efficient seed potato value chains that can improve production, distribution, use and yields (USAID, 2011). Public and private sector agencies' involvement in developing seed potato value chains is recommended in Ethiopia (Hirpa *et al.*, 2016). The diverse genetic resources in potato should be explored further to address the current climatic, food security and nutrition challenges. The introduction of orange-fleshed sweet potato has increased its production in Africa. This could provide a good case for future strategies to increase its production. It is important to make healthy seed material that can reduce diseases and improve resistance to droughts, accessible to smallholders (Karanja *et al.*, 2014). There are many techniques including the rapid seed multiplication that can help to shorten the duration of certified seed potato (CSP) production and thereby increase the supply of seed material to smallholders as observed in Kenya (Okello *et al.*, 2017). CSP can have a positive effect on both yield and food security. Other methods such as tissue culture in the form of in vitro micro and mini-tuber production are popular in South America as important sources of potato seed material (Huarte, 2006; Hirpa *et al.*, 2016).

Starting with *soil fertility and water management*, smallholders need training to optimize input usage that can increase yields and profits. Evaluation of potato cultivars with high Nitrogen and Phosphorous efficiency is essential for sustainable production of the crop (Gebru *et al.*, 2017b). Yield response variation of potato varieties can differ in their nutrient uptake and use efficiencies, and some of the improved potato varieties can respond better and give higher yields. The varieties that respond well, and have market preference are to be promoted as part of the VC improvement. Adjusting planting dates, managing soil water supply through irrigation, and drought resistant varieties can help potato farmers to adapt to climate change (FAO, 2008). In Kenya, potato farmers use crop diversification and off-season approaches to adapt to rainfall variability including irrigation in the area. In many situations, improving micronutrient availability in the soil and resistance to heat and drought can significantly increase productivity (Thiele *et al.*, 2010).

Kromann *et al.* (2014) find that the most promising measures for *potato late blight disease*, are to integrate the use of resistant cultivars, fungicides (for late blight) and capacity building of farmers to manage the disease. For other pests, cultural management practices and biological control could help smallholders to reduce the pest damage. Participatory variety selection can be an effective approach where scientists and farmers have the opportunity to cooperate and choose the potato varieties that address the local problems that vary from one region to another even within one country (Kolech *et al.*, 2015).

Enabling institutions and policies

Smallholders need better and timely *access to microfinance institutions* to purchase farm inputs and increase productivity (Assam *et al.*, 2012). This is a major constraint for smallholders who find it difficult to purchase certified seed material on time from the formal sources. There is a need for innovative extension services and farmer trainings, accompanied with improved access to credit in order to improve potato production efficiency of smallholders (Nyagaka *et al.*, 2009). Sebatta *et al.* (2014) recommend that potato farmers should be sensitized on investment of external incomes into potato production and the village market collection centers should be promoted to provide farmers better access to markets. Market linkages to support potato farmers are comparatively well established in Asia, especially in China, India, Vietnam, Thailand etc. Lessons from Asia can help boost potato production in Africa and elsewhere, especially the potato farmer collectives which are well organized to exploit the marketing opportunities. This requires policies that promote collective action among smallholders because it eases access to production and marketing information as well as cheaper inputs. According to Karanja *et al.* (2014), policies that spur investment in public infrastructure, private investments and collective organization can address the problems of high transaction costs to investors, and reduce the risks faced by farmers (Kuyiah, 2007).

National Extension Services should integrate themselves well into the VC development, and at the same time engage in farmer capacity building on VC enhancement, including through Farmers Field Schools. Good examples of VC enhancement from other fruit products such as Avocado, Mango can be used in potato. It is important that Extension Services, while designing future programs, consider the underlying social structures, in particular, the gender relations in potato farming that influence variety selection, crop management and post handling operations (Kolech *et al.*, 2015). Dissemination of technologies for better understanding and efficient use of resources should preferably be in local languages and through mass media targeting both men and women farmers (Okonya and Kroschel, 2014).

Empowerment of women farmers through training on market choices, home economics, entrepreneurship skills, decision making and leadership capacity are potential areas for reducing gender differences (Kuma and Limenih, 2015). Without targeting women, potato production on smallholder farms cannot be improved, given the increasing role they will play in farming in rural areas. Experiences from dairy and horticulture projects show that public and private sector initiatives can successfully involve women and increase production, income and the stock of household assets (Quisumbing *et al.*, 2015). Future adaptive measures need to encourage gender-equitable outcomes if the potato VCs have to be improved in smallholder settings. In particular, policies should support women's acquisition and ownership of the physical assets required to expand production or enter other nodes of the value chain.

Market value addition: Farmers have to be trained in quality grading of potatoes, not just based on color and shape but also the variety, that is often not done on small farms. This would fetch them a better price from intermediaries and other buyers. Establishing community owned warehouses close to farms can be useful to store the produce and sell when the market is conducive. In some regions, road networks to production areas have to be improved for easier movement of the produce and linkage to markets. In the African context, often the small-scale and local institutional innovations motivate smallholders to participate in the VC improvement. The State has not been able to provide adequate investments in public, as well as in private sectors nor the policy support necessary for the desired VCs improvements in potato. Hence, public–private partnerships at different scales are necessary to ensure demand-led VCs and sustainable intensification of potato production.

Conclusions and recommendations

The demand for potato is likely to grow in the coming years. It is time for governments to invest in improving the potato VCs, make quality seed material available for smallholders, provide adequate support and quality education to farmers for improving adoption of more effective, efficient and sustainable agronomic practices, reduce gender constraints, and ensure stable prices for the produce. Experiences from various regions emphasize good management practices and intensification, need-based application of nutrients, and improving the entire VC including post-production processes and market operations. Future strategies aiming at sustainable intensification of potato production should ensure increasing productivity and minimizing the environmental impact. Low-input, stable-yielding potato varieties will enable smallholder farmers to take advantage of market opportunities.

To fully tap the potential of potato production, thereby improve food and nutrition security and livelihoods of smallholder farmers, interventions are needed in technology, institutions and policies that can improve potato production in developing countries.

Technologies

The Vietnam experience in no-till potato cultivation with IPM practices using the FFS approach has shown how potato production can become more profitable and environmentally sustainable. Lessons from the Vietnam case can be useful to other regions where potato is becoming an important food and cash crop to smallholders impacted by similar challenges especially in parts of Africa. In addition to this:

- *Healthy seed material*: The demand for potato is likely to grow in the coming years. It is time for governments and private sector to facilitate provision of better access to quality and affordable seed material for smallholders through rapid seed multiplication that shortens the duration of certified seed potato production or using tissue culture methods such as in vitro micro and mini-tuber production.

- *Soil fertility management*: A need-based nutrient management and measures that promote organic potato farming is necessary to address the smallholder problems and enable sustainable intensification of potato production while minimizing the adverse environmental impacts.
- *Disease and pest control*: Integrated pest management is necessary to address the smallholder problems related to potato pest and diseases. This practice includes the use of resistant cultivars, responsible use of fungicides (for late blight) and capacity building of farmers on pest and diseases control, preferably through Farmers Field Schools. Cultural practices and biological control could also help smallholders to reduce pest damage. FAO and CIP have developed an Ecological Guide for Potato IPM and an associated FFS Manual with relevant learning exercises for farmers (www.vegetableipmasia.org/documents/type/1 under the title *An Ecological Guide to Potato Integrated Crop Management: Farmer Field School for Potato Integrated Pest Management: Facilitator's Guide.*

Market and institutions

- *Post-harvest handling operations*: Establishing community owned warehouses close to farms can be useful to store the produce and sell when the market is conducive, provided that the road networks to production areas are easier for movement of the produce and linkage to other market outlets.
- *Market value addition*: improving the entire VC including post-production processes and market operations through potato products transformation (to highly priced products), commanding price premium (ensuring stable prices for the produce), niche marketing and offering services.
- *Gender mainstreaming*: Potato production on smallholder farms cannot be improved without empowering women, which is possible through training on potato market choices, home economics, entrepreneurship skills, decision making and leadership capacity. These are potential areas for reducing gender differences.

Policy interventions

- *Market linkages*: Lessons from Asia with regard to polices that promote collective action among smallholders can help boost potato production in Africa and elsewhere because it eases access to production and marketing information as well as cheaper inputs. These policies address the problems of high transaction costs to investors, and reduce farmers' risks by increasing investment in public infrastructure and private sectors, and forming collective farmer organizations.

- *Support for women farmers*: Policies should support women's acquisition and ownership of the physical assets required to expand potato production in different nodes of the value chain. Development strategies and innovations to be promoted through government and private sector extension systems should also aim to reduce labor and drudgery in potato production and agriculture in general, in line with a greater recognition of the key role that women play in contemporary agriculture as the Vietnam case study clearly shows.

References

Assam, M.M., Edriss, A.K. and Matchaya, G.C. (2012) Unexploited profit among smallholder farmers in Central Malawi: What are the sources? *International Journal of Applied Economics*, 9 (2), pp. 83–95.

Bac Giang Plant Protection Sub Department (2016) *Report on Minimum Tillage Potato Production (FAO project): Towards a non-toxic Southeast Asia*, Hanoi.

Bruun, E.W., Müller-Stöver, D., Ambus, P. and Hauggaard-Nielsenet, H. (2011) Application of biochar to soil and N2O emissions: Potential effects of blending fast-pyrolysis biochar with anaerobically digested slurry. *European Journal of Soil Science*, 62 (4), pp. 581–589.

Consultative Group on International Agricultural Research (CGIAR) (2017) Understanding gender roles in Uganda's potato and cooking banana value chains. Available at: www.rtb.cgiar.org/blog/2016/07/25/understanding-gender-roles-ugandas-potato-cooking-banana-value-chains/ (accessed 10 September, 2017).

Devaux, A., Kromann, P. and Ortiz, O. (2014) Potatoes for sustainable global food security. Available at: https://link.springer.com/article/10.1007/s11540-014-9265-1 (accessed 09 September 2017).

Finckh, M.R., Schulte-Geldermann, E. and Bruns, C. (2006) Challenges to organic potato farming: Disease and nutrient management. *Potato Research*, 49 (1), pp. 7–42.

Firman, D.M., Potato Council Research Project R269 (2008) *Factors Affecting Tuber Numbers per Stem Leading to Improved Seed Rate Recommendations: Introduction to the project and key findings from analysis of historic data.* Cambridge, Oxford: Potato Council Research Project R269.

Food and Agriculture Organization of the United Nations (FAO) (2008) International Year of the Potato. Available at: www.fao.org/potato-2008/en/events/book.html (accessed 09 September 2017).

Food and Agricultural Organization (FAO) (2009) Sustainable potato production guidelines for developing countries. Available at: www.fao.org/3/a-i1127e.pdf (accessed 15 September 2017).

Food and Agricultural Organization (FAO) and Common Fund for Commodities (CFC) (2010) Strengthening potato value chains: Technical and policy options for developing countries. Available at: www.fao.org/docrep/013/i1710e/i1710e.pdf (accessed 12 September 2017).

Food and Agriculture Organization of the United Nations (FAO) (2016) *Save and Grow in Practice – Maize, Rice, Wheat: A guide to sustainable cereal production.* Rome: FAO.

Gebru, H., Mohammed, A. and Belew, D. (2017a) Assessment of production practices of smallholder potato (*Solanum tuberosum* L.) farmers in Wolaita zone, southern Ethiopia. *Agriculture and Food Security*, 6, p. 31.

Gebru, H., Nigussie, D., Ali, M. and Derbew, B. (2017b) Nitrogen and phosphorus use efficiency in improved potato (*Solanum tuberosum* L.) cultivars in southern Ethiopia. Available at: https://link.springer.com/article/10.1007%2Fs12230-017-9600-6 (accessed 09 September 2017).

Hao, T.Q. (2017) Technique for processing potato tubers before growing. Available at: http://nongnghiep.vn/cach-bo-va-dat-cu-khoai-tay-giong-post2660.html (accessed 10 August 2017).

Hirpa, A., Meuwissen, M.P.M., Lommen, W.J.M., Oude Lansink, A.G.J.M., Tsegaye, A. and Struik, P.C. (2016) Improving seed potato quality in Ethiopia: A value chain perspective. Available at: www.wageningenacademic.com/doi/abs/10.3920/978-90-8686-825-4_5 (accessed 12 September 2017).

Huarte, M. (2006) Seed potato strategies in Latin America. Available at: www.potatocongress.org/wp-content/uploads/2012/03/Dr_Marcelo_Huarte.pdf (accessed 12 January 2018).

Independent. Global rice shortage caused by El Niño threatens rice crisis. Available at: www.independent.co.uk/news/world/asia/global-rice-crisis-el-nino-prices-increase-a7012526.html (accessed 23 March 2017).

International Potato Center (CIP) (2017) Potato facts and figures. Available at: http://cipotato.org/potato/facts/ (accessed 12 February 2017).

Independent (2016) Global rice shortage caused by El Niño threatens rice crisis. Available at: www.independent.co.uk/news/world/asia/global-rice-crisis-el-nino-prices-increase-a7012526.html (accessed 23 March 2017).

Karanja, A.M., Shisanya, C. and Makokha, G. (2014) Analysis of the key challenges facing potato farmers in Oljoro-Orok Division, Kenya. Available at: http://file.scirp.org/Html/4-3000628_48689.htm (accessed 06 September 2017).

Kolech, S.A., Halseth, D., Perry, K., De Jong, W., Tiruneh, F.M. and Wolfe, D. (2015) Identification of farmer priorities in potato production through participatory variety selection. Available at: https://link.springer.com/article/10.1007/s12230-015-9478-0 (accessed 10 September 2017).

Kromann, P., Mietbauer, T., Ortiz, O. and Forbes, G.A. (2014) Review of potato biotic constraints and experiences with integrated pest management interventions. Available at: https://link.springer.com/chapter/10.1007/978-94-007-7796-5_10 (accessed 09 September 2014).

Kuma, B. and Limenih, B. (2015) Women farmers in practices: Opportunities and challenges in accessing potato production technologies in Welmera Ethiopia. *Asian Journal of Agricultural Extension, Economics & Sociology*, 6 (3), pp. 149–157.

Kuyiah, J.W. (2007) *Economic analysis of smallholder agricultural production under conditions of risk: The case of Vihiga and Kilifi Districts in Kenya*. MSC Thesis, Egerton University, Kenya.

Lutaladio, N., Ortiz, O., Haverkort, A. and Caldiz, D. (2009) *Sustainable Potato Production: Guidelines for developing countries*. Rome: Food and Agriculture Organization of the United Nations.

McKenzie, B. (2013) Soil platforms for potatoes. Available at: https://potatoes.ahdb.org.uk/publications/r467-soil-platforms-potatoes (accessed 15 September 2017).

Maranzani, B. (2013) After 168 years, potato famine mystery solved. Available at: www.history.com/news/after-168-years-potato-famine-mystery-solved (accessed 17 September 2017).

MARD (Ministry of Agriculture and Rural Development) Vietnam (2013) Decision No: 204/QD-TT-CLT acknowledging "Technical protocol for minimum tillage production of potato covered with rice straw in northern provinces" as a technical advance. Available at: www.ppd.gov.vn/index.php?language=vi&nv=news&op=Ung-dung-chuyen-giao-TBKT/Quyet-dinh-204QD-TT-CLT-ve-viec-cong-nhan-Quy-trinh-ky-thuat-trong-khoai-tay-bang-phuong-phap-lam-dat-toi-thieu-co-phu-rom-ra-tai-cac-tinh-phia-Bac-la-tien-bo-ky-thuat-ke-web-tphc-320 (accessed 17 February 2017).

Miura, Y. and Kanno, T. (1997) Emissions of trace gases (CO_2, CO, CH_4, and N_2O) resulting from rice straw burning. *Soil Science Plant Nutrition*, 43 (4), pp. 849–854.

Mudege, N.N., Kapalasa, E., Chevo, T., Nyekanyaka and Demo, P. (2015) Gender norms and the marketing of seeds and ware potatoes in Malawi. *Journal of Gender, Agriculture and Food Security*, 1 (2), pp. 18–41.

Nyagaka, D.O., Obare, G.A. and Nguyo, W. (2009). Economic efficiency of smallholder Irish potato producers in Kenya: A case of Nyandarua North District. Available at: http://ageconsearch.umn.edu/bitstream/49917/2/CCONTRIBUTED_PAPER_98.pdf (accessed 06 September 2017).

Okello, J.J., Kwikiriza, N., Ogutu, S., Barker, I., Schulte-Geldermann, E., Atieno, E. and Ahmed, J.T. (2017) Productivity and food security effects of using of certified seed potato: the case of Kenya's potato farmers. *Agriculture and Food Security*, 6, pp. 25.

Okonya, J.S. and Kroschel, J (2014) Gender differences in access and use of selected productive resources among sweet potato farmers in Uganda. Available at: https://agricultureandfoodsecurity.biomedcentral.com/articles/10.1186/2048-7010-3-1 (accessed 10 September, 2017).

Plant Protection Department (PPD)(2012) *Report on Minimum Tillage Potato Production in Hanoi and Thai Binh* (FAO project): Towards a non-toxic Southeast Asia, Hanoi.

Plant Protection Department (PPD) (2015) *Report On Minimum Tillage Potato Production* (FAO project): Towards a non-toxic Southeast Asia, Hanoi.

Quisumbing, A.R., Rubin, D., Manfre, C., Waithanji, E., van den Bold, M., Olney, D., Johnson, N. and Meinzen-Dick, R. (2015) Gender, assets, and market-oriented agriculture: Learning from high-value crop and livestock projects in Africa and Asia. Available at: https://link.springer.com/article/10.1007/s10460-015-9587-x (accessed 17 September 2017).

Romasanta, R., Sander, B.O., Gaihre, Y.K., Alberto, M.C., Gummert, M., Quilty, J., Nguyen, V.H., Castalone, A.G., Balingbing, C., Sandro, J., Correa, T. and Wassmann, R. (2017) How does burning of rice straw affect CH_4 and N_2O emissions? A comparative experiment of different on-field straw management practices. *Agriculture, Ecosystems and Environment*, 239, pp. 143–153.

Rosen, C.J. (2016) Nutrient management for potato production. Available at: www.hort.cornell.edu/expo/proceedings/2016/Potato.Nutrient%20management%20for%20potato%20production.Rosen.pdf (accessed 15 September 2017).

Sebatta, C., Mugisha, J., Katungi, E., Kashaaru, A. and Kyomugisha, H. (2014) Smallholder farmers' decision and level of participation in the potato market in Uganda. *Modern Economy*, 5 (8), pp. 12.

Thiele, G., Theisen, K., Bonierbalel, M. and Walker, T. (2010) Targeting the poor and the hungry with potato science. *Potato Journal*, 37 (3–4), pp. 75–86.

The Crop Trust (2017) The Crop Trust. *Potato*. Available at: www.croptrust.org/crop/potato/ (accessed 12/02/17).

Truc, N.T.T., Sumalde, Z.M., Espaldon, M.V.O., Pacardo, E.P., Rapera, C.L. and Palis, F.G. (2012) Farmers' awareness and factors affecting adoption of rapid composting in Mekong Delta, Vietnam and Central Luzon, Philippines. *Journal of Environmental Science and Management*, 15, pp. 59–73.

Uddin, M.A., Yasmin, S., Rahman, M.L., Hossain, S.M.B. and Choudhury, R.U (2010) Challenges of potato cultivation in Bangladesh and developing digital databases of potato. *Bangladesh Journal of Agricultural Research*, 35 (3), pp. 453–463.

UN (United Nations–Vietnam) (2013) Greenhouse gas emissions and options for mitigation in Viet Nam, and the UN's responses. Available at: www.un.org.vn/en/publications/cat_view/130-un-viet-namjoint-publications/209-climate-change-joint-un-publications.html (accessed 15 March 2016).

USAID (2011) Roadmap for investment in the seed potato value-chain in Eastern Africa. Available at: http://cipotato.org/wp-content/uploads/2014/05/CIP_Roadmap-Final.pdf (accessed 12 September 2017).

Wang, J., Pan, X., Liu, Y., Zhang, X. and Xiong, Z. (2012) Effects of biochar amendment in two soils on greenhouse gas emissions and crop production. *Plant Soil*, 360, pp. 1–2.

Wurr, D.C.E., Fellows, J.R., Akehurst, J.M., Hambidge, A.J. and Lynn, J.R. (2001) The effect of cultural and environmental factors on potato seed tuber morphology and subsequent sprout and stem development. *Journal of Agricultural Science*, 136, pp. 55–63. Available at: http://wrap.warwick.ac.uk/797/1/WRAP_Wurr_cultural_environmental.pdf.

6 Pulses–millets crop diversification by smallholders and their potential for sustainable food and nutrition security

Mehreteab Tesfai, Udaya Sekhar Nagothu and Asfaw Adugna

Introduction

The current global agricultural landscape is dominated by cereals-based mono-cropping systems (particularly wheat, rice, maize) and potato in response to 'feeding the world' narratives. Hence, a few crops dominate the food crops produced worldwide and they constitute the staple diets for the majority of the world population. However, these specialized crops do not meet the required proteins, vitamins and minerals contents necessary for a healthy life. On top of that, crop monocultures have resulted in negative outcomes on the environment and ecosystem services (IPES-Food, 2016).

Shifting to diverse farming systems may be needed in the future as they promote sustainable intensification and provide multiple services and benefits from the use of the same piece of land. The services entail provision of diverse food sources for human nutrition, agro-biodiversity conservation, climate resilience, improving soil fertility, increasing smallholder income while reducing risks. Crop diversification by growing millets and pulses through crop rotations or intercropping cropping can provide these services and benefits on a truly sustainable footing. However, millets (including sorghum) and pulses have not been given enough attention and are even neglected in areas where they were first domesticated and cultivated for decades, despite their superior nutritional value, adaptability to climate change (CC) and their ecosystem services (Kajuna and Mejia, 2001). Lack of awareness, inadequate policy support, shortage of quality seeds and low prices are some of the main bottlenecks that discourage smallholders from growing pulses and millets.

Millets and pulses could represent important crops for future food and nutrition security (FNS) and human well-being considering the increasing world population and food demand on one hand and climate change effects of dwindling water supplies and soil resources on the other (ICRISAT & FAO, 1996). There is a growing interest in exploring the potential of the millets–pulses and/or cereal–pulses based cropping systems for sustainable food and nutrition security and ecosystem services. In general, data on global trends of individual millets is scarce.

The objectives of this chapter therefore are to: i) review and describe the state of millets and pulses crop production systems; ii) assess the contributions of millets–pulses cropping systems to FNS and ecosystem services; and iii) suggest possible changes in technology, institutions and/or policies that can help smallholders to adopt diverse cropping systems. The information in this chapter highlights the role of pulses and millets to achieve some of the UN Sustainable Development Goals (SDGs) and the changes needed in technology, institutions and policies to realize the potentials of these future crops.

The chapter is organized into four main sections. Section 1 provides an introduction to the global FNS challenges, and the need for crop diversification. This is followed by describing the state of millets and pulses and factors hindering their production and developments. Section three evaluates the contributions of millets–pulses cropping systems through the lens of SDGs and addressing environmental/climate challenges. Finally, concluding remarks and future directions with regard to the changes needed in technology, socio-economics, institutions and policies to unlock the potentials of millets–pulses cropping systems, are discussed.

Pulses and millets

'Pulse' is a collective term referring to the plants whose fruit is enclosed in a pod. Pulses are a subgroup of the legume family (Fabaceae) that are harvested solely as dry seed crops and that have edible seeds of leguminous crops. They include crops such as dried beans, lentils and peas, chick peas, pigeon peas, black gram, green gram. Whereas the term 'millets' includes tall grasses (classified under Gramineae family) with heads of small seeds that are grown for food, feed, forage and fuel in harsh environments where other crops generally fail (Kothari *et al.*, 2005). The major millets are sorghum and pearl millet, and minor millets include finger millet, foxtail millet, proso millet, kodo millet, barnyard millet, tef and little millet. Some of the advantages of pulses and millets are mentioned in Box 6.1. Both pulses and millets are staple crops for many of the poorest people in the semi-arid tropics of Asia and Africa. Millets in particular are regarded as a subsistence product and generally perceived as a famine crop for the poor (Kajuna and Mejia, 2001).

Box 6.1 Advantages of pulses and millets

Pulses

- Inexpensive source of plant-based protein, vitamins and minerals for the poor. Their consumption can help to manage obesity, diabetes and coronary conditions;
- Improve soil fertility via biologically fixing nitrogen and free soil-bound phosphorous (Hauggaard-Nielsen *et al.*, 2009);

(continued)

(continued)

- Ability to reduce greenhouse gases (GHG) emissions and enhance soil carbon sequestration (Zander *et al.*, 2016);
- Reduced fossil energy consumption in plant production (Jensen *et al.*, 2010);
- Efficient usage of residual moisture in the soil and thus withstands drought (FAO, 2016);
- Disrupt the life cycle of weeds, pests and disease agents (FAO, 2016);
- Have a broad genetic diversity from which climate-resilient varieties can be selected and/or bred (Calles, 2016).
- Pulses are low-fat, low-sodium, good source of iron, excellent supplier of fibre, excellent source of folate, good supplier of potassium, low glycemic index, and cholesterol-free (FAO, 2016).

Millets

- Highly drought resistant, resilient and adaptable crops to harsh, hot and dry environments (Gupta *et al.*, 2017);
- Grow in nutrient poor soils, making them an attractive crop for marginal farming environments (Padulosi *et al.*, 2015);
- Require low external inputs that fit very well to smallholder farmers' production systems (Esele, 2002) and principles of sustainable intensification;
- Offer multiple benefits for smallholders in terms of diverse diets that are rich in minerals and micronutrients (Singh, 2016);
- Millets are non-acid forming food, gluten free and easy to digest (Singh and Raghuvanshi, 2012);
- Source of feed for livestock, fuel and building materials (Tadele, 2016);
- Fast growing, produce abundant seed and grown as weed-supressing cover crop (Tadele, 2016); and
- Several types of traditional foods and beverages are made from millets (Tadele, 2016).

Source: Authors' own compilation from several sources.

Globally, the average grain yield of pulses per hectare was rather steady between 1961 and 1981 and increased to 1000 kg/ha in 2013 but then decreased slightly (Figure 6.1). In general, the yield of pulses per hectare plus the area under cultivation has not increased significantly at the global level (Figure 6.2).

Millets have been underutilized in general, some minor millets are under recognized, and their importance in diversification and complementing other food items are underestimated. Globally, the productivity of millets

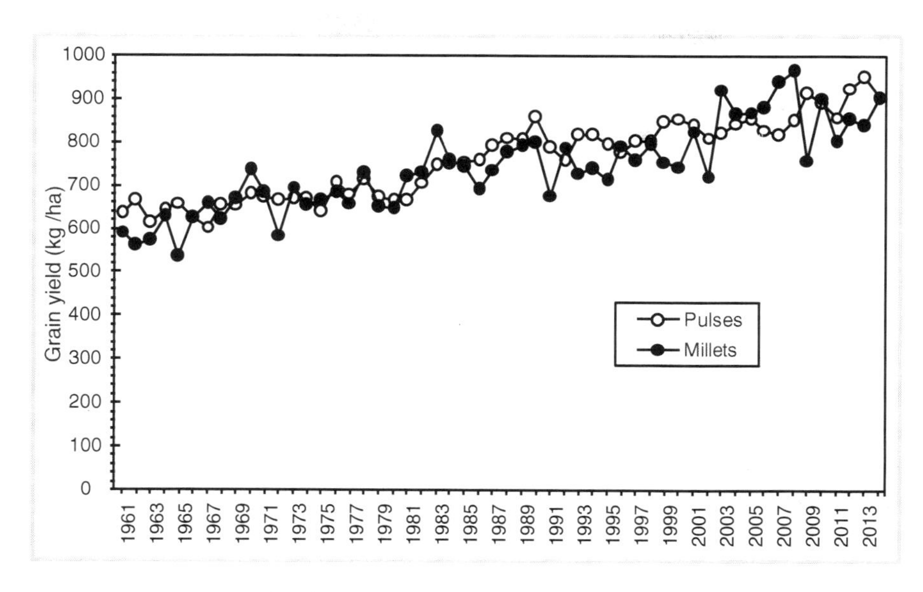

Figure 6.1 Average yield of coarse grain crops (millets in general) and pulses in the world during 1961–2014

Source: Authors' own compilation based on FAOSTAT data base: www.fao.org/faostat/en/#data/QC.

per hectare has increased during the last five decades perhaps due to improved varieties released from research institutions. Although millet intensification has increased, the area under cultivation of millets has not increased over the last five decades globally (Figure 6.2).

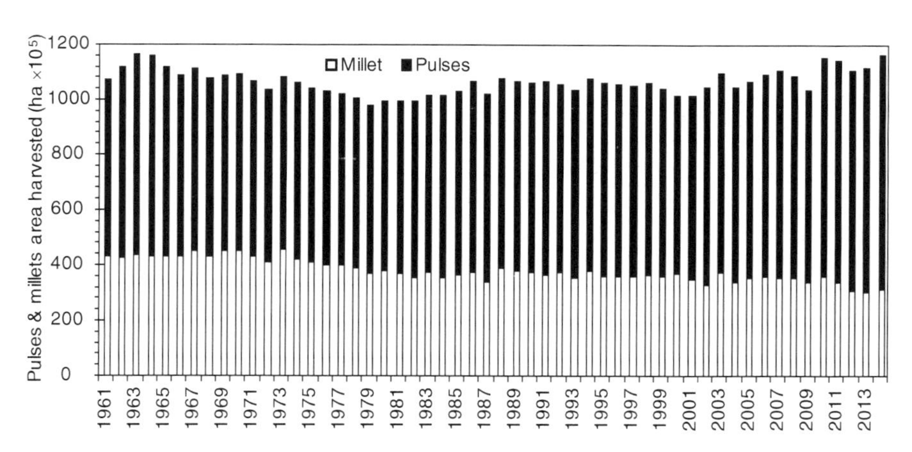

Figure 6.2 Total cultivation area of pulses and millets harvested during 1961–2014 in the world

Source: Authors' own compilation based on FAOSTAT data base: www.fao.org/faostat/en/#data/QC.

It is noteworthy to mention that there are some limitations and risks associated with growing pulses or millets throughout the value chains. For instance inclusion of grain pulses in the farming system could increase risk of nitrate leaching from soils (Urbatzka *et al.*, 2009) though it can be reduced by efficient rotation design, cover crops/catch crop management (Plaza-Bonilla *et al.*, 2015), intercropping pulses with millets (Hauggaard-Nielsen *et al.*, 2003) or sowing winter grain pulses, where possible. There are also some constraints associated with the growing of pulses and millets. Pulses are also commonly hard to cook and thereby require more energy or fuelwood (Graham and Vance, 2003). But, soaking helps the removal of the seed coat and thereby reduces the cooking time of pulses (Tharanathan and Mahadevamma, 2003). With regard to millets, seeds shatter easily and plants tend to lodge. The nature of their small seed size and lack of access to improved technology means there is more drudgery in millet cultivation. A majority of the millets have anti-nutritional factors such as polyphenols, phytates and tannins (Singh, 2016). The fluctuating and low prices discourage smallholders from growing millets and pulses and hence they are mostly consumed by poor people and they do not pay off the production costs of smallholders.

Pulses production system

This chapter focuses on pulses that are harvested for their dry grains such as lentils or chickpeas, thereby excluding legume crops that are harvested as greens (green peas, green beans, etc.), also crops used mainly for oil extraction (soybean and groundnuts) and seeds of clover and alfalfa that are exclusively used for fodder and green manuring. Pulses are mostly grown and consumed in developing countries, and in fact the largest pulse producing countries are found in the developing world where they contribute about 74 per cent of the total global pulse production. The major pulse producing countries are India (contributing nearly 25 percent), followed by china (10 per cent), Brazil (5 per cent), Canada (5 per cent), Myanmar (4 per cent) and Australia (4 per cent). And it is mostly smallholders in the developing countries that are involved in the production of pulses. The year 2016 declaration as the International Year of Pulses was rightly timed to heighten public awareness of the role of pulses and provide support to the farmers (FAO, 2016). Below is provided a brief description of the most common pulses.

Common beans, dry (*Phaseolus vulgaris L.*): The common beans category includes the haricot (white) beans, kidney beans, black beans, and numerous green beans (Murphy, 2007). Globally, the common bean species is the most cultivated (32 per cent of global production). India, Myanmar, Brazil are the major producing countries (FAO, 2014). The main production constraints are pests and diseases, and low prices.

Peas, dry (*Pisum sativum L.*): Peas are the pulses with second highest world production (i.e. 17 per cent of the total). They are one of the world's oldest domesticated crops, dating as far as back 10,000 years (Warkentin *et al.*, 2016). Canada, China, Russia are the largest pea producers (FAO, 2014). The main production constraints are pests and diseases, and low price.

Chickpea (*Cicer arietinum L.*): Chickpeas are the pulses with the third highest global production (i.e. 16 per cent of the total). Chickpea is listed as one of the first domesticated pulses and originated in the Middle East. The major producing countries are India, Australia, Pakistan (FAO, 2014). The main production constraints include pests and diseases and low market prices.

Broad bean (*Vicia faba L.*): Broad beans are the pulses with the fourth highest global production (i.e. 8 per cent of the total). Broad bean is also an ancient crop together with chickpea, peas and lentils. China, Ethiopia and Australia are the largest producers of broad beans (FAO, 2014). Pests and diseases are the main production constraints.

Cowpeas, dry (*Vigna ungiculata L.*): Cowpeas are the pulses with the fifth highest world production (i.e. 6 per cent of the total) and believed to have originated in Africa (Smartt, 1990). Nigeria, Niger, Burkina Faso are the largest cowpea producers (FAO, 2014). Pests and diseases are the main production constraints.

Lentils (*Lens culinaris L.*): Lentils are the pulses with the sixth highest global production (i.e. 6 per cent of the total) and were domesticated in the Middle East. Like chickpeas and peas, lentils were one of the first domesticated pulses (Stefaniak and McPhee, 2016). Canada, India, Australia are the largest lentil producers (FAO, 2014). Pests and diseases are the main production constraints.

Pigeon peas (*Cajanus cajan L.*): Pigeon peas are the pulses with the seventh highest global production (i.e. 4 per cent of the total). Pigeon peas are currently cultivated in Africa, Asia and the Americas. India, Myanmar, Malawi are the largest pigeon pea producers (FAO, 2014). The main production constraints are pests and diseases.

Lupines (*Lupinus albus L.*): The lupines category is aggregated into several species of the genus *Lupinus*. Lupines are the pulses with the eighth highest world production. Australia, Poland, Russia are the largest producers of lupines in the world (FAO, 2014). Pests and diseases are the main production constraints.

Common vetch (*Vicia sativa L.*): Vetches are the pulses with the ninth highest world production. The common vetch is native to southern Europe (Frame, 2005). This species is mainly used for animal nutrition. Ethiopia, Russia and Mexico are the largest producers of common vetches (FAO, 2014). Frost and severe winter are the main production constraints.

Bambara beans/Bambara groundnuts (*Vigna subterranean*): Bambara beans are the pulses with the tenth highest world production. It is generally accepted that Bambara bean is native to the African continent (Heller *et al.*, 1997) specifically to West Africa. Bambara bean is mainly cultivated in Africa where Mali, Burkina-Faso and Cameroon are the largest producers (FAO, 2014). Pests and diseases are the main production constraints.

Pulses as sources of nutrition

Pulses are important food crops that can play a major role in addressing future global FNS and climate challenges while providing multiple ecosystem services (FAO, 2016). They occupy an important place in human nutrition, especially in the dietary pattern of low-income groups of people in developing countries. In India, for example, they constitute the major source of protein. Pulses are often called 'the poor man's meat' for their protein source and their rich content of minerals, especially iron and zinc, excellent supplier of fibre, and vitamins (Tharanathan *et al.*, 2003). Except for carbohydrate, the nutrient composition of pulses in general is higher than rice and wheat (Table 6.1).

Millets production system

This chapter focuses not only on the most important millets, namely sorghum (*Sorghum bicolor*), pearl millet (*Pennisetum glaucum*), foxtail millet (*Setaria italica*), finger millet (*Eleusine coracana*) and tef (*Eragrostis teff*), but also gives insights on the minor millets that include little millet, kodo millet,

Table 6.1 Nutrient composition of pulses per 100 g edible portion of fresh weight basis

Pulse	*Energy* (kJ)*	*Protein (g)*	*Fat (g)*	*Carbohydrate* (g)*	*Dietary fibre (g)*	*Iron (mg)*	*Zinc (mg)*	*Folate (mcg)*
Chickpeas	1430	21.2	5.4	45.5	12.4	5.4	3.2	557
Cowpeas	1330	21.2	1.3	47.2	15.3	7.3	4.6	417
Lentils	1420	25.4	1.8	49.3	10.7	7.0	3.9	295
Pigeon peas	1260	18.4	1.5	43.2	20.2	4.7	2.0	456
Broad bean	1260	26.1	1.8	31.7	26.3	6.1	3.1	423
Lupines	1490	36.2	9.7	21.5	18.9	4.4	4.8	355
Common vetch	1470	28.3	1.6	43.5	5.0	0.04	3.6	n.a.
Bambara beans	1632	20.8	6.55	n.a.	10.3	8.8	1.9	n.a.
Wheat	1105	11.8	2.0	71.0	2.0	5.3	3.5	n.a.

Source: Authors' own compilation based on data from USDA (2015); Huang *et al.* (2017)
* metabolizable energy calculated from the energy producing food components (n.a. – not available).

barnyard millet, proso millet and fonio millet. Sorghum and pearl millet are traditionally known as major millets whereas the rest of the small millets are called minor millets because of the limited research investments given so far (Padulosi *et al.*, 2009). Due to this, minor millets are also classified under 'neglected and underutilized crops' (e.g. Dutta *et al.*, 2007).

In general, millets rank sixth among the world's most important cereal grains, sustaining more than one-third of the world population. They are one of the main staples for the world's poorest and most food-insecure people (ICRISAT & FAO, 1996). Asia and African countries are the biggest millet producers (Figure 6.3). Millets are the principal food cereals in the Sahelian zone of Africa, except in Senegal where rice is more important (Schaffnit-Chatterjee, 2014). A brief description of the major millets is provided in Figure 6.3.

Millets are reported to be the first crops to have been domesticated by humans more than 5000 years ago (Habiyaremye *et al.*, 2017). Millets including sorghum are C_4 crops that are climate change compliant. Different species of millets are adapted to different agricultural systems, but in general, most of them are reported to do well in nutrient poor soils and low moisture environments that are not suitable for other cereals (e.g. Chandel *et al.*, 2014). They require as little a rainfall as 200-500 mm for their production (Gupta *et al.*, 2017).

Sorghum (*Sorghum bicolor L. Moench*): Sorghum is cultivated over a wide range of agro-climatic zones in the tropics and believed to have been

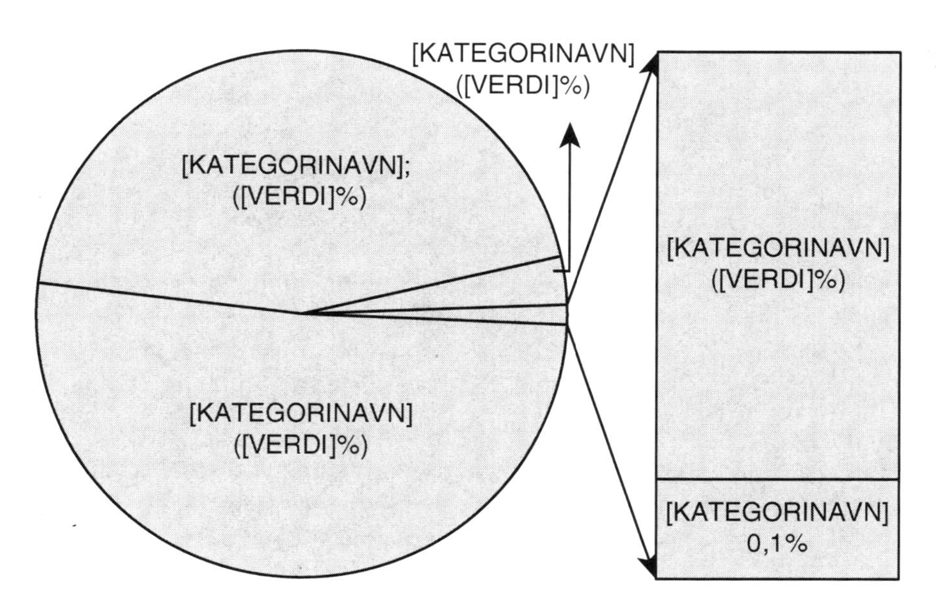

Figure 6.3 Worldwide millet production by region in 2014

Source: Adapted from Habiyaramye *et al.* (2017).

originated in the north quadrant of Africa (Dogget, 1988) more specifically, the region extending from Lake Chad to Ethiopia (Aldrich and Doebley, 1992). It is the fifth most important cereal grain in the world and serves as a staple for millions of people in Africa and Asia (Ejeta and Grenier, 2005). USA, India, China/Nigeria are the major producing countries in the world. The main production constraints of sorghum include birds' damage, weeds (Striga sp.), pests and diseases, changing food preferences, and low market prices (Thilakarathna and Raizada, 2015). Sorghum has versatile uses but mainly as food, feed for animals, construction purposes for houses and fences and fuel. It has been proposed as a model for designing C4 grass bioenergy crops (e.g., Mullet *et al.*, 2014).

Pearl millet (*Pennisetum glaucum, L.*): Pearl millet, also called bulrush millet, cattail millet, candle millet (Baker, 1998), is the millet with the second highest global production. It is a staple for over 400 million people in the semi-arid regions of the world living mainly in Asia and Africa (M'Khaitir and Vanderlip, 1992). It is adapted to marginal environments such as poor, droughty and infertile soils (Tadele, 2016). The best soils are well-drained light loamy to sandy soils, but pearl millet can tolerate acid subsoils as low as pH 4 (NRC, 1996). Pearl millet could have a great potential for industrial purposes as feedstock, for fuel, ethanol production in dry areas where maize and even sorghum are not able to cope (Wu *et al.*, 2006). The top three pearl millet producers are India, China, and Nigeria. Pests and diseases are the main production constraints of pearl millet.

Foxtail millet (*Setaria italic, L.*): It is the oldest of all the cultivated millets, and probably originated in China more than 4000 years ago (Oelke *et al.*, 1990). It ranks third in the total millet global production (Baltensperger, 2002), of which China produces more than 90 per cent and it is used both for food and feed. The top three foxtail millet producers are China, India and the USA. It has been proposed that *S. italica* and *S. viridis* together with sorghum (*Sorghum bicolor*) could serve as models for studying the C_4 photosynthesis and for engineering C_4 traits into important C_3 crops such as rice and wheat in the long run. The primary limitation of foxtail millet production is pests and diseases. In addition, foxtail millet continues to suffer from limited research and market developments (Baltensperger, 2002).

Finger millet (*Eleusine coracana, L.*): In Africa, finger millet (also called *ragi* in India) is mostly grown in areas with low soil fertility. Swinton *et al.* (1984) reported that 87 per cent of millets area in Niger is intercropped with pulses, notably pigeon pea, soybeans and field beans. In southern India, finger millet is intercropped with beans and pigeon pea (Hegde and Gowda, 1989). The top three finger millet producing countries are India, Ethiopia and Nepal. It has been recently named as a 'certain' crop for an 'uncertain' future (Gupta *et al.*, 2017). Diseases, especially blast, weeds, drudgery harvesting and threshing practices and lack of adequate marketing avenues are the main production constraints of finger millet.

Little millet (*Panicum miliare, L.*): Little millet is considered of Indian origin. It has short duration, withstands both drought and waterlogging conditions. India is the top producing country of little millet. The main production constraints include low productivity due to poor agricultural practices, inadequate research and extension services.

Kodo millet (*Paspalum scrobiculatum, L.*): Kodo millet is a native to tropical Africa and believed to have been domesticated in India about 3000 years ago. In West Africa, kodo millet frequently infests rice fields as a weed, but farmers tolerate it as they consider it as an insurance crop (NRC, 1996). India and West African countries are the major growers of kodo millet. The main production constraints include low market prices, low awareness of kodo poisoning and its precautionary treatment measures (Deshpande *et al.*, 2015).

Barnyard millet (*Echnochloa frumentacea, L.*): Barnyard millet, also called Japanese millet (Baker, 1998) is the fastest growing crop with 45–60 days maturity and voluminous fodder (Padulosi *et al.*, 2009). The top three barnyard millet producers are India, Japan and China. The main production constraints include low productivity.

Proso millet (*Panicum miliaceum L.*): Proso millet, also called broomcorn millet, hog millet and Hershey millet, is a short season crop with a growing season of 60 to 75 days and low water requirement (Baker, 1998). Its unique characteristics of drought and heat tolerance make proso millet a promising alternative cash crop. It is a highly nutritious crop used for human consumption, bird feeding, in cattle-fattening rations, and/or ethanol production. It is one of the oldest crops, and serves as food and in culture in China where it is believed to have been domesticated (Bettinger *et al.*, 2007; Wang *et al.*, 2016). The top proso millet producing country is USA (Habiyaremye, et al., 2017). The main production constraints are low market prices attributed by no value addition.

Fonio millet (white fonio (acha)-*Digitalis exilis* (Kippist) Stapf; black fonio (iburu)-*Digitaria iburua* (Stapf)): Fonio millet, also known as findi, fundi, Asian millet and hungry rice (NRC, 1996) is a fast maturing crop. The top three fonio millet producers are Guinea, Nigeria and Mali. Seed shattering and drudgery are the main production constraints (Jideani and Jideani, 2011).

Tef (*Eragrostis teff, Zucc.*): Tef is a versatile crop that grows under both low moisture and waterlogged conditions. It is a staple crop in Ethiopia and in the Eritrean highlands. The primary use of tef grain is for the preparation of the most popular flat bread locally known as *injera* (see Box 6.2). It is also used for making porridge and local alcoholic beverages, known as *tella* and *katikala* in Ethiopia (Teklu and Tefera, 2005). Moreover, its grain can be stored for many years without insect damage. Cattle prefer to feed on tef straw than any other cereal straw (Ketema, 1993). Tef straw is also used for construction purposes to reinforce mud for plastering walls of houses and other household items. The top three tef producers are Ethiopia, Eritrea and the USA. Production constraints include lodging and drudgery.

Box 6.2 The traditional food 'injera'

'*Injera*' is the popular flat circular bread (usually around 50 cm diameter) consumed in Ethiopia and Eritrea. It is made mostly from *tef* grain but can also be prepared from millets. To make injera, the grain is first debranned then ground into a fine flour using a disk mill. The flour is mixed with water to a 50:50 ratio and kneaded to form a dough. Starter culture from a previous fermentation is added to initiate fermentation. The dough is covered and allowed to stand for 2–3 days. After fermenting, hot water is added to the fermented batter, thinning it a great deal. Then, the mixture is allowed to sit for up to 8 hours. When ready to cook, the batter foams greatly and is poured onto a heated round clay griddle (locally called *mitad*) and a lid is placed over it. Good *injera* is soft and can be folded or rolled. *Injera* is served with other dishes such as meat, pulses or vegetables, thus making it a complete healthy meal.

Source: Authors' own interpretation.

Millets as sources of nutrition

Millets, in general, are a rich source of proteins, fibre, minerals, iron and calcium when compared to rice and wheat (Table 6.2). Finger millet, for example, has 7.6 times more Ca than rice while some of the other millets contain more Ca compared to rice and wheat (Table 6.2). Therefore, consuming millets-based diets can alleviate the malnutrition that affects millions of people in Africa and Asia.

Multi-dimensional benefits of pulses–millets cropping system

This section describes the multi-dimensional benefits delivered by a millets–pulses cropping system that are essential for human well-being. These include food and nutrition provision that improves human health, adaptation to a changing climate, improving soil fertility, enhancing agrobiodiversity, contributing to reducing GHG emissions, and providing socioeconomic benefits and cultural values for smallholder farmers and for the society at large.

Food and nutrition provision

One of the benefits provided by both pulses and millets is better food and nutrition. The nutrient composition of pulses and millets varies among

Table 6.2 Nutrient composition of millets as compared to rice and wheat grains per 100 g edible portion

Crop	*Protein (g)*	*Carbohydrate (g)*	*Crude fibre (g)*	*Iron (mg)*	*Calcium (mg)*
Sorghum	7.9	73.0	2.3	6.6	27.0
Pearl millet	10.6	67.0	1.3	16.9	38.0
Finger millet	7.3	72.0	3.6	3.9	344.0
Foxtail millet	12.3	60.9	8.0	5.0	31.0
Little millet	7.7	67.0	7.6	6.0	17.0
Kodo millet	8.3	65.0	9.0	12.0	27.0
Barnyard millet	6.2	65.5	10.1	15.0	11.0
Proso millet	12.5	70.4	2.2	10.0	14.0
Fonio millet	9.0	81.8	3.3	8.5	44.0
Tef	9.6	83.0	3.0	5.8	159.0
Rice	6.8	78.2	0.2	1.8	45.0
Wheat	11.8	71.2	1.2	3.5	41.0

Source: Authors' own compilation based on data from Habiyaremye *et al.* (2017).

different species and even cultivars as shown in Table 6.3. The variation in nutrient composition could be influenced by environmental conditions, storage, processing and genetics. This has led pulses and millets crop researchers to improve varieties through biofortification – the process of incorporating nutrient enhancing genes into food crops through conventional breeding and/or modern biotechnology.

Pulses are N (protein)-rich foods, used for livestock feeds, and green manures (Voisin *et al.*, 2014). Pulses contain on average 27 per cent protein, with over 30 per cent in some newly developed varieties (USDA, 2015) which is more than three times the amount of protein found in cereal grains such as wheat (Table 6.3). Pulses are a significant source of minerals, notably Fe, Mg, K, P and Zn and are a good source of B-vitamins such as Folate, which all have beneficial effects on human health. High levels of micronutrients were observed in pulses such as common beans (e.g. Blair, 2013). Due to their high nutritional value, pulses can improve the diet of the poorest who cannot rely on a diversified diet enriched by animal-derived food (such as meat, eggs, fish, milk) consumption. Besides, pulses are used for the provision of livestock and poultry feed. In this regard, lupins tend to be the most widely used pulse crops followed by field peas. The nutritional quality of pulses makes them particularly helpful in the fight against some non-communicable diseases such as heart disease, strokes and type two diabetes. Eating pulses as a replacement for some animal protein also helps limit the intake of saturated fats and increases the intake of fibre.

On the other hand, millets provide multiple securities (food, fodder, health, nutrition, livelihood and ecological benefits). Millets are higher in

Table 6.3 Nutrient compositions of pulses and millets per 100 g edible portion versus wheat*

Components	*Unit*	*Pulses*		*Millets*		*Wheat*
		Mean	*Range*	*Mean*	*Range*	*Mean*
Energy**	kJ	1350	1210–1490	1341	1142–1541	1105
Protein	g	27.3	18.4–36.2	9.9	7.3–12.5	11.8
Dietary fibre	g	18,5	10.7–26.3	5.1	1.2–9.0	2.0
Fat	g	5.1	0.53–9.74	3.1	1.1–5.0	1.1
Carbohydrate***	g	35.9	21.5–50.2	66.5	61–72	71.0
B-vitamins, Folate	µg	420	216–625	n.a.	n.a.	n.a.
Iron, Fe	mg	6.0	4.4–7.6	4.9	0.5–9.3	5.3
Calcium, Ca	mg	153	110–197	206	14–398	41
Magnesium, Mg	mg	4.5	4.2–4.8	154	137–171	3.6
Phosphorus, P	mg	277	251–303	242	188–296	400
Potassium, K	mg	1077	874–1280	366	314–418	330
Zinc, Zn	mg	3.5	1.96–5.0	2.6	1.4–3.7	3.5

* air-dry basis at 12 per cent moisture content ** metabolizable energy calculated from the energy producing food components. *** Available carbohydrate.

Source: Authors' own compilation based on data from Obilana (2003); Iqbal *et al.* (2006); Bergamini *et al.* (2013); FAO (2014); USDA (2015); Singh (2016).

calorific value, fibre, fat and minerals such as calcium and magnesium when compared to other cereals such as wheat (Table 6.3). However, millets contain lower nutritional values than pulses, except in carbohydrates, calcium and magnesium. Millets are generally gluten free crops, which can help people with gluten intolerance and they are easy to digest. They are also rich in micronutrients and the B vitamins such as niacin, vitamin B6 and folic acid (Pathak, 2013). Micronutrient deficiency is one of the widespread problems causing health issues, affecting billions of people in developing countries. Studies by Velu *et al.* (2006) at ICRISAT (in India) showed a large variability among the pearl millet genotypes in micro-nutrient contents for, e.g. Fe (30.1–75.7 mg/kg on dry weight basis) and Zn (24.5–64.8 mg/kg). Fonio millets proteins have relatively higher sulphur amino acid content (methionine and cystine) than rice proteins, which are crucial for proper heart function and nerve transmission (Jideani and Jideani, 2011). Cystine is a rare amino acid in cereals (NRC, 1996). A number of value added food products such as biscuits, cakes and beverages can be made from millets grains (e.g. Gaffa *et al.*, 2004).

Jayne *et al.* (2009) suggested that yields of sorghum, millet, groundnut and cowpeas could be easily increased by more than 300 per cent with rotation or intercropping pulses. More specifically, increased production and greater food utilization of mixed pulses and millets are important to achieve several of the UN's Sustainable Development Goals (SDGs), in particular SDGs 1, 2, 3, 12, 13, and 15 (Table 6.4).

Climate change adaptation and mitigation

The climate change resilience services offered by pulses and millets include adaptation to drought and climate resilience and climate mitigation in terms of emission reduction in N (including N_2O, NH_3, NO) and reduce loss of NO_3^- via leaching.

Drought and heat tolerance

Pulse species have a broad genetic diversity from which improved varieties can be selected and/or bred – an attribute that is particularly important for adapting to climate change because more climate-resilient varieties can be developed from this broad diversity. Since climate experts suggest that heat stress will be the biggest threat to pulse production (particularly to beans) in the

Table 6.4 Pulses and millets contributions to SDGs and targets

SDGs#	*Targets#*
1	1.2: Reduce at least by half the proportion of men, women and children of all ages living in poverty
2	2.2: End all forms of malnutrition and address the nutritional needs of adolescent girls, pregnant and lactating women and older persons
3	3.4: Reduce by one-third premature mortality from non-communicable diseases and promote mental health and well-being
12	12a: Support developing countries to strengthen their capacity to move towards more sustainable patterns of consumption and production
	12.2: Achieve the sustainable management and efficient use of natural resources
13	13.1: Strengthen resilience and adaptive capacity to climate-related hazards in all countries
	13.2: Integrate CC measures into national policies, strategies and planning
	13.3: Improve education, awareness-raising and capacity on CC mitigation, adaptation
15	15a: Increase financial resources to conserve and sustainably use biodiversity and ecosystems.
	15.3: Combat desertification, restore degraded soil, and strive to achieve a land degradation neutral world
	15.4: Ensure the conservation of mountain ecosystems including their biodiversity
	15.5: Take action to reduce the degradation of natural habitats, loss of biodiversity and extinction of threatened species

Source: Authors' own compilation based on information from UN (2015).

coming decades, these improved pulse varieties will be of critical importance, especially for low-input agricultural production systems. Further to that, pulses' ability to grow on residual soil moisture makes them climate resilient crops in water scarce areas (FAO, 2016).

Millets are also so hardy that they have always been grown in situations where there is a risk of famine because they offer a more reliable harvest relative to other cereal crops (Kajuna, 2001). Millets need very little water for their growth and production compared to other cereal crops. The common characteristics of all the millets is their adaptation to low moisture that is inadequate for other cereals (e.g. Tadele, 2016). For instance, pearl millet can grow in areas where annual rainfall is as low as 200 mm (M'Khaitir and Vanderlip, 1992; Baltensperger, 2002). Pearl millet and sorghum are the most climate resilient crops due to their drought and heat tolerance.

Greenhouse gases emission reductions

Emission reduction in N: Pulses have the potential to directly and indirectly reduce emissions of Greenhouse gases (GHGs), notably nitrous oxides and loss of plant nutrients such as nitrates and phosphorus. The manufacturing of synthetic fertilizers (such as DAP or Urea) is energy intensive and emits N_2O into the atmosphere, thus their overuse contributes to global warming and climate change. Pulses mitigate emissions of N_2O (a potent GHG) by reducing dependency on synthetic N fertilizers through supplying N via symbiotic fixation. This reduces combustion of fossil fuel that is needed to produce N fertilizers (Crews and Peoples, 2004). Pulses can provide inorganic N up to 4–11 fold greater than obtained in DAP or urea fertilizers through biologically fixing nitrogen (Table 6.5). Consequently, farmers can apply less N fertilizer to their soils when growing pulses in rotations or intercropping with millets or other cereals which in turn lowers fertilizer costs and reaps higher yields.

Pulses indirectly reduce emissions of carbon dioxide as Biological Nitrogen Fixation (BNF) saves the fossil energy resources required for manufacture of synthetic N fertilizers. Compared to cereals or pastures fertilized with N, pulses and pulses-based pastures can reduce fossil energy use by 35 to 60 per cent (Jensen *et al.*, 2012). Using a life cycle assessment approach, the introduction of pulses into crop rotations has been estimated to reduce the emissions of NH_3 by about 25 per cent and N_2O and NO by about 10 per cent (Nemecek and Baumgartner, 2006). Jensen *et al.* 2010) calculated average N_2O emissions from grain pulses of 1.23 kg N_2O–N/ha compared to 2.71 kg N_2O–N/ha from annual non-legume crops using data from 71 site-years of crop experiments. Aside from this, many pulses also promote higher rates of accumulation of soil carbon than cereals or grasses and thus contribute to improving the carbon sequestration of agro-ecosystems.

Millets also sequestrate carbon and thereby reduce the burden of GHGs (NAAS, 2013). Millets contain phytolith occluded carbon which is an

Table 6.5 Estimates of nitrogen fixed (kg N/ha) by some pulses and their contributions to urea/DAP

Pulse	*Nitrogen fixation**	*Contributions to Urea/DAP***	
	Range (kg N/ha)	*Urea (46% N)*	*DAP (18% N)*
Chickpea	23–97	2.1 fold	5.4 fold
Common bean	3–57	1.2 fold	3.2 fold
Cowpea	9–125	2.7 fold	6.9 fold
Lentil	35–100	2.2 fold	5.6 fold
Pea	46	1.0 fold	2.6 fold
Pigeon pea	4–200	4.3 fold	11.1 fold

* Source: Adapted from Wani and Lee (1992). ** Authors' own analysis.

important long-term terrestrial carbon reservoir that has a major role in the global carbon cycle (XinXin and Yuan, 2011). Millet fields in general emit less GHG when compared to other cereal cropped fields (Tongwane *et al.*, 2016).

Emission reduction in NO_3^- leaching: Emissions of reactive N to water through nitrate leaching and to the atmosphere through emissions of NH_3, N_2O and NO cause nutrient and acid accumulation in vulnerable ecosystems (Galloway *et al.*, 2004). In arable systems, contribution to nitrate leaching by pulses is conditioned by the crop management, the rotation design and the synchrony of crop N supply from pulses and the demand from following crops. Rotation design, cover crops (Plaza-Bonilla *et al.*, 2015) and inter-cropping of pulses with catch crops (Hauggaard-Nielsen *et al.*, 2003) are measures to reduce nitrate leaching. Deep rooted perennial pulses reduce the risk of groundwater contamination (caused by NO_3^- leaching) or development of dryland salinity and reduce eutrophication (Angus *et al.*, 2001).

Soil fertility improvements

Pulses have the potential to increase soil organic matter, decrease soil pH, improve soil structure and improve soil porosity (McCallum *et al.*, 2004) through their deep rooting system (for e.g. lupin). Nutrient recycling and soil formation are improved through the pulses' abilities to fix nitrogen and free phosphorous that can be used by plants, while improving soil fertility. Pulses are able to solubilize soil phosphorus through releasing root exudates that contain up to eight carboxylic acids (Egle *et al.*, 2003). This can improve the phosphorus (P) uptake of a cereal grown intercropped in a mixture (Li *et al.*, 2007) and also in cereals grown after a pulse crop (Nuruzzaman *et al.*, 2005). Switching from cereal monoculture to a rotation with pulse crops is reported to stimulate the accumulation of 0.5–1.0 t/ha of soil organic carbon annually, with the pulse component of the cropping sequence contributing up to 20 per cent of the carbon gain (Wu *et al.*, 2003).

Millets are more resilient on nutrient poor soils when compared to other cereals and reclaim soil quality in degraded lands. All millets are known to respond well to nitrogen and phosphorus fertilizers (Baker, 1998). Millet glume or chaff (residues left after threshing of millet) represents a potential source of reusable organic material when applied with N and P fertilizers. Millet glume-based compost ameliorates soil fertility and improves crop productivity because it contains macro- and micro-nutrients (Issoufa, 2015).

Biodiversity enhancement

Cultivating pulses with cereals (including millets, maize, rainfed rice) increases soil microbial biomass and its activity which enriches agrobiodiversity. Millets are not just crops but are cropping systems inherently biodiverse. In most millet fields, several landraces are planted on the same plot of land at the same time. This increases diversity in cropping systems and soil fauna.

i) *Diversity in cropping systems*: Pulses have a multiple role in promoting living organisms and ecological complexity to re-establish the natural good functioning of ecosystems. A high soil biodiversity not only provides ecosystems with greater resistance and resilience against disturbance and stress, but also improves the ability of ecosystems to suppress diseases. When used in multiple cropping systems, pulses help to curb and control pests and diseases. For instance, intercropping of sorghum with cowpeas and greengram (*Vigna radiate*) greatly reduces striga infestation (Khan *et al.*, 2007). Diversification of crop species via rotations reduces the requirement for pesticides and other agrichemicals, encouraging systems resilience and biodiversity (Jensen and Hauggaard-Nielsen, 2003; Köpke and Nemecek, 2010). Pulse production increases the diversity of cropping systems as pulses are minor crops and enable temporal and spatial diversification of the agro-ecosystem at the field and landscape level (Peoples *et al.*, 2009). Pulses are able to increase biodiversity as they are able to fix their own nitrogen into the soil through symbiosis with certain types of bacteria, namely *Rhizobium* and *Bradyrhizobium*.

ii) *Diversity in soil fauna*: Increased crop diversity supports increased above and below ground biodiversity (Köpke and Nemecek, 2010) including decomposer invertebrates such as earthworms and Collembola (Sabais *et al.*, 2011) and pollen- and nectar-gathering wild and domesticated bees as well as bumblebees that are attracted by the crop's mass-flowering habit (Köpke and Nemecek, 2010). The mass flowering habit of pulses is an excellent source of food for insect species and pulses improve soil biodiversity (Chaer *et al.*, 2011). Pulses enhance pollination by providing feed to pollinators and beneficial insects (Köpke and Nemecek, 2010). Some varieties of pulses are able to nurture the development of those organisms that are responsible for promoting soil structure and nutrient availability by freeing soil-bound phosphorous and/or increasing vital microbial biomass and activity in the soil.

Conclusions and future directions

In spite of all the positive qualities and benefits provided by pulse–millets (as described in the preceding sections), they are often viewed as less important when compared to cereal crops such as wheat, maize and rice. The current state of the art of pulse–millets cultivation systems varies between regions and countries. Worldwide, the area under pulses and millets production has been shrinking and the consumption of these crops has declined particularly since the Green Revolution. Unless special technological, market, institutional and policy improvements are taken, pulses and/or millets might disappear from the agrarian landscape of many countries over the next 50 years.

Technological improvements

There are a range of technology-related constraints that have to be addressed in order to improve millets–pulses production on smallholder farms.

i) *Improved seed varieties*: This includes better access to quality seeds of improved varieties that are rich in the essential micronutrients, minerals and vitamins for human nutrition and livestock feeding. The possible measures could be an increase in investments in research to develop and produce improved seed varieties that are climate resilient and tolerant to abiotic and biotic stress through participatory plant breeding programmes.

ii) *Improved cultivation practices including input management (crop, fertilizers, water and machinery).* There is a need to intensify crop diversification in smallholders' farming systems through crop rotations or intercropping of pulses with millets or other suitable cereals to maximize their synergetic effects on provision of food and nutrition, climate change adaptation and mitigation, soil fertility and biodiversity improvements, and their contribution to the 2030 Agenda of SDGs. To reduce post-harvest yield losses at all levels, develop low-cost pulse/millets seed harvesting/processing machines and storage warehouses/storage units that are accessible and affordable by smallholders.

Market value chain improvements

Smallholders in the developing world are the largest producers and, at the same time, consumers of pulses and millets. However, these smallholders are exposed to a range of socioeconomic constraints and risks.

i) *Measures to improve market facilities and strengthen value chains*: Smallholders lack access to up-to-date information on pulse and millets grains markets, e.g. in relation to market outlets, prices fluctuations.

They sell their products to nearby markets whatever price they get, often at low price through middlemen who exploit their weak points. In addition, low public awareness and changing food habits to animal products have contributed to a declining demand for pulses and millets. In most pulses and millet growing countries, the value chain (VC) actors are poorly organized and/or even not organized at all, in some countries. In these countries, value addition of pulse and millet products is non-existent and income generation from these crops is curtailed. Hence, pulses and millets growers are positioned at low economic competitiveness when compared to cereal farmers. Making pulses–millets marketing systems accessible and attractive and responding to the aspirations of consumers implies taking the following actions:

- investing in value added product innovations (for e.g. developing pulse/millet-based food recipes) gives the opportunity to diversify their use;
- promoting access to markets by establishing effective pulse/millet networks that connect the different VC actors and enhance public–private partnerships;
- delivering high standard quality grains at food marketing centres to retailers/traders;
- controlling the role of middlemen to make prices more consumer- and producer-friendly by establishing minimum support prices and weather based price insurance schemes.

Institutional and policy factors

The national and international research institutions have developed and released a number of agricultural technologies on pulses and millets production systems. These technologies entail improved seeds (that are resistant to abiotic and biotic stress and higher yielding), and modern cultivation practices related to crops, soil and water management. However, in general the adoption and diffusion of these technologies by smallholders is low mainly due to weak extension/advisory services and lack of policy support.

i) *Measures to improve extension and advisory services*: The services provided by private and public agencies are targeted towards only the production aspect of pulses and millets. In other words, the processing, marketing and consumption aspects of these crops are not address properly mainly due to low awareness and lack of capacity. Henceforth, there is no link between agriculture and nutrition/health. This is where the role of pulses and millets comes in to bridge the gap between agriculture and nutrition. The possible measures could be:

- an awareness-raising promotional campaign particularly targeting women, children and youth on the health and nutrition benefits of pulses and millets. The campaign should address the entire food systems approach (i.e. from the field to the fork) by undertaking life cycle assessment;

- providing customized trainings on seed production, multiplication, storage and consumption of pulses with millets;
- stimulating the development of agribusiness services to support smallholders' access to inputs and services, e.g. by supporting pulses–millets seed systems (such as community seed banks) that empower smallholders and improve the availability and accessibility of varieties that are suited to local conditions; and
- promoting research on innovative ways to make the cooking process of pulses easier (in particular) to increase the use of pulses among city-based consumers.

ii) *Enabling policies and institutional support*: Policies that are barriers to the development of pulses and millets cultivation include insecure land tenure systems (where land is owned by the state and the user does not possess long-term ownership as in the case of most African countries) and absence of clear strategy/incentives that promote pulse and millet cultivation, not only for increasing production but also for increasing smallholders' income and livelihood.

In cognizance of the advancing climate change and food insecurity challenges faced by smallholders, the policy environment in pulses and millets growing countries should be conducive and supportive. In this regard, some progress has been made by countries such as India where local governments have initiated a Public Distribution System to provide poor households with food items such as pulses at subsidized prices. This helps both producers and consumers to grow and consume pulses. With regard to millets, a country-wide millets network has been established and a series of policy papers on the role of millets in nutritional security have been produced in India (NAAS, 2013). There are also national and international initiatives under way (e.g. the 2016 international year of pulses by the UN) that are aimed at raising public awareness, and the pulses and millets improvements programme run, e.g. by ICRISAT.

Yet, to fully tap the potentials of pulses and millets sustainably and raise the profile of these climate-smart cropping systems and multipurpose crops, specific measures should be taken targeting the smallholders' needs and capacities.

References

Aldrich, P.R. and Doebley, J. (1992) Restriction fragment variation in the nuclear and chloroplast genomes of cultivated and wild *Sorghum bicolor*. *Theory of Applied Genetics*, 85, pp. 293–302.

Angus, J.F., Gaul, R.R., Peoples, M.B., Stapper, M. and van Hawarden, A.F. (2001) Soil water extraction by dryland crops, annual pastures and lucerne in south-eastern Australia. *Australian Journal of Agricultural Resource*, 52, pp. 183–192.

Baker, R.D. (1998). *Millet Production. Guide A-414*: Cooperative Extension Service, College of Agriculture and Home Economics, New Mexico University, Las Cruces, USA.

Baltensperger, D.D. (2002) *Progress with Proso, Pearl and Other Millets*. Alexandria, VA: ASHS.

Bergamini, N., Padulosi, S., Ravi, S.B. and Yenagi, N. (2013) Minor millets in India: A neglected crop goes mainstream. In: Fanzo, J., Hunter, D., Borelli, T. and Mattei, F. (eds) *Diversifying Food and Diets: Using agricultural biodiversity to improve nutrition and health*. New York, NY: Biodiversity International, Routledge, pp. 312–325.

Bettinger, R.L., Barton, L., Richerson, P.J., Boyd, R., Wang, H. and Choi, W. (2007) The transition to agriculture in northwestern China. In: Marsden, D.B., Chen, F.H. and Gao, X. (eds) *Developments in Quaternary Science, 9, Late Quaternary Climate Change and Human Adaptation in Arid China*. Amsterdam: Elsevier, pp. 83–101.

Blair, M.W. (2013) Mineral biofortification strategies for food staples: The example of common bean. *Journal of Agriculture and Food Chemistry*, 61, pp. 8287–8294.

Calles, T. (2016) The International Year of Pulses: What are they and why are they important? Available at: www.fao.org/3/a-bl797e.pdf (accessed 05 December 2016).

Chaer, G.M., Resende, A.S., de Balieiro, F.C., Boddey, R.M. (2011) Nitrogen-fixing legume tree species for the reclamation of severely degraded lands in Brazil. *Tree Physiology*, 31, pp. 139–149.

Chandel, G., Meena, R.K., Dubey, M. and Kumar, M. (2014) Nutritional properties of minor millets: Neglected cereals with potentials to combat malnutrition. *Current Science*, 107(7), pp. 1109–1111.

Crews, T.E. and Peoples, M.B. (2004) Legume versus fertilizer sources of nitrogen: Ecological tradeoffs and human needs. *Agriculture, Ecosystems and Environment*, 102, pp. 279–297.

Deshpande, S.S., Mohapatra, D., Tripathi, M.K. and Sadvatha, R.H. (2015) Kodo millet nutritional value and utilization in Indian foods. *Journal of Grain Processing and Storage*, 2 (2), pp. 16–23.

Doggett, H. (1988) *Sorghum*, 2nd edn. London, UK: Longman.

Dutta, M., Phogat, B.S. and Dhillon, B.S. (2007) Genetic improvement and utilization of major underutilized crops in India. In: Ochat, S. and Jain, S.M. (eds) *Breeding of Neglected and Underutilized Crops, Spices and Herbs*. Enfield, NH: Science Publishers, pp. 251–298.

Egle, K., Romer, W., Keller, H. (2003) Exudation of low molecular weight organic acids by Lupinus albus L., Lupinus angustifolius L. and Lupinus luteus L. as affected by phosphorus supply. *Agronomie*, 23, pp. 511–518.

Ejeta, G. and Grenier, C. (2005) Sorghum and its weedy hybrids. In: Gressel, J. (ed.) *Crop Ferality and Volunteerism*. Boca Raton, FL: Taylor & Francis, pp. 123–135.

Esele, J.P. (2002) Diseases of finger millet: A global overview. In: Leslie, J.F. (ed.) *Sorghum and Finger Millet Diseases*. Iowa, IA: Iowa State Press, Blackwell, pp. 19–26.

FAO (Food and Agriculture Organization of the United Nations) (2014) FAOSTAT: Statistics for the year 2014. Available at: http://faostat3.fao.org/home/E (accessed 15 March 2016).

FAO (Food and Agriculture Organization of the United Nations) (2016): FAO Committee on Agriculture, Twenty-fifth Session. The International Year of Pulses: Nutritious seeds for a sustainable future. COAG/2016/3. Available at: www.fao.org/3/a-mr021e.pdf.

Frame, J. (2005) *Forages Legumes for Temperate Grasslands*. Rome: Food and Agriculture Organization of the United Nations.

Gaffa T., Jideani I.A. and Nkama, I. (2004) Chemical and physical preservation of kunun zaki: A cereal based non-alcoholic beverage in Nigeria. *Journal of Food and Science Technology*, 41, pp. 66–70.

Galloway, J.N., Dentener, F.J., Capone, D.G., Boyer E.W., Howarth, R.W., Seitzinger, S.P., Asner, G.P., Cleveland, C.C., Green, P.A., Holland, E.A., Karl, D.M., Michaels, A.F., Porter, J.H., Townsend, A.R., and Vorosmarty, C.J. (2004) Nitrogen cycles: Past, present, and future. *Biogeochemistry*, 70, pp. 153–226.

Graham, H.P. and Vance, C. (2003) Legumes: Importance and constraints to greater use. Available at: www.plantphysiol.org/cgi/doi/10.1104/pp.017004.

Gupta, S.M., Arora, S., Mirza, N., Pande, A., Lata, C., Puranik, S., Kumar, J. and Kumar, A. (2017) Finger millet: A 'certain' crop for an 'uncertain' future and a solution to food insecurity and hidden hunger under stressful environments. *Frontiers in Plant Science*, 8, pp. 643.

Habiyaremye, C., Matanguihan, J.B., Guedes, J.D. Ganjyal, G.M., Whiteman, M.R., Kidwell, K.K. and Murphy, K.M. (2017) Proso millet (Panicum miliaceum L.) and its potential for cultivation in the Pacific Northwest, U.S.: A review, *Frontiers in Plant Science*, 7:1961. doi: 10.3389/fpls.2016.01961.

Hauggaard-Nielsen, H., Ambus, P. and Jensen, E.S. (2003) The comparison of nitrogen use and leaching in sole cropped versus intercropped pea and barley. *Nutrient Cycling in Agroecosystems*, 65, pp. 289–300.

Hauggaard-Nielsen, H., Gooding. M., Ambus, P., Corre-Hellou, G., Crozat, Y., Dahlmann, C., Dibet, A., von Fragstein, P., Pristeri, A., Monti, M. and Jensen, E.S. (2009) Pea–barley intercropping for efficient symbiotic N2-fixation, soil N acquisition and use of other nutrients in European organic cropping systems. *Field Crop Research*, 113, pp. 64–71.

Hegde, B.R. and Gowda, B.K.L. (1989) Cropping systems and production technology for small millets in India. In: Seetharam, A. *et al.* (eds) *Small Millets in Global Agriculture*. Delhi, India: Oxford & IBH, pp. 209–235.

Heller, J., Begemann, F. and Mushonga, J. (1997) *Bambara Groundnut. Vigna subterranea (L.) Verdc.* Gatersleben, Germany: Institute of Plant Genetic and Crop Plant Research (IPK) and Rome: International Plant Genetic Resources Institute (IPGRI).

Huang, Y.F., Gao, X.L., Nan, Z.B. and Zhang, Z.X. (2017) Potential value of the common vetch (Vicia sativa L.) as an animal feedstuff: A review. *Journal of Animal Physiology and Animal Nutrition*, 101 (5), pp. 807–823.

ICRISAT & FAO (1996) *The World Sorghum and Millet Economies*. Patancheru, India/ FAO, Rome: ICRISAT.

IPES-Food (2016) From uniformity to diversity: A paradigm shift from industrial agriculture to diversified agro-ecological systems, International Panel of Experts on Sustainable Food Systems. Available at: www.ipes-food.org (accessed on 15 November 2016).

Iqbal, A., Khalil, I.A., Ateeq, N. and Khan, M.S. (2006) Nutritional quality of important food legumes. *Food Chemistry*, 97, pp. 331–335.

Issoufa, B.B. (2015) Composting millet glume for soil fertility improvement and millet/cowpea productivity in semi-arid zone of Niger. PhD thesis, University of Science and Technology, Kumasi, Ghana.

Jayne, T., Rashid, S., Minot, N. and Kasule, S. (2009) Promoting fertilizer use in Africa: Current issues and empirical evidence from the COMESA region.

Presented at the COMESA African Agricultural Markets Programme Policy Conference, June 15–16, in Livingstone, Zambia. Available at: http://tinyurl.com/fertilizer-africa-comesa.

Jensen, E.S., Peoples, M.B. and Hauggard-Nielsen, H. (2010) Faba bean in cropping system. *Field Crops Research*, 115, pp. 203–216.

Jensen, E.S., Peoples, M.B., Boddey, R.M., Gresshoff, P.M., Hauggaard-Nielsen, H., Alves, B.J.R. and Morrison M.J. (2012) Legumes for mitigation of climate change and the provision of feedstock for biofuels and biorefineries: A review. *Agronomy for Sustainable Development*, 32 (2), pp. 329–364.

Jideani, I.A. and Jideani, V.A. (2011) Developments on the cereal grains Digitaria exilis (acha) and Digitaria iburua (iburu). *Journal of Food Science Technology*, 48 (3), pp. 251–259.

Kajuna, S.T.A.R. and Mejia, D. (ed.) (2001) MILLET: Post-harvest Operations Organization. INPhO-Post-harvest Compendium. Sokone University of Agriculture (SUA). Available at: www.suanet.ac.tz.

Ketema, S. (1993) *Tef (Eragrostis tef): Breeding, Genetic Resources, Agronomy, Utilization and Role in Ethiopian Agriculture*. Addis Ababa: Institute of Agricultural Research.

Khan, Z.R., Midega, C.A.O., Hassanali, A., Pickett, J.A. and Wadhams, L.J. (2007) Assessment of different legumes for the control of Striga hermonthica in maize and sorghum. *Crop Science* 47: 728–734.

Kothari, S.L., Kumar, S., Vishnoi, R.K., Kothari, A. and Watanabe, K.N. (2005) Applications of biotechnology for improvement of millet crops: Review of progress and future prospects. *Plant Biotechnology*, 22, pp. 81–88.

Köpke, U. and Nemecek, T. (2010) Ecological services of faba bean. *Field Crops Research*, 115, pp. 217–233.

Li, L., Li, S., Sun, J., Zhou, L., Bao, X., Zhang, H. and Zhang, F (2007) Diversity enhances agricultural productivity via rhizosphere phosphorus facilitation on phosphorus-deficient soils. *Proceedings of the National Academy of Science USA*, 104: 11192–11196.

McCallum, M.H., Kirkegaard, J.A., Green, T., Cresswell, H.P., Davies, S.L., Angus, J.F. and Peoples, M.B. (2004) Improved subsoil macroporosity following perennial pastures. *Australian Journal of Experimental Agriculture*, 44, pp. 299–307.

M'Khaitir, Y.O. and Vanderlip, R.L. (1992) Grain sorghum and pearl millet response to date and rate of planting. *Agronomy Journal*, 84, pp. 579–582.

Mullet, J., Morishige, D., McCormick, R., Truong, S., Hilley, J., McKinley, B., Anderson, R., Olson, S.N. and Rooney, W. (2014) Energy sorghum: A genetic model for the design of C4 grass bioenergy crops. *Journal of Experimental Botany*, 65(13): 3479–3489.

Murphy, D.J. (2007) People, plants, and genes. In: *The Domestication of Non-cereal Crops*. New York, NY: Oxford University Press, pp. 96–105.

NAAS (National Academy of Agricultural Sciences) (2013) Role of millets in nutritional security of India. Policy Paper No. 66, New Delhi.

Nemecek, T. and Baumgartner, D.U. (2006) *Environmental Impacts of Introducing Grain Legumes into European Crop Rotations and Pig Feed Formulas: Concerted action GL-Pro*. Zürich, Switzerland: Agroscope Reckenholz-Tänikon Research Station (ART).

NRC (National Research Council) (1996) *Lost Crops of Africa. Vol. 1: Grains*. Washington, DC: National Academy Press.

Nuruzzaman, M., Lambers, L., Bolland, M.D.A. and Veneklaas, E.J. (2005) Phosphorus uptake by grain legumes and subsequently grown wheat at different levels of residual phosphorus fertilizer. *Australian Journal of Agricultural Research*, 56, pp. 1041–1047.

Obilana, A.B. (2003) Overview importance of millets in Africa. Available at: www.afripro.org.uk/papers/paper02obilana.pdf (accessed 10 January 2017).

Oelke, E.A., Oplinger, E.S., Putnam, D.H., Durgan, B.R., Doll, J.D. and Undersander, D.J. (1990) Millets. In: *Alternative Field Crops Manual of Agriculture and Home Economics*. Las Cruces, NM: New Mexico University.

Padulosi, S., Mal, B., King, O.I. and Gotor, E. (2015) Minor millets as a central element for sustainably enhanced incomes, empowerment, and nutrition in rural India. *Sustainability*, 7, pp. 8904–8933.

Padulosi, S., Mal, B., Ravi, S.B., Gowda, J., Gowda, K.T.K., Shanthakumar, G., Yenagi, N. and M Dutta, M. (2009) Food security and climate change: Role of plant genetic resources of minor millets. *Indian Journal of Plant Genetic Resources*, 22 (1), pp. 1–16.

Pathak, H.C. (2013) *Role of Millets in Nutritional Security of India*. New Delhi: National Academy of Agricultural Sciences, pp. 1–16.

Peoples, M.B., Brockwell, J., Herridge, D.F., Rochester, I.J., Alves, B.J.R., Urquiaga, S., Boddey, R.M., Dakora, F.D., Bhattarai, S., Maskey, S.L., Sampet, C., Rerkasem, B., Khan, D.F., Hauggaard-Nielsen, H. and Jensen, E.S. (2009) Review article: The contributions of nitrogen-fixing crop legumes to the productivity of agricultural systems. *Symbiosis*, 48, pp. 1–17.

Plaza-Bonilla, D, Nolot, J-M., Raffaillac, D. and Justes, E. (2015) Cover crops mitigate nitrate leaching in cropping systems including grain legumes: Field evidence and model simulations. *Agriculture, Ecosystems and Environment*, 212: 1–12.

Sabais, A.C.W., Scheu, S. and Eisenhauer, N. (2011) Plant species richness drives the density and diversity of Collembola in temperate grassland. *Acta Oecologica*, 37, pp. 195–202.

Schaffnit-Chatterjee, C. (2014) Agricultural value chains in sub-Saharan Africa: From a development challenge to a business opportunity. *Current Issues, Emerging Markets*. Deutsche Bank AG, Deutsche Bank Research, 60262 Frankfurt am Main, Germany. Print: ISSN 1612-314X / Internet/E-Mail: ISSN 1612-3158.

Singh, S.E. (2016) Potential of millets: Nutrients composition and health benefits. *Journal of Scientific and Innovative Research*, 5 (2), pp. 46–50. Available at: www.jsirjournal.com.

Singh, P. and Raghuvanshi, R.S. (2012) Finger millet for food and nutritional security. *African Journal of Food Science*, 6 (4), pp. 77–84.

Smartt, J. (1990) *Grain Legumes: Evolution and genetic resources*. Cambridge, UK: Cambridge University Press.

Stefaniak, T.R. and McPhee, K.E. (2016) Lentil. In: de Ron, A.M. (ed.) *Grain Legumes*. New York, NY: Springer, pp. 111–140.

Swinton, S.M., Numa, G. and Samba, L.A. (1984). Les cultures associées en milieu paysan dans deux regions du Niger: Filingue et Madarounfa. In: *Proceedings of the Regional Workshop on Intercropping in the Sahelian and Sahelo–Sudanian Zones of West Africa, November 7–10, 1984*. Niamey, Niger, Bamako, Mali: Institute du Sahel (in French), pp. 183–194.

Tadele, Z. (2016) Drought adaptation in millets. In: Shanker, A.K. and Shanker, C. (eds) *Abiotic and Biotic Stress in Plants: Recent advances and future perspectives*. Rijeka, Croatia: In Tech, pp. 639–662.

Teklu, Y. and Tefera, H. (2005) Genetic improvement in grain yield potential and associated agronomic traits of tef (*Eragrostis tef*). *Euphytica*, 141: 247–254

Tharanathan, R.N. and Mahadevamma, S. (2003) Grain legumes a boon to human nutrition. *Trends in Food Science and Technology*, 14, pp. 507–518.

Thilakarathna, M.S. and Raizada, M.N. (2015) A review of nutrient management studies involving finger millet in the semiarid tropics of Asia and Africa. *Agronomy*, 5, pp. 262–290.

Tongwane, M. Mdlambuzi, T., Moeletsi, M., Tsubo, M., Mliswa, V. and Grootboom, L. (2016) Greenhouse gas emissions from different crop production and management practices in South Africa. *Environmental Development*, 19, pp. 23–35.

UN (United Nations) (2015) Sustainable Development Goals. Available at: www.un.org/sustainabledevelopment/sustainable-development-goals/ (accessed 15 March 2017).

Urbatzka, P., Graβ, R., Haase, T., Schüler, C. and Heβ, J. (2009) Fate of legume-derived nitrogen in monocultures and mixtures with cereals. *Agriculture, Ecosystems and Environment*, 132, pp. 116–125.

US Department of Agriculture (USDA), (2015) USDA National Nutrient Database for Standard Reference, Agricultural Research Service, Nutrient Data Laboratory. Release 28. Version current: September 2015. Available at www.ars.usda.gov/nea/bhnrc/ndl.

Velu, G., Rai, K.N., Muralidharan, V., Kulkarni, V.N., Longvah, T. and Raveendran, T.S. (2006) Rapid screening method for grain iron content in pearl millet. *ISMN*, 47. Available at: http://oar.icrisat.org/1143/1/ISMN-47_158-161_2006.pdf (accessed 10 May 2017).

Voisin, A-S., Guéguen, J., Huyghe, C., Jeuffroy, M-H., Magrini, M-B., Meynard, J-M., Mougel, C., Pellerin, S. and Pelzer, E. (2014) Legumes for feed, food, biomaterials and bioenergy in Europe: A review. *Agronomy for Sustainable Development*, 34, pp. 361–380.

Wang, R., Hunt, H.V., Qiao, Z., Wang, L. and Han, Y. (2016) Diversity and cultivation of broomcorn millet (*Panicum miliaceum* L.) in China: A review. *Economic Botany*, 70, pp. 332–342.

Wani, S.P. and Lee, K.K. (1992) Role of biofertilizers in upland crop production. In: Tandon, H.L.S. (ed.) *Fertilizers Organic Manures Recycle Wastes and Biofertilizers*. Bhanot Corner, New Delhi: Fertilizer Development and Consultation Organization.

Warkentin, T.D., Smýkal, P., Coyne, C.J., Weeden, N., Domoney, C., Bing, D-J., Leonforte, A., Xuxiao, Z., Dixit, G.P., Boros, L., McPhee, K.E., McGee, R.J., Burstin, J. and Ellis, T.H.N. (2016) Pea. In: de Ron, A.M. (ed.) *Grain Legumes*. New York, NY: Springer, pp. 37–83.

Wu, T., Schoenau J.J., Li, F., Qian, P., Malhi, S.S. and Shi, Y. (2003) Effect of tillage and rotation on organic carbon forms of chernozemic soils in Saskatchewan. *Journal of Plant Nutrition and Soil Science*, 166, pp. 328–335.

Wu, X., Wang, D., Bean, S. R. and Wilson J. P. (2006) Ethanol Production from Pearl Millet Using Saccharomyces cerevisiae. *Cereal Chemistry*, 83 (2), pp. 127–131.

Xinxin, Z. and Yuan, L.H. (2011) Carbon sequestration within millet phytoliths from dry-farming of crops in China. *Chinese Science Bulletin*, 56, pp. 3451–3456.

Zander, P., Amjath-Babu, T.S., Preissel, S., Reckling, M., Bues, A., Schläfke, N., Kuhlman T., Bachinger, J., Uthes, S., Stoddard, F., Murphy-Bokern, D. and Watson, C. (2016) Grain legume decline and potential recovery in European agriculture: A review. *Agronomy for Sustainable Development*, 36, pp. 1–26.

7 Global challenges in today's horticulture and the prospects offered by protected vegetable cultivation

Hugo Despretz, Warwick Easdown and Mansab Ali

Introduction

Recent decades have seen positive developments in the global fight against undernourishment. According to a report by Meerman, Carisma and Thompson (2012), 'estimates of undernourishment prevalence have decreased fairly consistently, albeit slightly, since 1990 [. . .] and underweight and stunting have both decreased, although not fast enough to meet the MDG target of halving malnutrition rates by 2015'. In some parts of the world, especially in developing countries (World Bank, 2016), malnutrition appears in the form of deficiencies and remains a major problem that causes long term negative impacts on health, education and consequently on the economy (Alderman, Hoddinott and Kinsey, 2006; Caulfield *et al.*, 2006; Europarliament, 2014). Conversely, malnutrition can take the form of obesity or result in diabetes, which also affects the economy (INSPQ, 2014; Popkin, Linda and Ng, 2012). According to the WHO (2002), low intake of fruit and vegetables ranks as the sixth most important risk factor for mortality in the world. These global issues have accelerated the emphasis on production and consumption of healthy vegetables, as promoted by the World Vegetable Center, in order to meet the United Nations' Sustainable Development Goals, more specifically goal N°2 'End hunger, achieve food security and improved nutrition and promote sustainable agriculture' (UN, 2018). Numerous papers support this strategy to mitigate both malnutrition and poverty (Gibson, Perlas and Hotz, 2006; Goldman, 2003; Savage and Burgess, 1993; Susilowati and Karyadi, 2002) and the global horticultural sector is experiencing consistent growth (Ali, 2008). Nonetheless, the production of vegetables faces an array of local as well as global limitations and threats.

Climate change and more extreme weather

Climate change will have considerable impacts upon horticultural processes and productivity (Dixon, Collier, and Bhattacharya, 2014). It will affect plant physiology and soil microbial activity, and vital resources such as water and nutrients may become scarce in some regions, reducing the opportunities for growing horticultural crops. Climate change has been linked to the increased frequency of damaging extreme weather events such as droughts leading to water scarcity, storms or heavy rains (IPCC, 2012). The observed shift in the timing of seasonal events, precipitation patterns and the length of seasons influences the growing season and crop productivity (Gornall *et al.*, 2010). Climate change may also exacerbate desertification through the alteration of spatial and temporal patterns in temperature, rainfall, solar radiation and winds (Sivakumar, 2007). As a consequence, the land area suitable for agriculture, the length of growing seasons and yield potential, particularly in the dry tropics, are expected to decrease.

Major pests and diseases

Globalization, trade and climate change are breaking down the original biogeographical boundaries of agricultural pests (Loope and Howarth, 2003; Nelson *et al.*, 2009), exposing farmers to new biotic risks that they have to mitigate. Transboundary plant pests and diseases can easily spread to several countries and reach epidemic proportions. For instance, the threat represented by the rapid spread of South American tomato pinworm *Tuta absoluta* in Afro-Eurasia over the last decade is particularly alarming and this pest is likely to have a great impact on world tomato production (Desneux *et al.*, 2011). Such pest outbreaks can cause huge losses to crops and pastures, threatening the livelihoods of vulnerable farmers and the food and nutrition security of millions. This has been the case of the extraordinary upsurge of the desert locust in West Africa between 2003 and 2005 that disrupted agricultural production in areas already sensitive to food security (Ceccato *et al.*, 2006). Furthermore, some cosmopolitan pests such as whiteflies or aphids pose a systemic problem to farmers, especially horticultural crops, as they are not only vectors of incurable plant viruses, but also tend to adapt to pest management strategies and develop pesticide resistances (Denholm *et al.*, 1998; Prasad Singh, Srinivasa Rao and Shivashankara, 2013). Some farmers consequently react by over-spraying their crops with pesticides resulting in health impacts for themselves and consumers, in addition to the higher production cost incurred.

Stricter regulation of pesticides and health issues

The misuse of pesticides in food production and the growing concern of civil society about their environmental and health impacts encourage public national institutions to regulate their marketing and utilization. Major food importers such as the USA or European countries have set strict limitations on pesticide types and on residue quantities in products imported. Export-oriented producers have to find alternative strategies to protect their crops as they cannot use the pesticides that are banned by the EU or the USA.

Organic agriculture with an emphasis on pest management strategies such as biological control, good cultural practices (such as rotation) and mechanical barriers is also encouraging reduced reliance on pesticides. The trade in organic products from developing to developed countries is currently growing at over 20 per cent per year (Raynolds, 2004).

More scarce and more expensive resources

Rapidly growing urbanization together with soil pollution and a loss of soil fertility decrease the amount of fertile land available for food production worldwide. The shrinking amount of land per capita and more urban competition for resources, such as water and energy, contribute to increasing costs of production in agriculture (FAO, 2011), leading to a need for an increase in the productivity of horticulture per unit area (Andrews-Speed *et al.*, 2014). As a consequence, the concept of vertical farming to grow vegetables is gaining momentum in the newly constructed residential areas of major cities.

Growing economies and changing diets

Total fruit and vegetable consumption tends to increase with income (Hall, Moore and Lynch, 2009). As economies grow, especially in developing countries, incomes increase, middle classes expand and significant changes in diets and eating habits take place (Popkin *et al.*, 2012). As suggested by Diop and Jafee (2004), this intensifies the demand for horticultural products, since trade in fruit and vegetable is one of the most dynamic and fast-growing areas of national and international agricultural trade.

The production of high value agricultural products such as fruits and vegetables is also helping smallholder farmers to generate decent incomes, to create employment and limit rural emigration, and to ensure local food and nutrition security (Ibeawuchi *et al.*, 2015; Temu and Temu, 2005; Weinberger and Lumpkin, 2007). The production of vegetables has a comparative advantage over cereals where arable land is scarce and labour is abundant. On an average, it provides twice the amount of employment per hectare of production compared to cereal crop production (Ali, Farooq and Shih, 2002).

Vegetable production has great economic and social importance for both producers and consumers. Producers, especially smallholders need to produce crops economically and to mitigate the adverse impacts caused by biotic and abiotic factors on their crops, and to ensure that crops reach consumers with minimal postharvest losses. Consumers want high value and healthy vegetables at affordable prices. This chapter focuses on protected cultivation as a practical solution with a potential to meet the needs of both producers and consumers.

Protected cultivation of vegetables: a worldwide growing interest

State of the art of PC worldwide

Protected cultivation (PC) has existed in some countries, especially in Europe and China, for centuries; the practice of growing plants in environmentally controlled areas has existed since Roman times (Wittwer and Castilla, 1995). It can be defined as a combination of technologies and practices that aim to create a more conducive environment for plant growth. PC is usually divided into three different types, depending how the physical barriers are placed in relation to the crop (Castilla, 2007):

i Windbreaks if protection is only provided at the side of a field,
ii mulch, when only the soil is covered and
iii enclosed protective structures, ranging from floating covers to greenhouses.

PC is financially much more suitable for vegetable cropping than for cereal production because high value vegetables can return the cost of investment associated with improving yields from a small area, especially for smallholders. Most vegetables grown through PC are solanaceous and cucurbit crops because of their large market demand, adaptability to cultivation in unheated shelters and to long distance transportation, and because they allow complementary crop rotations that make most use of the shelter.

Kacira (2017) in his study presents the spread of the greenhouses in various countries. China in fact tops the list, followed by Spain and South Korea. Recent figures showed that the 'Global Greenhouse Vegetable Area Increased by 14 per cent over 2015' (*Hortidaily*, 2016). Yet, the expansion of PC worldwide is a two-speed process. On one hand, some developed countries, such as the Netherlands, Israel or Scandinavia, focus on expensive high-tech infrastructure to grow high-value vegetables, fruits and flowers year-round while protecting them from pests and diseases. This highly Controlled Environment Agriculture (CEA) is high-yielding but also highly energy-consuming. On the other hand, China, Southern European countries or North African states rely more on low-tech infrastructure with lower yields and lower energy needs. The trend in these countries is to improve yield with moderate investments (Baeza, Stanghellini and Castilla, 2013).

Among the leading countries in PC across the world, the Netherlands has a long tradition of PC under climate-controlled glasshouses (often soilless cultivation) for growing vegetables and flowers year-round with the most advanced and automated technologies in a naturally ambient cool climate. The Netherlands are today a world-leading exporter of horticultural products. This position can partly be explained by the fact that the Netherlands have the second highest private R&D investment rate, as a percentage of GDP in horticultural foodstuff in Europe (Netherlands Enterprise Agency, 2013).

Similarly, Israel relies on PC as the principal way to ensure a constant, year-round supply of high-quality products for the domestic market as well as for export, while minimizing the use of inputs. Unlike the Netherlands, Israeli growers face a dry and hot climate and shortages of water that push them to resort to a set of specific technologies (shading screens, pad and fan cooling, forced ventilation, precision-irrigation) that contribute to better greenhouse operations.

The dry Almeria province in Spain has risen from a low economic ranking among Spanish provinces to the third highest in the country in only 20 years, thanks to protected horticulture (Pardossi, Tognoni and Incrocci, 2004). Cantliffe and Vansickle (2012) explain that quality control, food safety and pesticide residues are major concerns for producers from these regions since they export a part of their production to other European countries. Almeria has become very competitive because it is able to sell high quality products at relatively low prices. This successful conversion to PC provides hope for rural development by means of economic growth and new job opportunities in other parts of the world in which governments, progressive farmers and private companies are willing to coordinate their efforts.

Although China has started soliciting the Netherlands for cooperation to develop high-tech greenhouse production, it has historically focused on low-cost PC and is now a leader in this field (Jiang, Qu and Mu, 2004). According to Kang *et al.* (2013), tunnels (low and high) and solar greenhouses are the two main types of protected structures in China, with most of the area under PC located in the temperate northern part of the country.

The use of protected structures for cultivation of crops as a commercial venture is a recent phenomenon in the tropics (Sidhu, 2014). Nowadays, there is increasing interest in the advantages of PC in tropical and subtropical regions as evidenced by its expansion and evolution in Asia, Mexico and the Middle East. Historically, most research in PC has been carried out to provide protection against cold and pests under temperate conditions. This is less relevant in warmer regions and there are visible gaps not only in technology but also in its proper implementation and in the understanding of crop economics (Singh *et al.*, 2012).

Protected cultivation to improve productivity, quality and economic viability

Protected cultivation is assumed to increase yields, product quality and water efficiency while reducing pesticide use (Castilla, 2013; Reddy, 2015). It does so by 'buffering' various stresses induced by biotic and abiotic factors on protected crops. The primary reasons for PC in the tropics and subtropics are for pest exclusion, protection from extreme solar radiation, heavy rain and winds. By reducing the impact of adverse environments, the growing season for vegetables can be extended, farmers can produce higher quality produce during the off-seasons for open field production and crops can be grown in locations unsuitable for open-field cultivation.

Box 7.1 Creating congenial conditions for plant growth

The beneficial effects of protective structures on the growth and yield of vegetable crops has been well documented. Depending on the design and materials of the structure, protective enclosed structures can:

1. produce a greenhouse effect by trapping solar radiation, increasing both air and soil temperature and relative humidity, which is beneficial for crops in the cooler parts of the tropics (high altitudes) or during winter time or at night in the subtropics;
2. limit water vapour losses coming from the plant and the soil, which is particularly important when potential evapotranspiration is high;
3. improve water-use efficiency (less water used per unit of biomass produced);
4. protect the plant from scorching radiations through reduced transmittance by the use of covers or a glazing material;
5. protect the plants and soil from heavy rains (lodging, diseases spreading and run-offs), winds (lodging, desiccation and aeolian erosion) or cold damages (snow and frost for instance);
6. enable efficient and high quality nursery production by protecting the seedlings from virus vectors and pests as well as hybrid seed production of high purity by hindering cross-pollination;
7. exclude pests – thus preventing spreading of diseases and reducing insecticides costs – as well as birds;
8. limit the infestation of the plot by weeds, by the use of covers blocking the dissemination of wind-borne weed seeds;
9. reduce chemical drift between the protective structure and neighbouring fields;
10. facilitate the entrapment of carbon dioxide, thus enhancing photosynthesis.

Source: Al-Helal and Abdel-Ghany (2011), Abou Hadid (2013), Baeza *et al.* (2013) and Singh *et al.* (2014).

The use of PC in tropical and subtropical regions is rapidly growing to meet the needs for off-season production (Gonzaga *et al.*, 2013). Although some massive high-tech operations are developing, their designs are often too expensive for smallholder farmers. More affordable and context-appropriate greenhouses can improve livelihoods for smallholder farmers and entrepreneurs while fostering food security (Pack and Mehta, 2012). As Groener *et al.* (2015) suggest, there is a need for more low cost PC structures in tropical and subtropical developing countries to improve farmers' income and to increase food security and food safety.

The economic viability of low-tech PC systems has been reported, notably in Benin where a cost–benefit analysis of the protected cropping of cabbage by smallholder farmers showed that insect proof net tunnel farming generated a threefold higher margin compared to open-field production relying on insecticides. Vidogbena *et al.* (2015) concluded from this study that netting technology led to less variation in costs, yields and cabbage quality than relying on insecticides. The economic feasibility of protected horticultural systems has also been confirmed by Armenia *et al.* (2015) in the Philippines where PC is used as a means to protect crops against heavy rains, which regularly damage vegetable productivity. Farmer skill levels (including crop selection) were important in contributing to productivity and revenue in such systems. Singh *et al.* (2009) also concluded from a study of naturally ventilated greenhouse production of hybrid tomatoes in the northern plains of India that the combination of minimized pest damages, off-season production and higher yields resulted in a high net return per unit area and made the investment worthwhile.

Box 7.2 Urban and peri-urban horticulture

According to Paroda (2014), PC is fully adapted to peri-urban and urban areas as it allows a better use of the limited arable land in those areas and helps growers fetch high returns due to close market proximity, meaning less need of intermediaries, shorter transportation and lower postharvest losses. Moreover, rooftop and backyard production is often feasible, offering direct exposure to sun and relative isolation from pests. For instance, a study conducted by Astee and Kishnani (2010) suggested that growing crops on available rooftops in Singapore would result in a 700 per cent increase in domestic vegetable production – satisfying 35.5 per cent of domestic demand, reducing food imports and decreasing Singapore's carbon footprint by 9052 tonnes of CO_2 annually. Nowadays, there is a growing trend of greenhouses being built on flat rooftops and of vertical farms within buildings in more and more Western and Asian cities (Specht *et al.*, 2014).

Source: Paroda (2014); Specht *et al.* (2014); Astee and Kishnani (2010).

The main financial difference between protected production and open-field production is due to the investment in the protective structure and its maintenance and operational costs. Those expenses vary according to the design, materials and technologies used.

Current technologies and management in the tropics and subtropics

In the tropics and subtropics, a critical balance must be maintained between protection against specific pests and control over the microclimate within the protective structure (Boulard, 2012). Both economic and management issues affect the degree of environmental control that is desirable and practical for vegetable production.

The adopted PC structure is dependent upon judicious choices based on the specific climate conditions, the funds available and the quality of environmental control (Castilla and Hernandez, 2007). Smallholder farmers in the tropics and subtropics using inexpensive low-tech protective technologies can only provide passive modification of the growing environment. Such greenhouses or tunnels are usually made of locally available materials such as bamboos, 'greenwood' or finished lumber poles that are cheaper to purchase or construct than metal-framed structures but less robust and long-lasting, mainly due to degradation by termites and fungi. A widely used design comprises insect proof nets (permeable to fluids) as side walls and a polyethylene film or polycarbonate panels (impermeable to fluids) on the roof. A roof vent is usually required for this kind of enclosed structure to enable a proper ventilation of the inside environment. According to Hoffmann and Waaijenberg (2002), the covering material properties are particularly important in subtropical regions: the cover should ideally allow the combination of near infrared radiations reflection during the day to reduce temperatures inside the greenhouse during periods of high irradiation, with the interception of far infrared radiations during the night to prevent heat losses especially during cold clear nights.

The most common types of enclosed protective structures found in the tropics and subtropics are presented below:

> Shade nets provide a practical way to decrease heat stresses of crops during hot and dry conditions. Set up in the form of an enclosed house, within or above a greenhouse, or as a mere horizontal cover above the crop in the open-field, they are particularly well adapted to arid and intensely sunlit regions. Al-Helal and Abdel-Ghany (2011) list the benefits of shading screens as follows:
>
> i reduction of the energy consumption for cooling the environment;
> ii reduced crop transpiration and irrigation water consumption (confirmed by Tanny, Haijun and Cohen, 2006, in Israel);
> iii fewer pests and less pesticide use (still under research); and
> iv diffusion of solar radiation, allowing its better use by crops.

Shade nets differ in light transmission rates and colours. However, the effects of coloured shading nets are varied and plant responses may be different even among cultivars. Much additional research is needed to understand the effects of coloured shade nets on the crop production (Stamps, 2009).

Nethouses or screenhouses can be seen as a type of shade houses. Net houses are usually defined as structures with perforated screens, primarily to act as a barrier to pest entry. Their design varies greatly, e.g. gable-, hut-, Gothic arch, Quonset- or saw-tooth shape. The screen is usually made of woven or knitted (or both) plastic threads. Depending on its properties (mesh size and geometry, thread diameter, colour, etc.), it can prevent the entry of specific pests and influences the fluid exchanges between the inside and the outside: bigger mesh sizes allow better ventilation and a more conducive nethouse microclimate, but are more prone to pest entry. They are usually naturally ventilated zero-energy structures, as the insect-proof nets allow natural ventilation. Desmarais (1996) demonstrated that the architectural shape, followed by screen mesh size and screen colour, had the largest influences on temperature, ventilation and the heat transfer properties of screenhouses.

Polyhouses can be similar to nethouses in architectural shape, but they are fully covered with an impermeable plastic film, mostly polyethylene. If there is no mechanical (also called 'forced') ventilation to renew the inside air, a roof vent has to be set up. If the sides are clad with screens, the shelter gets called a polynet house. This polyethylene film is impermeable and translucent, thus blocking the entry of rain, direct sunlight and limiting the penetration of dust, weeds seeds, insects and wind.

Tunnel farming has become incredibly popular in some countries, especially in Asia. Tunnels, whether they are covered with net or plastic film, are quite easy to build and vary in dimensions according to the crop and the farmer's investment capacity. High tunnels (height >3 m) offer enough room for indeterminate varieties and tall crops (cucumber, tomatoes, etc.), walk-in tunnels (1.5–3 m high) can accommodate semi-indeterminate crops and low tunnels (height <1.5m) are best suited for determinate varieties and short crops. Low tunnels are more difficult for labour to access but are cheaper, quick to build, maintain and dismantle (Lodhi, Kaushal and Singh, 2014). This enables farmers to change their location from season to season, therefore minimizing damage by soil-borne pests and diseases (Talekar *et al.*, 2003) which is less feasible with larger fixed structures.

Box 7.3 The advantages of naturally ventilated systems

The leading advantages of natural ventilation systems are:

i reduced costs of ventilation and cooling equipment;
ii reduced electrical operation and maintenance costs; and
iii avoiding problems created by irregular or absent power supplies often found in rural areas of developing countries.

As Buffington *et al.* (2016) observe, natural ventilation systems become more desirable as the cost of energy and the likelihood of power failures increase. Recent research by McCartney and Lefsrud (2014) in Barbados on a Natural Ventilation Augmented Cooling (NVAC) greenhouse design is offering promising prospects for the future of PC in warm humid climates.

Source: Buffington *et al.* (2016); McCartney and Lefsrud (2014).

Obviously, there is a need to create suitable greenhouse or tunnel designs that are customized to local agroclimatic factors. Although generally higher than for open-field systems, yields obtained in protected systems are also highly dependent on crop management, including the choice of cultivar, irrigation management, nutrition and pest control (Gonzaga *et al.*, 2013). So, regardless of the design of the protective structures, success depends on how the entire cropping system is managed. A combination of economic research, agronomy and engineering is needed for the development of context-appropriate and cost-effective protected cropping systems (Armenia *et al.*, 2015).

Research perspectives of PC in the tropics and subtropics

Developing a protected agrosystem that meets farmers' needs and suits the local environmental conditions requires a good understanding of the protected agrosystem components, the local context and their interactions.

The cultivars grown in enclosed spaces and the cultural practices that they require need to be well adapted to the specific conditions induced by this confined environment, usually higher temperatures, relative humidity, and consequently greater fungal pressure than in the open field. Grafting technologies, soil-less cultivation in bags filled with a sterilized medium, hydroponics, suitable crop rotations that are economically viable, and 'moveable' protective structures are useful additional management strategies to minimize the build-up of pests and disease problems under intensive PC production conditions.

The general trend in PC goes towards the use of a more advanced technology for better climate control and lower use of farm inputs, especially water and agrochemicals (Pardossi *et al.*, 2004). There is a serious need for more data on abiotic and biotic stresses under different types of naturally ventilated zero-energy structures in the tropics and subtropics. The decreasing cost and size of sensors and microcontrollers offers the opportunity to collect such data and to enable farmers to use them in order to optimize the effectiveness of their low cost PC production systems (Groener *et al.*, 2015).

Modelling the bio-physical phenomena within protected agrosystems can help to optimize their management, productivity and profitability.

The processes in action within the different components (namely the soil, atmosphere, plants and other living organisms) and the interactions between the components in such agrosystems are numerous and complex to study as a whole. Therefore, scientists try to develop mathematical models that are focusing on a few components (climate models, crop models, etc.) and that can later be linked and combined together to explain the functioning of an agrosystem, through the use of modular digital frameworks such as STICS[1] or DSSAT.[2] Besides their cognitive aspect, those models help to predict the values of variables of interest such as, for instance, the air temperature in a greenhouse, the residual nitrogen in the soil or the crop yield, without having to resort to expensive and time-consuming experimental approaches. The simulation accuracy of these models for specific conditions can be improved by confronting their outputs with data collected from experiments carried out in those same specific conditions.

On the one hand, global greenhouse-climate models have been produced for different designs of greenhouses in different climates to help optimize the use of such structures (Luo *et al.*, 2005; Vanthoor *et al.*, 2011). Computational Fluid Dynamic methods (CFD) are currently used for mapping the distributed climate inside the greenhouse, along with new experimental methods such as geostatistics or sonic anemometry (Boulard, 2012). As an example of their usefulness, Teitel *et al.* (2009) found ways to improve ventilation in screenhouses through judicious screen inclination using CFD approaches and experimental studies. On the other hand, there exist crop models capable of simulating the growth and development of specific crops in particular soil and climatic conditions such as the generic and adaptable-to-different-crops CROPGRO model. The combination of the different modelling approaches can be used to fine-tune management and crop production in different soil and climate conditions.

Box 7.4 The emergence of PC in sub-Saharan Africa

The African continent is about to witness a spectacular demographic explosion. Already affected by global warming, it has to adapt to rapid structural shifts in the agri-food industry, from the production systems to the food habits (Binswanger-Mkhize, 2009; Jayne, Mather and Mghenyi, 2010). There is growing emphasis on traditional vegetables such as Amaranths (Amaranthus spp.), the spider plant (Cleome gynandra) or the African eggplant and nightshade (Solanum spp.) to help take up those challenges (Dinssa *et al.*, 2015). Yet, it is estimated that production losses due to crop pests in Western and Eastern Africa exceed 50 per cent (Oerke, 2006). The frequent outbreaks and the transcontinental progression of exotic pests such as Tuta absoluta causing up to 100 per cent losses in crops in sub-Saharan Africa stress the need to adopt new practices and technologies to sustainably protect the crops (Brévault *et al.*, 2014; Tonnang *et al.*, 2015; Zekeya *et al.*, 2016). However, this part of the continent lags behind the rest of the world regarding the development of protected cultivation of horticultural crops, in terms of area under PC as well as of adoption rate of physical protection technologies (Castilla, 2013). PC in Africa is mostly the preserve of large greenhouse operations targeting niche or export markets in the flower and vegetable industry. Context-appropriate innovations are popping up here and there, such as Kenya's young entrepreneurs' smart and affordable greenhouses, and larger funds are invested into PC-related extension and research projects to promote PC among smallholders. The increasing establishment of international agribusiness companies specialized in greenhouse farming, flower or seed production, and farm input distributions contributes to boosting the development of protected cultivation, especially on the mid- and highlands of Eastern Africa. Nevertheless, to increase the adoption rate of PC by small vegetable growers, there is an urgent need to adapt the designs sold by greenhouse suppliers to the local context (climatic and economic), to train farmers to protective structure management, to give them access to financial help (subsidies, loans, micro-credits), and to help them identify and supply profitable market sectors.

Source: Adapted from several sources.

Development of PC of vegetables in South Asia: an inspiring case

A challenging context

Eight countries form what is geographically known as South Asia, is one of the most dynamic regions of the world, with India as its largest economy. With a population density of around 366 inhabitants/km^2 (Worldbank, 2016), it faces enormous challenges such as – ensuring food security and nutrition, social stability, environmental preservation, and water and energy supplies – unrivalled elsewhere in the world (Palit and Spittel, 2012). The claims of a mostly rural population, mainstreaming gender, a growing demand for fresh vegetables and a low-capital based agriculture on shrinking arable lands provide real challenges to increase productivity, and improve food quality and the incomes of farmers.

This region is the second largest vegetable producing area in the world after China, with India harvesting around 170 million tons of vegetables from roughly 9.5 million hectares in 2014–2015 (APEDA, 2016). With most of the region in the tropics and subtropics there are high pest and disease pressures with often harsh climatic conditions. Producing vegetables in the open field in such environments is challenging and there is a growing interest in the use of PC to make it easier and more economic.

Efforts to promote PC in India

The use of PC in India grew by 50 to 60 per cent between 2007 and 2012. Protected cultivation has been incorporated into national and state agricultural development policies that are supported by various subsidies. As Singh *et al.* (2009) note, 'The economic liberalization during the early 90's gave a boost to greenhouse expansion [and PC in general] by way of liberated import/export and project funding'. However, by 2012, the area used for PC was still only 0.23 per cent of the total area used for horticultural crops – roughly 25,000 ha (Singh *et al.*, 2014). The major crops grown under PC were tomato, capsicum, cucumber and melons (Sidhu, 2014). While a majority of research efforts so far promoted PC for growing crops in winter or cold areas of the country, it is now developing faster in the warmer and more humid regions of the country.

Paroda (2014) explains that the government of India has adopted two main strategies to promote PC:

1. *Information and training services*: Public organizations such as the National Horticulture Mission, Rashtriya Krishi Vikas Yojana and National Horticulture Board provide a large amount of horticultural information online, handy for literate farmers with access to the internet. They provide training for extension workers as well and support for field demonstrations. Many public or semi-public research and training centres focus on PC promotion, including:
    - The 'School on Protected Cultivation' at CCS HAU (Hisar, Haryana)
    - The 'Centre for Protected Cultivation Technology' (IARI, Pusa, Bihar)
    - Training from the 'Institute of Horticulture Technology' (Noida, Delhi)
    - 'Precision Farming Development Centres' throughout India (belong to the National Committee on Plastic Applications in Horticulture) and
    - International cooperation projects (with the Dutch at 'Durgapura Horticulture Farm', Jaipur, or the Israelis through the 'Centres of Excellence' across India).
2. *Financial assistance for farmers*: This covers PC components such as mulch, different types of PC structures, and irrigation. The pattern of assistance in 2015 was given by the Indian Ministry of Agriculture as: 'Credit linked back-ended subsidy at 50% of the total project cost limited to 5.6 million Indian Rupees per project as per admissible cost norms for green houses, shade net house, plastic tunnel, anti-bird /hail nets & cost of planting material etc.' The Indian government offers huge subsidies for PC in order to encourage farmers to adopt these new

technologies. This strategy works when farmers know about subsidies and are skilled and motivated enough to access them. However, excessive and complex bureaucracy makes it hard to access the financial support. This has encouraged the development of NGOs to help farmers to make it through the administrative procedures to obtain subsidies.

Private companies providing PC-related equipment or planting material are also accelerating PC adoption in India. The Indian Petrochemical Corporation Ltd (IPCL) contributed significantly to developing and promoting early PC adoption in India by providing covers, claddings (UV-stabilized LPDE) and frames for protective houses (Singh, 2015). Nowadays, there are many Indian and international companies competing to sell a wide array of PC supplies. Still, improvement is needed in terms of material cost and quality, particularly regarding the covers such as nets: heterogeneity in the mesh size of nets and short lifespans of covers sometimes appear dissuasive or unacceptable to farmers.

Both the public and private sectors are actively involved in developing PC and developing more collaborative approaches as the economics of protected cultivation continue to be attractive. Paroda (2014) advocates promoting PC using a 'cluster approach' to facilitate PC understanding by groups of farmers and to help solve field-related technical and marketing problems more efficiently.

Pakistan bets on PC of vegetables

Agriculture in Pakistan represents almost 20 per cent of GDP and employs 42.3 per cent of the population (GOP, 2016). There is a growing development of profitable high value vegetable production, as typified by protected cultivation of cucumber to meet a growing urban demand for the crop (Balal *et al.*, 2015). This increase of interest in PC has been stimulated by the work of the National Agricultural Research System (NARS) and international organizations through projects such as the 'Agricultural Innovation Programme' (2013–2017). In Punjab, a province of Pakistan, tunnel farming has increased in popularity, and the Punjab provincial government provides 50 per cent subsidy on the installation of tunnel farming technologies (*Hortidaily*, 2017). Started in the early 2000s, low-tech PC in Pakistan has quickly developed and there are currently more than 45,000 acres of tunnel farms in Punjab alone according to the GOP (2016). They are mainly used to produce off-season vegetables or to extend the crop production period during colder seasons, enabling producers to fetch higher prices (Saeed, 2015).

Protected cultivation of vegetables usually entails the setup of a drip irrigation system to facilitate the precision watering of crops within protective structures and to limit weed infestation, drudgery and pest entrance by farmers' coming and going in the greenhouse with watering cans.

Table 7.1 Vegetables yield with reduced usage of water and fertilizer (drip vs furrow irrigation) in Pakistan

Location	*Crop*	*Irrigation*			*Fertigation*			*Yield*		
		Drip (Litres)	*Furrow (Litres)*	*Saving (%)*	*Drip (Kgs)*	*Furrow (Kgs)*	*Saving (%)*	*Drip (Kgs)*	*Furrow (Kgs)*	*Additional yield/tunnel (%)*
Mingora	Cucumber	87468	145636	40	5	7	29	1171	855	37
Sheikhupur		63043	95326	34	9	14	39	2308	1832	26
Gojra		66325	88600	25	10	17	40	2530	1772	43
Quetta		87216	144352	40	13	21	39	2750	1925	43
DI Khan		74673	135326	45	12	18	34	1708	1350	27
Mingora	Tomato	67468	108546	38	7	12	39	1245	975	28
Islamabad		52613	88635	41	6	9	33	1168	861	36
Noorpur Thal		54097	546200	90	18	127	86	1580	1195	32
Quetta		94970	148628	36	13	20	34	2647	1895	40
Gojra	Bitter gourd	68250	112568	39	10	16	42	1967	1644	20
DI Khan		74673	135326	45	12	18	34	1958	1475	33
Mingora	Vegetable marrow	90443	133383	32	6	10	37	1870	1540	21
Quetta	Onion	195000	251000	22	8	14	43	1734	1468	18

Source: Authors' own compilation based on data collected by AIP–WorldVeg Staff, Pakistan.

As part of the Agricultural Innovation Program for Pakistan (2013–2017), drip irrigation systems were installed at 30 locations across Pakistan in Khyber Pakhtunkhwa (KP), Punjab and Balochistan to demonstrate the effect of drip and furrow irrigation on water and fertilizer efficiency as well as crop growth in 45 tunnels (of 252 m^2) covered with insect proof net. Respectively, 25–45 per cent and 29–40 per cent of water and fertilizer were saved with an increased yield of 27–43 per cent at five locations by using drip irrigation as compared to furrow irrigation for cucumber cropping. For tomato cropping, 36–41 per cent of water and 33–39 per cent of fertilizer was saved with an additional yield of 28–40 per cent at four locations (Table 7.1). The biggest savings of water (90 per cent) and fertilizer (85 per cent) were achieved with increase in yield (26 per cent) in the sandy soil of Noorpur Thal. Bitter gourd yield increased from 20–33 per cent by utilizing 39–45 per cent and 34–42 per cent less water and fertilizer respectively at two locations. Vegetable marrow yielded 21 per cent more as compared to crop irrigated through the furrow system by saving 32 per cent in water and 37 per cent in fertilizer at Mingora-KP.

Drip irrigation systems not only save water and fertilizer but also help to reduce the cost of production and minimize the usage of weedicides and fungicides under plasticulture. Onion crop cultivated in 500 m^2 in open field under a drip irrigation system in Quetta, Balochistan, yielded 18 per cent more and saved 22 per cent of water and 43 per cent of fertilizer as compared to crop irrigated through furrow irrigation. Results from the project showed that the adoption rate of drip irrigation in Pakistan was encouraging and that farmers were becoming aware of its manifold benefits in comparison with traditional irrigation.

Box 7.5 A success story of vegetables tunnel farming in Pakistan

World Vegetable Center (WorldVeg) scientists introduced the opportunities provided by protected cultivation of vegetables to poor subsistence wheat farmers in the Pothwar region, Punjab, Pakistan. In this part of the country, crops can fail due to untimely rains. To address this situation, the WorldVeg team in Pakistan conducted surveys across the Rawalpindi district and found that the highly fertile land (sandy loam and moderately well-drained soil) and limited water resources could be much better utilized for vegetable production than cereals. A cluster of 15 interested farmers was initially formed. In 2013, WorldVeg introduced the concept of tunnel farming to grow off-season vegetables to one of the farmers, Mr Bhatti, through the USAID funded 'Agricultural Innovation Program for Pakistan' project. The farmer decided to invest in two

(continued)

(continued)

small walk-in tunnels for growing indeterminate tomatoes. As the pioneer for trying the PC system in his district, he earned good profits and decided to expand into high tunnels in 2014. With continuous and additional training provided by WorldVeg, the farmer was able to set up the netting structures with a drip irrigation system for PC over longer periods. He earned about US$ 1,250 from 250 m² tunnels during 2014–2015 and realized that there was more opportunity under the PC system. After receiving training in healthy seedling production, the farmer started to promote the concept of healthy and quality seedlings to other farmers. The farmer used the readily available plastic plug trays and high quality compost to raise vegetable seedlings. He turned this into a business and earned US$ 1,057 from selling the seedlings. An additional income of US$ 961 was generated during 2015–2016, from selling the side shoots of improved varieties of tomato as planting materials. In the process, Mr Bhatti has become a progressive lead farmer and recently appeared on air along with representatives of USAID and WorldVeg. He has become a model to fellow farmers in the district to shift from subsistence wheat production to growing off-season vegetables under protected cultivation. His latest foray is into growing winter crops such as coriander and spinach under green shade nets in summer when the prices for the products are very high. New ideas have changed his life and he has become an inspiration to many farmers that farming in the Pothwar region has a brighter future than just subsistence.

Source: Authors' own contribution.

Khan *et al.* (2011) emphasize the need for Pakistani farmers to have a good understanding of markets, varieties and their performance, as well as a good knowledge of how to manage protected cultivation for the successful production of high value crops around the year. Increasing urbanization in Pakistan as in other south Asian countries, is increasing the demand for vegetables in urban centres and guarantees success to innovative farmers and entrepreneurs shifting into PC a bright future.

Efforts to promote new agricultural practices must be tailored to reflect the particular conditions of individual locales. A study carried out by Coelli and Battese in 1996, using data collected in Indian villages showed that the age, level of education of farmers and farm size had an effect on the technical inefficiency of Indian farmers. For Paroda (2014), education is the principal solution for raising awareness among farmers about the potential of PC and to overcome technical inefficiency.

The main reasons behind the low adoption of PC

Although the PC area in India was expected to rise by over 84 per cent for the period between 2013 and 2017, primarily because of government support schemes (Paroda, 2014), the targets could not be met as small-scale farmers were left out. There is wide variation in the progress and government commitment towards PC across Indian states (Singh *et al.*, 2014). States such as Jammu and Kashmir (through the DIHAR), Haryana and Karnataka are pioneers in India in PC, whereas, states such as Odisha or Jharkhand lag behind.

Chatterjee *et al.*, (2013) claim that the major limitations and failures for PC adoption are mainly due to economic reasons – the additional investment required for protected structures, lack of knowledge and skills about PC technologies, and the maintenance of protective structures. Singh *et al.* (2012) state that there is lack of knowledge among both farmers and agricultural extension centres. There is an overall lack of expertise in state departments and State Agricultural Universities in PC.

The absence of suitable assistance coupled with the facts that the entire developmental pattern of PC is subsidy-driven and that investments in larger structures (usually more than 250 m^2) are encouraged, implies that farmers are often left to their own devices once the greenhouses have been constructed. To manage crops under PC requires a different mindset in comparison to traditional open-field agricultural systems. Even if they are willing to invest in PC, farmers face the additional hurdles of poor storage and marketing support and access to reliable marketing networks. With an absence of such networks to pay for their higher quality produce, it may prove difficult to recoup their investment costs.

The World Veg Center focuses on providing vegetable producers in the tropics that have limited resources with good agricultural practices and sustainable technologies. Complex and expensive technologies that require larger investments for the acquisition and maintenance of equipment and that demand a lot of energy and considerable know-how can be out of reach for most small farmers. Nonetheless, the attraction for the complex persists. For instance, a recent publication in the *Indian Journal of Applied Research* entitled 'Protected Cultivation of Vegetables – Present Status and Future Prospects in India' by Nair and Barche (2014) emphasizes the need to develop high-tech greenhouses in India but neglects the advantages of low-cost structures. The use of low-cost systems that farmers can readily understand is needed to enhance the shift from conventional agriculture to PC. Naturally ventilated plastic walk-in tunnels or greenhouses might offer a good compromise for many producers.

Conclusion

Today's world is not only exposed to climate change and transboundary pest proliferation, but is also affected by a reduction in available arable land, water scarcity and increases in the cost of energy and agrochemicals.

Protected cultivation can form part of the solution to the challenges of increasing the efficiency of resource use and mitigating the harmful effects of biotic and abiotic factors on crop yields and quality.

Protected cultivation is gaining popularity across the world, especially for the purpose of vegetable production. As compared to conventional open-field farming, farmers properly using PC technologies and techniques are able to produce higher yields of high-quality foodstuff year-round and to reduce the effects of climatic and pest damage, thus increasing their income. However, the development of PC is bifurcating: on one hand, resourceful farmers are moving into expensive high-tech, high-yielding structures and, on the other, resource-poor smallholder farmers are resorting to more basic low-cost shelters with less control of their growing environment.

In the case of South Asia and especially India, Central or State Governments and other private or public institutions intend to facilitate the former type of innovative agriculture, supported with massive subsidies. However, the area under PC is not increasing as fast as expected, and the current system appears to be benefiting wealthier, more educated and well-informed farmers. In addition to the lack of capital, a greater constraint is the lack of knowledge among smallholder farmers and no effort to provide appropriate technologies to meet their needs. PC is therefore seen by some as beneficial but not accessible.

Yet, low-cost, rudimentary and easy-to-manage protective structures already exist and can be improved to meet regional environmental conditions. However, insufficient research has been conducted on low-cost structures in tropical and subtropical contexts so far. There is a need not only to improve the structures' designs, but also to recommend appropriate crop sequences and cultural practices to producers, and to breed or provide plant material fit for local agro-climatic conditions. A better understanding of the requirements of farmers in targeted regions would also help tailor such technologies to their financial and technical capacities. Formation of self-help groups and cluster approaches could help to tackle knowledge gaps, financial limitations and marketing issues.

Finally, systematizing data collection on farms equipped with low cost sensors and data loggers, experimenting with various protected agro-systems and modelling the interactions of climate, crops, pests and diseases within protected environments is a necessary step to optimize low-cost protected cropping systems and in particular soil and climatic conditions.

Notes

1 www6.paca.inra.fr/stics_eng/
2 http://dssat.net/

References

Abou Hadid, A.F. (2013) Protected cultivation for improving water-use efficiency of vegetable crops in the NENA region. In: FAO (eds) *Good Agricultural Practices for Greenhouse Vegetable Crops: Principles for Mediterranean climate areas*, Rome: FAO, pp. 137–148.

Alderman, H., Hoddinott, J. and Kinsey, W. (2006) Long term consequences of early childhood malnutrition. *Oxford Economic Papers*, 58 (4), pp. 450–474.

Al-Helal, A.M. and Abdel-Ghany, I.M. (2011) Measuring and evaluating solar radiative properties of plastic shading nets. *Solar Energy Materials & Solar Cells*, 95, pp. 677–683.

Ali, M. (2008) *Horticulture Revolution for the Poor: Nature, challenges and opportunities*. Background paper for the World Development Report 2008, Asian Vegetable Research and Development Center, Shanhua, Tainan, Taiwan.

Ali, M., Farooq, U. and Shih, Y.Y. (2002) Vegetable research and development in the ASEAN region: A guideline for setting priorities. In: Kuo, C.G. (ed.) *Perspectives of ASEAN Cooperation in Vegetable Research and Development*. Shanhua, Tainan, Taiwan: Asian Vegetable Research and Development Center, pp. 20–64.

Andrews-Speed, P., Bleischwitz, R., Boersma, T., Johnson, C., Kemp, G. and VanDeveer, S.D. (2014) *Want, Waste or War?: The global resource nexus and the struggle for land, energy, food, water and minerals*. Abingdon, Oxon/New York, NY: Routledge p. 240.

APEDA (2016) Fresh fruits and vegetables. Ministry of Commerce and Industry, Government of India. Available at: http://apeda.gov.in/apedawebsite/six_head_product/FFV.htm (accessed 3 February 2017).

Armenia, P.T., Menz, K.M., Rogers, G.S., Gonzaga, Z.C., Gerona, R.G. and Tausa, E.R. (2015) *Economics of vegetable production under protected cropping structures in the Eastern Visayas, Philippines*, Visayas State University, Philippines/Applied Horticultural Research, Australia.

Astee, L.J. and Kishnani, N.T. (2010) Building integrated agriculture: Utilising rooftops for sustainable food crop cultivation in Singapore. *Journal of Green Building*, 5 (2), pp. 105–113.

Baeza, E.J., Stanghellini, C. and Castilla, N. (2013) Protected cultivation in Europe. *Acta Hortic*, (ISHS) 987, pp. 11–27.

Balal, R.M, Akhtar, G., Khan, M.W. and Akram, A. (2015) Pakistan: Cucumber growers shift to tunnels to meet growing demand. *Hortidaily*, 9 August. Available at: www.hortidaily.com/article/20294/Pakistan-Cucumber-growers-shift-to-tunnels-to-meet-growing-demand (accessed 15 February 2017).

Binswanger-Mkhize, H. (2009) Challenges and opportunities for African agriculture and food security: High food prices, climate change, population growth, and HIV and aids. Paper presented at FAO Expert Meeting on *How to Feed the World in 2050*, 24–26 June 2009, available at: www.fao.org/3/a-ak542e/ak542e16.pdf.

Boulard, T. (2012) Recent trends in protected cultivations–microclimate studies: A review, *Acta Horticulturae*, (ISHS) 957, pp. 15–28.

Brévault, T., Sylla, S., Diatte, M., Bernadas, G. and Diarra, K. (2014) Tuta absoluta Meyrick (Lepidoptera: Gelechiidae): A New threat to tomato production in sub-Saharan Africa. *African Entomology*, 22 (2), pp. 441–444.

Buffington, D.E., Bucklin, R.A., Henley, R.W. and McConnell, D.B. (2016) *Greenhouse ventilation*. Agricultural and Biological Engineering Department, UF/IFAS Extension, September 1987 (revised March 2016). Available at: http://edis.ifas.ufl.edu/pdffiles/AE/AE03000.pdf.

Cantliffe, D.J. and Vansickle, J.J. (2012) *Competitiveness of the Spanish and Dutch Greenhouse Industries with the Florida Fresh Vegetable Industry*. Horticultural Sciences Department, Florida Cooperative Extension Service, Institute of Food and Agricultural Sciences, University of Florida, May 2003 (revised July 2009 and August 2012). Available online at: https://edis.ifas.ufl.edu/cv284.

Castilla, N. (2007) *Invernaderos de plástico: Tecnología y manejo*, 2nd edn, revised and improved. Madrid: Mundi-Prenso Libros, p. 462.

Castilla, N. and Hernandez, J. (2007) Greenhouse technological packages for high-quality crop production. *Acta Horticulturae*, 761, pp. 285–297.

Castilla, N. (2013) *Greenhouse Technology and Management*, 2nd edn. Wallingford, Oxon: CABI, p. 170.

Caulfield, L.E., Richard, S.A., Rivera, J.A., Musgrove, P. and Black, R.E. (2006) Stunting, wasting and micronutrients deficiency disorders. In: Jamison, D.T., Breman, J.G., Measham, A.R., Alleyne, G., Claeson, M., Evans, D.B., Jha, P., Mills, A. and Musgrove, P. (eds) *Disease Control Priorities in Developing Countries*, 2nd edn. New York, NT: Oxford University Press. Available at: www.ncbi.nlm.nih.gov/books/NBK11761/ (accessed 19 January 2018).

Ceccato, P., Cressman, K., Giannini, A. and Trzaska, S. (2006) The desert locust upsurge in West Africa (2003–2005): Information on the desert locust early warning system and the prospects for seasonal climate forecasting. *International Journal of Pest Management*, 53 (1), pp. 7–13.

Chatterjee, R., Sandip, H. and Pal, P.K. (2013) Adoption status and field level performance of different protected structures for vegetable production under changing scenario. *Innovare Journal of Agricultural Science*, 1 (1), pp. 11–13.

Coelli, T.J. and Battese, E.B. (1996) Identification of factors which influence the technical inefficiency of Indian farmers. *Australian Journal of Agricultural Economics*, 40 (2), pp. 103–128.

Denholm, I., Cahill, M., Dennehy, T.J. and Horowitz, A.R. (1998) Challenges with managing insecticide resistance in agricultural pests, exemplified by the whitefly Bemisia tabaci. *Philosophical Transactions of the Royal Society of London B: Biological Sciences*, 353 (1376), pp. 1757–1767.

Desmarais, G. (1996) Thermal characteristics of screenhouse configurations in a West-African tropical climate, PhD thesis, McGill University, Quebec, pp. 172–177.

Desneux, N., Luna, M.G., Guillemaud, T. and Urbaneja, A. (2011) The invasive South American tomato pinworm, Tuta absoluta, continues to spread in Afro-Eurasia and beyond: The new threat to tomato world production. *Journal of Pest Science*, 84, pp. 403–408.

Dinssa, F.F., Stoilova, T., Nenguwo, N., Aloyce, A., Tenkouano. A., Hanson. P., Hughes. J.d'A. and Keatinge, J.D.H. (2015) Traditional vegetables: Improvement and development in sub-Saharan Africa at AVRDC – The World Vegetable Center. *Acta Horticulturae*, 1102.

Diop, M. and Jaffee, S.M. (2004) Fruits and vegetables: Global trade and competition in fresh and processed product markets. In: *Global Agricultural Trade and Developing Countries*. Washington, DC: The International Bank for Reconstruction and Development/The World Bank: pp. 237–257.

Dixon, G.R., Collier, R.H. and Bhattacharya, I. (2014) An assessment of the effects of climate change on horticulture. In: Dixon, G.R. and Aldous, D.E. (eds) *Horticulture: Plants for people and places, Volume 2*, Dordrecht: Springer, pp. 817–857.

Europarliament (2014) *The Social and Economic Consequences of Malnutrition in ACP Countries*, Background Document of the European Union Parliament meetings. Available at: www.europarl.europa.eu/meetdocs/2009_2014/documents/acp/dv/background_/background_en.pdf.

FAO (2011) *Energy-smart food for people and climate, Issue Paper*. Rome: FAO.

Gibson, R.S., Perlas, L. and Hotz, C. (2006) Improving the bioavailability of nutrients in plant foods at the household level. *Proceedings of the Nutrition Society*, 56 (2), pp. 160–168.

Goldman, I.L. (2003) Recognition of fruits and vegetables as healthful: Vitamins and phytonutrients. *HortTechnology*, 13 (2), pp. 252–258.

Gonzaga, Z.C., Capuno, O.B., Loreto, M.B., Gerona, R.G., Borines, L.M., Tulin, A.T., Mangmang, J.S., Lusanta, D.C., Dimabuyu, H.B. and Rogers, G.S. (2013) Low-cost protected cultivation: Enhancing year-round production of high-value vegetables in the Philippines. *ACIAR Proceedings Series*, 139, pp. 123–137.

GOP (2016) *Economic Survey of Pakistan, 2015–2016*. Ministry of Finance, Government of Pakistan.

Gornall, J., Betts, R., Burke, E., Clark, R., Camp, J., Willett, K. and Wiltshire, A. (2010) Implications of climate change for agricultural productivity in the early twenty-first century. *Philosophy Transactions of The Royal Society*, 365 (1554).

Groener, B., Knopp, N., Korgan, K., Perry, R., Romero, J., Smith, K., Stainback, A., Strzelczyk, A. and Henriques, J. (2015) Preliminary design of a low-cost greenhouse with open source control systems. *Procedia Engineering*, 107, pp. 470–479.

Hall, J.N., Moore, S. and Lynch, J.W. (2009) Global variability in fruit and vegetable consumption. *American Journal of Preventive Medicine*, 36 (5), pp. 402–409.

Hoffmann, S. and Waaijenberg, D. (2002) Tropical and subtropical greenhouses: A challenge for new plastic films. *Proceedings of International Symposium on Tropical and Subtropical Greenhouses. Acta Horticulturae (ISHS)*, 578, pp. 163–169.

Hortidaily (2016) Global greenhouse vegetable area increased 14% over 2015. Available at: www.hortidaily.com/article/23258/Global-greenhouse-vegetable-area-increased-14-procent-over-2015 (accessed 19 January 2018).

Hortidaily (2017) Pakistan: Punjab announces subsidies on tunnel farms. Available at: www.hortidaily.com/article/32264/Pakistan-Punjab-announces-subsidies-on-tunnel-farms (accessed 19 January 2018).

Ibeawuchi, I.I., Okoli, N.A., Alagba, R.A., Ofor, M.O., Emma-Okafor, L.C., Peter-Onoh, C.A. and Obiefuna, J.C. (2015) Fruit and vegetable crop production in Nigeria: The gains, challenges and the way forward. *Journal of Biology, Agriculture and Healthcare*, 5 (2): 194–208.

INSPQ (2014) *The Economic Impact of Obesity and Overweight*, TOPO. Summaries by the Nutrition–Physical Activity–Weight Team, Institut National de Sante Publique du Quebec (INSPQ).

IPCC (2012) *Managing the Risks of Extreme Events and Disasters to Advance Climate Change Adaptation*. Cambridge, UK: Cambridge University Press.

Jayne, T.S., Mather, D. and Mghenyi, E. (2010) Principal challenges confronting smallholder agriculture in sub-Saharan Africa. *World Development, The Future of Small Farms*, 38 (10), pp. 1384–1398.

Jiang, W., Qu, D. and Mu, D. (2004) Protective cultivation of horticultural crops in China. *Horticultural reviews*, 30, pp. 115–162.

Kacira, M. (2017) Greenhouse structures and glazing. Presentation at the 2017 UA-CEAC Annual Greenhouse Crop Production and Engineering Design Short Course, Tucson, Arizona, USA, April 3–7.

Kang, Y., Yao-Chien, A.C., Hyun-Sug, C. and Mengmeng, G. (2013) Current and future status of protected cultivation techniques in Asia. *Proceedings of*

International Symposium on High Tunnel Horticultural Crop Production. Acta Horticulturae, 987, pp. 33–40.

Khan, A., Islam, M., ul-Haq, I., Ahmad, S, Abbas, G. and Athar, M. (2011) Technology transfer for cucumber (Cucumis sativus L.) production under protected agriculture in uplands Balochistan, Pakistan. *African Journal of Biotechnology*, 10 (69), pp. 15538–15544.

Lodhi, A.S., Kaushal, A. and Singh, K.G. (2014) Low tunnel technology for vegetable crops in India. In: *Research Advances in Micro-irrigation: Best Management Practices for Drip Irrigation Crops*, Oakville, ON/CRC Press: Boca Raton, FL: Apple Academic Press, pp. 3–4.

Loope, L.L. and Howarth, F.G. (2003) Globalization and pest invasion: Where will we be in five years?, *1st International Symposium on Biological Control of Arthropods*, Honolulu, Hawaii.

Luo, W., Stanghellini, C., Dai, J., Wang, X., Feije De Zwart, H. and Bu, C. (2005) Simulation of greenhouse management for the subtropics, part II: Scenario study for the summer season. *Biosystems Engineering*, 90 (4), pp. 433–441.

McCartney, L. and Lefsrud, M. (2014) Naturally ventilated augment cooling greenhouse: A step towards solving food security problems in Barbados, *Barbados Interdisciplinary Tropical Studies – Summer 2014*. Available at: www.mcgill.ca/bits/files/bits/naturally_ventilated_augment_cooling_greenhouse.pdf.

Meerman, J., Carisma, B. and Thompson, B. (2012) *Global, Regional and Subregional Trends in Undernourishment and Malnutrition*, report for SOFA. Rome: FAO.

Nair, R. and Barche, S. (2014) Protected cultivation of vegetables: Present status and future prospects in India. *Indian Journal of Applied Research*, 4 (6), pp. 245–247.

Nelson, G.C., Rosegrant, M.W., Koo, J., Robertson, R., Sulser, T., Zhu, T., Ringler, C., Msangi, S., Palazzo, A., Batka, M., Magalhaes, M., Valmonte-Santos, R., Ewing, M. and Lee, D. (2009) *Climate Change: Impact on Agriculture and Adaptation*. Washington, DC: Food Report Policy, International Food Policy Research Institute.

Netherlands Enterprise Agency (2013) Holland compared 2013 (p. 39). The Hague, the Netherlands: Netherlands Enterprise Agency. Available at: https://issuu.com/nvnom/docs/holland_compared_summe_2013_hr (accessed 19 January 2018).

Oerke, E.C. (2006) Crop losses to pests. *The Journal of Agricultural Science*, 144, pp. 31–43.

Pack, M. and Mehta, K. (2012) Design of affordable greenhouses for East Africa. *2012 IEEE Global Humanitarian Technology Conference*.

Palit, A. and Spittel, G. (2012) *South Asia in the New Decade: Challenges and prospects*. Singapore: World Scientific Publishing Company, p. 280.

Pardossi, A., Tognoni, F. and Incrocci, L. (2004) Mediterranean greenhouse technology. *Chronica horticulturae*, 44 (2), pp. 28–34.

Paroda, R.S. (2014) Strategies for promoting protected cultivation in India. In: Singh B., Singh, B., Sabir, N. and Hasan, M. (eds) *Advances in Protected Cultivation*, New Delh: New India Publishing Agency, pp. 1–9.

Popkin, B.M., Linda, S.A. and Ng, S.W. (2012) Now and then: The global nutrition transition: The pandemic of obesity in developing countries. *Nutrition Reviews*, 70 (1), pp. 3–21.

Prasad Singh, H.C., Srinivasa Rao, N.K. and Shivashankara, K.S. (2013) *Climate-Resilient Horticulture: Adaptation and mitigation strategies*. New Delhi: Springer India.

Raynolds, L.T. (2004) The globalization of organic agro-food networks. *World Development*, 32 (5), pp. 725–743.

Reddy, P.P. (2015) *Sustainable Crop Protection under Protected Cultivation*. Singapore: Springer.

Saeed, S. (2015) Rags to Riches: Tunnel farming: the swift money maker. *The Express Tribune*, June 7. Available at: https://tribune.com.pk/story/899495/rags-to-riches-tunnel-farming-the-swift-money-maker/.

Savage, K.F. and Burgess, A. (1993) *Nutrition for Developing Countries*, 2nd edn. Oxford, UK: Oxford University Press.

Sidhu, A.S. (2014) Protected cultivation of vegetables in Southern India. In: Singh B., Singh, B., Sabir, N. and Hasan, M. (eds) *Advances in Protected Cultivation*. New Delhi: New Indian Publishing Agency, pp. 47–50.

Singh, B., Singh, B., Kumar, R., Arora, S.K., Chadha, M.L. (2012) *Working Group Report on Development of Protected Cultivation in Haryana*, Haryana Kisan Ayog, CCS Haryana Agricultural University Campus, Hisar 125004, Government of Haryana. Available at: www.haryanakisanayog.org/Reports/Report_on_PC.pdf.

Singh, B., Singh, B., Sabir, N. and Hasan, M. (2014) *Advances in Protected Cultivation*, New Delhi: New India Publishing Agency, p. 235.

Singh, B. (2015) DRDO-Harbinger of protected cultivation of vegetable crops in India. In: Singh B. et al., *Advances in Protected Cultivation*. New Delhi: New India Publishing Agency, pp. 11–24.

Singh, S.K., Tiwari, P.S., Singh, G.R., Singh, B.R. and Singh, D. (2009) Economics of naturally ventilated greenhouse (SVBPU model) for tomato production in northern plain of India. *Progressive Agriculture*, 9 (1), pp. 130–133. Available at: www.indianjournals.com//ijor.aspx?target=ijor:pa&volume=9&issue=1&article=026.

Sivakumar, M.V.K. (2007) Climate and land degradation: An overview. In Sivakumar, M.V.K. and Ndiang'ui, N. (eds) *Climate and Land Degradation* (pp. 105–135), New York: Springer.

Specht, K., Siebert, R., Hartmann, I., Freisinger, U.B., Sawicka, M., Werner, A., Thomaier, S., Dietrich, H., Walk, H. and Dierich, A. (2014) Urban agriculture of the future: An overview of sustainability aspects of food production in and on buildings. *Agriculture and Human Values*, 31 (1), pp. 33–51.

Stamps, R.H., (2009) Use of colored shade netting in horticulture. *HortScience*, 44 (2), pp. 239–241.

Susilowati, D. and Karyadi, D. (2002) Malnutrition and poverty alleviation. *Asia Pacific Journal of Clinical Nutrition*, 11, pp. 323–330.

Talekar, N.S., Su, F.C. and Lin, M.Y. (2003) How to produce safer leafy vegetables in nethouses and net tunnels. The World Vegetable Center, Shanhua, Taiwan. Available at: www.avrdc.org/LC/cabbage/nethouse.pdf.

Tanny, J., Haijun, L. and Cohen, S. (2006) Airflow characteristics, energy balance and eddy covariance measurements in a banana screenhouse. *Agricultural and Forest Meteorology*, 139, pp. 105–118.

Teitel, M., Dvorkin, D., Haim, Y., Tanny, J. and Seginer, I. (2009) Comparison of measured and simulated flow through screens: Effect of screen inclination and porosity. *Biosystems Engineering*, 104, pp. 404–416.

Temu, A.E. and Temu, A.A. (2005) High value agricultural products for smallholder markets in sub-Saharan Africa: Trends, opportunities and research priorities. International workshop on *How can the poor benefit from the growing markets for high value agricultural products?*, October 2005, International Center for Tropical Agriculture, Cali, Colombia.

Tonnang, H.E., Mohamed, S.F., Khamis, F. and Ekesi, S. (2015) Identification and risk assessment for worldwide invasion and spread of Tuta absoluta with a focus on Sub-Saharan Africa: Implications for phytosanitary measures and management. *PloS one*, 10 (8): e0135283.

United Nations (2018) *Sustainable Development Goals: SDG N°2*. Available at: www.un.org/sustainabledevelopment/hunger/.

Vanthoor, B.H.E., Stanghellini, C., Van Henten, E.J. and De Vieser, P.H.B. (2011) A methodology for model-based Greenhouse design: Part 1, a greenhouse climate model for a broad range of designs and climates. *Biosystems Engineering*, 110 (4), pp. 363–377.

Vidogbena, F., Adegbidi, A., Assogba-Komlan, F., Martin, T., Ngouajio, M., Simon, S., Tossou, R. and Parrot, L. (2015) Cost:Benefit analysis of insect net use in cabbage in real farming conditions among smallholder farmers in Benin. *Crop Protection*, 78, pp. 164–171.

Weinberger, K. and Lumpkin, T.A. (2007) Diversification into horticulture and poverty reduction: A research agenda. *World Development*, 35 (8): 1464–1480.

Wittwer, S.H., and Castilla, N. (1995) Protected cultivation of horticultural crops worldwide. *HortTechnology*, 5 (1), pp. 6–23.

Worldbank (2016) Data: Population density (people per sq. km of land area) from 1961 to 2015 – South Asia. Food and Agriculture Organization and World Bank population estimates. Available at: http://data.worldbank.org/indicator/EN.POP.DNST?locations=8S.

World Health Organization (WHO) (2002) *The World Health Report 2002: Reducing Risks, Promoting Healthy Life*. Geneva, Switzerland: WHO.

Zekeya, N., Chacha, M., Ndakidemi, P.A., Materu, C., Chidege, M. and Mbega, E.R. (2016) Tomato leafminer (Tuta absoluta Meyrick 1917): A threat to tomato production in Africa. *Journal of Agriculture and Ecology Research International*, 10 (1), pp. 1–10.

8 Sustainable crop–livestock intensification in sub-Saharan Africa

Improving productivity through innovative adaptation

Donald M.G. Njarui, Mupenzi Mutimura, Elias M. Gichangi and Sita R. Ghimire

Introduction

Integrated crop–livestock systems are sources of livelihood and poverty alleviation to millions of smallholder farmers in arid and semi-arid lands (ASAL) across sub-Saharan Africa (SSA). These systems represent a key solution for enhancing agricultural productivity, ensuring food and nutritional security and safeguarding the environment by means of prudent and efficient use of natural resources. There is generally a strong interaction between livestock and crops in ASAL. Livestock provide manure and draft power in crop production (AGRA, 2014). In return, crop residues and forage on fallow lands provide feed for livestock (Williams, Hiernaux and Fernández-Rivera, 2010). Livestock are a primary source of food for millions, a valuable means for capital accumulation and generator of employment. Livestock also reduce the risks resulting from seasonal crop failures, as they add to the diversification of production (Sansoucy *et al.*, 1995). Whereas, livestock are relatively resilient to drought and low rainfall, thus cushioning farmers from food insecurity, crops are more susceptible to drought.

The chapter will begin with a literature review of the integrated crop–livestock farming systems and crop–livestock intensification as a means of ensuring food and nutritional security (FNS) while sustaining natural resources in SSA. This is followed by a brief account of farmers' innovative practices in respect of crop–livestock production systems and sustainable intensification that offer opportunities for increasing productivity and environmental sustainability in fragile agro-ecological systems. In the later part, the chapter briefly presents case studies of successful crop–livestock production systems from eastern, western, central and southern regions of SSA. The case studies focus on innovations that were successful in mitigating climate change and intensification of livestock-agriculture in ASAL. The factors affecting the sustainability of crop–livestock farming systems have been examined in the context of land tenure, population growth, climate change, infrastructure, institutional framework and policies. The potential options to overcome the constraints are provided towards the end.

Sub-Saharan Africa: biophysical environment, land use and agriculture

The total land area of SSA is 2,455 million hectares, and about a quarter (173 million hectares) of potential arable land is under cultivation (FAO, 2002). SSA has a diversified agro-ecology with ASAL being the most extensive, covering over 43 per cent of the land area; the dry sub-humid zone accounts for 13 per cent and the moist sub-humid and humid zones jointly cover 38 per cent (Dixon, Gulliver and Gibbon, 2001). The ASAL covers most of northern and central Senegal, parts of central Mali, northern and southern Burkina Faso and Niger, northern Nigeria and Chad. It extends eastward across Africa and covers large parts of Sudan and extends into Rift Valley in Ethiopia, northern and eastern parts of Kenya into Somalia and central Tanzania. It is also widespread in southern Mozambique, Zimbabwe and Botswana, and extends into southern and eastern Zambia and southern Angola. It is dominated by zones of lowland basins, highland plateaus and tropical savanna. The Sahel, a tropical savanna zone found north of the Equator, along the southern border of the Sahara Desert is the dominant physical feature of North Africa. The extensive Rift Valley is characterized by mountains, valleys and freshwater lakes. To the extreme south of the continent are the Namib and Kalahari deserts.

There is a wide range of climatic variation within SSA, but all of the areas within it observe the general categories of wet and dry season. The annual rainfall varies between 300 and approximately 600 mm in the ASAL region. In the western and southern regions, rainfall is unimodal while in the eastern regions it is bimodal in distribution, which influences crop growth (Williams *et al.*, 2010). Temperatures are generally high in ASAL (22–33°C), but are comparatively moderate around the Equator and rise up to 44 degrees Celsius in the Sahel. There is considerable variation in soils: alfisols and dunes sands are found in parts of West Africa, vertisols are found around and along the Sahel region and the Horn of Africa, including Ethiopia, Somalia and the Sudan, and oxisols and entisols can be located in southern Africa.

Agriculture is the backbone of the economy of SSA countries (Thornton and Herrero, 2015), which is reflected by its large contribution to gross domestic product (GDP) ranging from 20–50 per cent (World Bank, 2008). For example, in Burundi, the Democratic Republic of the Congo (DRC), Ethiopia, Sudan and Tanzania agriculture accounts for more than 50 per cent of GDP, while in Kenya and Eritrea, it accounts for less than 30 per cent of GDP (Waithaka *et al.*, 2013). Smallholder farmers account for 80 per cent of all farmers in SSA and contribute up to 90 per cent of food supply in some SSA countries (Wiggins, 2009; Wiggins and Sharada, 2013). In Botswana, for example, 76 per cent of the population depends on smallholder subsistence agriculture; in Kenya, 85 per cent; in Malawi, 90 per cent; and in Zimbabwe, 70–80 per cent (Ngigi, 2011; Rockström, 2000), thus underscoring the importance of the sector in job creation and poverty reduction.

Despite the huge agricultural potential in SSA, overall performance has been poor and unable to meet its burgeoning population (Waithaka *et al.*, 2013). Over the past decades, the region has become a net food importer. Crop yield levels remain low compared with other regions of the world (Chauvin, Mulangu and Mulangu, 2012). On an average, a farmer in SSA produces only one ton of cereal per hectare, which is less than half of what an Indian farmer produces and a quarter of what an average Chinese farmer produces (World Bank, 2007). These large differences can mainly be attributed to limited access to improved technologies and agronomic practices. The agricultural outputs of African farmers are constrained by inherently low soil fertility, poor access to inputs including improved crop seeds and affordable fertilizers, poor infrastructure and storage, and weak linkages to institutions and markets (Salami, Kamara and Brixiova, 2010). In recent years, agriculture in SSA has been compounded by the volatile food and energy prices and even more recently by the global financial crisis (Dixon, Tanyeri-Abur and Wattenbach, 2003). Furthermore, agriculture is also vulnerable to climate change (higher temperatures, shifting seasons, more extreme weather events, flooding and drought) due to its dependence on rains (IFAD, 2011; Rockström, 2003). About 90 per cent of the population of SSA depends on rain-fed agriculture for food production (FAO, 2006; Rockström *et al.*, 2004). In such vulnerable agro-ecological settings, an integrated crop–livestock system becomes quite relevant to practice.

Context of crop–livestock production systems

Integrated crop–livestock systems can be defined as farming systems that to some degree integrate crop and livestock production activities so as to gain benefits from the resulting crop–livestock interactions (Sumberg, 2003) and the production activities are managed by the same economic entity as the household (Williams *et al.*, 2010). Integrated crop–livestock systems are widespread in SSA and existing diversity in the region has led to evolvement of the different types of systems. This diversity includes variations in agro-ecological conditions, population densities, economic opportunities and institutional frameworks (Williams *et al.*, 2010). About 15 farming systems were identified in SSA (Dixon *et al.*, 2001; Garrity, Dixon and Boffa, 2012). The farming systems were classified based on: (a) the natural resource base; (b) the dominant livelihoods (main staple and cash income source – a balance between crops, livestock, fishing, forestry and off-farm activities); (c) the degree of crop–livestock integration; and (d) the scale of operation. The vast majority of mixed systems are rain-fed, and cover large areas of the ASAL and humid–sub-humid zones from Senegal in the west to Ethiopia in the east, and down the eastern side of the continent to South Africa (Thornton and Herrero, 2015).

Several drivers that are important for the development of smallholder crop–livestock production system were discussed in detail by Schiere *et al.* (2006). These drivers are exogenous and endogenous biophysical factors such as climate change, population pressures, market prices and soil types. As land becomes scarce, due to an expanding population, farmers are forced to keep animals together with cultivated crops. Furthermore, farmers tend to opt for mixed enterprises when they want to save resources because the system permits wider crop rotations and thus reduces dependence on agro-chemicals and allows diversification for better risk management.

There is a wide variation in the type of crops grown in different systems. The types of crops vary according to soil types, topographical positions and distance from the household compound (Salami *et al.*, 2010). Nevertheless, even at the individual household or farm level, there is a considerable diversity in the crops grown and livestock reared. Most smallholder operations occur in farming systems in which the family is at the centre of planning, management, decision making and implementation (Dixon *et al.*, 2003).

The World Bank's Rural Strategy defines smallholders as those with a low asset base, operating less than 2 hectares of cropland (World Bank, 2003). However, the smallholder generally differs in individual characteristics, resource distribution between food and cash crops, livestock and off-farm activities, their use of external inputs and hired labour, the proportion of food crops sold, and household expenditure patterns. The major characteristics of small farms in SSA are weathered soils of low inherent soil fertility, declining soil fertility and a minimal use of external input resulting in low yields. Crop diversity is high and farmers cultivate several crops in a mixture depending on soil type, topography and climatic condition. The ruminant component of an integrated crop–livestock system is the dominant system in terms of household livelihoods (Ellis and Freeman, 2004; McDermott *et al.*, 2010). The relative importance of livestock species varies in different regions. In eastern, central and southern Africa, cattle are regarded as the most important component of the farming systems followed by sheep and goats, poultry, horses, donkeys, mules and lastly pigs. By contrast, sheep and goats are ranked highest, followed by poultry and cattle, then horses, donkeys, mules and pigs in West Africa. The species of animal kept depends on adaptation to the local climate, relevance to the household economy, religion and cultural aspects.

Crop–livestock intensification for improved food security and livelihoods

The crop–livestock integration allows for more efficient use of resources than specialized systems and the synergies offer various opportunities for increasing productivity (Thornton and Herrero, 2001). The integrated systems play a major role in the livelihoods of millions of people in developing countries while providing significant quantities of both livestock and crop food products and increasing flexibility to cope with socio-economic and climate variability (Herrero *et al.*, 2010; Tarawali *et al.*, 2011). The benefits, both positive and negative, are summarized in Table 8.1.

Table 8.1 Positives and negatives of mixed crop–livestock systems

Factor	*Positive aspect*	*Negative aspect*
Trade and price fluctuations	Act as a buffer	Need high levels of management skill ('double expertise') Fewer economies of scale
Weather fluctuations	Buffer against weather fluctuations	Increase risk of disease and crop damage
Erosion	Control erosion by planting forages	Cause erosion through soil compaction and overgrazing
Nutrients	Improved nutrient cycling because of direct soil–crop–manure relations	Increased nutrient losses through intensive recycling
Draught power	Allow larger areas to be cultivated and more flexible residue management Allow more rapid planting	Extra labour (often by women) required for weeding increased area
Labour	Increase labour efficiency	Continuous labour requirements
Income	Diversified income sources More regular income streams	Risks due to climate change and disease outbreaks
Investment	Provide alternatives for investment	Require capital
Crop residues	Provide alternative use for low-quality roughage If mulched, controls weeds and conserves water	Feeding competes with other uses of crop residues (e.g. mulching, construction, nutrient cycling)
Security and savings	Provide security and a means of savings	Require investments on a regular basis
Social function	Confer prestige on the household	Cause of social conflicts

Source: Adapted from Thornton and Herrero (2015).

Economic and social benefits

In mixed crop livestock systems, livestock production contributes to poverty reduction in various ways. Livestock and livestock products are the most important sources of cash income in many smallholder mixed-farming systems in SSA since they can be easily sold to generate cash. The systems enable smallholder farmers to diversify farm production, increase cash income, and improve quality and quantity of food produced and exploitation of unutilized resources. For example, the monogastrics (pigs and poultry) convert by-products from grain into high value foods (Schiere *et al.*, 2006). The ruminants (cows, sheep or goats) provide added value by converting fibrous feeds such as straw. The nutrients and value of these products would largely be lost if they were not consumed by livestock. Thus, livestock serve to transform feed into food and marketable products, adding value to farming enterprises, increasing income, and enhancing the biophysical and economic viability of agriculture.

Livestock serve as a reserve for farmers, readily convertible to cash. Overall, 50 per cent or greater of the cash income earned by farmers engaged in mixed crop–livestock production in tropical Africa are obtained from livestock products. In Mali, for example, approximately 78 per cent of the total farm cash income of crop–livestock farms was generated from livestock sales (Debrah and Sissoko, 1990).

Environmental sustainability and ecological benefits

The success in long-term agricultural production in smallholder farming systems relies on the efficiency with which nutrients are conserved and recycled. In SSA, the depletion of soil nutrients is severe due to continuous cultivation with near universal negative soil–nutrient balances (Shepherd and Soule, 1998). The cycling of plant biomass through animals into manure that fertilizes the soil is an important linkage between livestock and soil productivity in many farming systems of SSA and plays a critical role in improving soil nutrient balances and sustaining crop production (McDermott *et al.*, 2010). Animal manure, if properly used can influence nutrient availability by: (i) the total nutrients added; (ii) controlling net mineralization–immobilization patterns; (iii) serving as a source of carbon and energy to soil microbes; (iv) functioning as precursors to soil organic matter fractions; and (v) interactions with the mineral soil in complexion of toxic cations and reducing the phosphorous sorption capacity of the soil (Gichangi, Mnkeni and Brookes, 2010; Palm, Myers and Nandwa, 1997; Schlecht, Fernández-Rivera and Hiernaux, 1998). However, the composition of manures varies depending largely on the animals' diet, type of animals (Probert, Okalebo and Jones, 1995) and in the ways manure is collected, stored and applied in the field (Gichangi, Karanja and Wood, 2006; Giller *et al.*, 1997; Karanja, Gichangi and Wood, 2005). Most efficient use of animal manures requires critical improvement in handling and storage, and in synchrony with mineralization of crop uptake (Rufino *et al.*, 2006).

In most farms, manure is not adequate and generally of poor quality for crop production while inorganic fertilizer is not readily accessible to most farmers due to high cost. Studies in the regions have identified integrated soil fertility management interventions that entail use of mineral fertilizer with organics to mitigate food insecurity and improve resilience of the soil productive capacity (Bationo *et al.*, 2003). The integration improves the agronomic efficiency of the external inputs used, reduces the risks of acidification and provides a more balanced supply of nutrients.

The interactions between crop resources and livestock revolve around the supply of nutrients and energy in feed with several key interactions between the various crop and livestock components of the system as illustrated in Figure 8.1. Plant materials, such as green manure or crop residues if used directly or after composting as nutrient inputs will, after decomposition, be taken up by crops to produce grain or farmers may decide to feed plant materials to livestock, which generally increases the nutrient concentration (Rufino *et al.*, 2006). Thus, the integration of crop and livestock production offers possibilities of reducing risk (Ellis, 2000) that makes these systems more adapted to semi-arid conditions where great seasonal and annual fluctuations in feed availability and quality affect the type and numbers of livestock that farmers keep, manure availability and quality, and the impact of manure on crop production (Powell *et al.*, 1993; Rufino *et al.*, 2007).

Across SSA, the production of crops is directly and indirectly related to livestock production. A direct relationship arises from the need for manure to sustain crop yields (Vanlauwe and Giller, 2006). Indirect relationships arise from the competition for biomass to feed growing livestock numbers or to restore degraded agricultural soils (Rufino *et al.*, 2011).

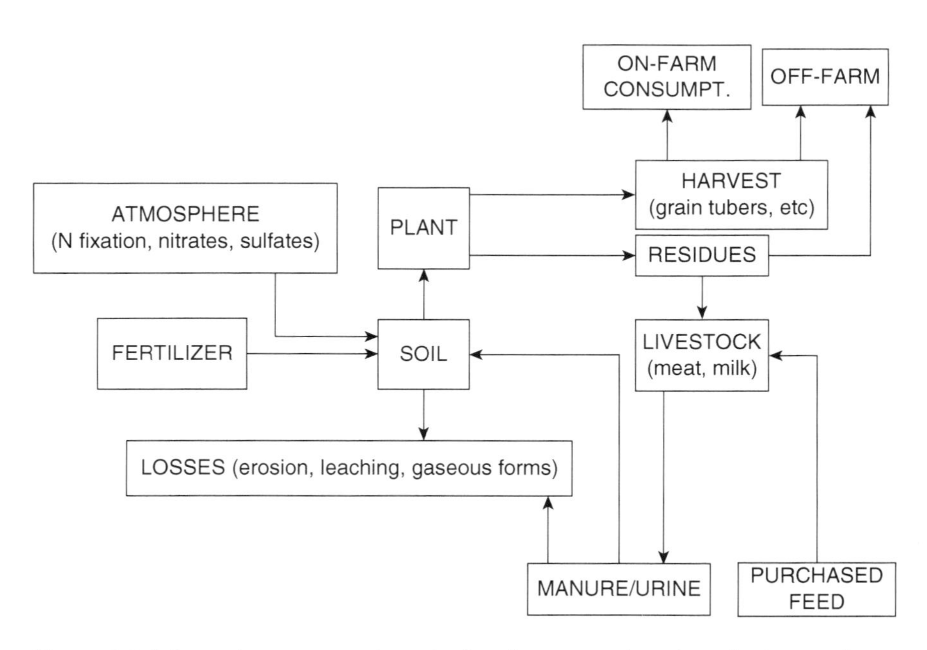

Figure 8.1 Schematic representation of a farming system based on the integration of crops and livestock

Source: Adapted from Stangel (1995).

A number of studies have confirmed that integrated systems tend to be more sustainable, use less energy per unit area and have higher energy efficiency than specialized crop or livestock systems (Entz, Bellotti and Powell, 2005). Incorporating pastures and animals in rotation with crops cultivated in no-tillage systems optimizes the beneficial characteristics, particularly via the capacity of pastures to sequester carbon (Landers, 2007), but also by increasing biodiversity via the attributes of organic matter provided by pastures (Gichangi *et al.*, 2016) and improving nutrient cycling. The resulting microbial diversity changes the soil and its physio-chemical properties (Carvalho *et al.*, 2010).

Crop residues largely from cereals play an increasingly important role as livestock feed in the crop–livestock systems when fodder and natural pastures are scarce in the semi-arid regions, particularly in the dry spells. For example, in a survey of 12 locations in nine countries across sub-Saharan Africa and South Asia, Valbuena *et al.* (2012) reported that crop residues accounted for up to 60 per cent of the total livestock diet in crop–livestock systems. In the semi-arid areas of eastern Kenya, crop residues, mainly maize stovers, are the main livestock feeds during the peak of the long dry season, either grazed in situ or gathered for stall feeding with over 60 per cent of farmers using them during the months of September and October (Njarui *et al.*, 2016).

The predicted increase in demand for livestock products will further fuel feed demand and increase the usage of crop residues for livestock feeding. Increased crop residue yield will facilitate partitioning of crop residue between livestock and soil improvement and improved crop residue fodder quality will support intensification of livestock production. Research has shown that the integration of forage legumes with cereal crops (e.g. hedgerow intercropping, under-sowing, rotations) can lead to sustainable increases in feed and soil productivity. Forage legume–cereal intercropping often increases the quantity and feeding value of crop residues, but decreases the yield of the companion cereal crop (Kouamé *et al.*, 1993; Mohamed-Saleem, 1985). However, grain yields are improved when cereals are rotated with short-term fallows of forage legumes (Mohamed-Saleem and Otsyina, 1986). Grain legumes (cowpeas, pigeon peas, groundnuts, etc.) are already an integral part of many mixed-farming systems. Their residues are valuable diet supplements for livestock and income sources for rural households. There are opportunities to add biological nitrogen fixation via the growth of herbaceous or woody legumes. Many grain legumes, however, shed their leaves upon grain maturity which greatly reduces their feeding value but this can be overcome by planting dual-purpose grain legumes, e.g. cowpea (*Vigna unguiculata*) and pigeon pea (*Cajanus cajan*) that retain their leaves, thus providing both food and feed. In semi-arid Kenya, maize (*Zea mays*) is commonly intercropped, or rotated with common bean (*Phaseolus vulgaris.*), pigeon pea or cowpea, although the relative proportion of the legume in mixed systems is small.

In general, where land is scarce, fodder production can be combined with soil conservation and stall feeding or tethering.

Crop–livestock farming systems and food and nutritional security

Mixed crop–livestock farming systems provide most of the staples consumed, thus, are critical for future food security in SSA. Between 41 and 86 per cent of maize, rice, sorghum and millet, 90 per cent of milk and 80 per cent of meat is produced from integrated systems (Herrero *et al.*, 2010). Livestock in mixed systems can increase local food supplies, supply inputs and services for crop production. Meat and milk are important sources of protein for rural communities and smallholder households. Crop–livestock systems also provide food for urban populations. The effect of manure application to increase yields of different crops is well documented. In nutrient deficient conditions, livestock dung and urine play a critical role in improving soil nutrient balances and sustaining crop production (McDermott *et al.*, 2010; Rufino *et al.*, 2006; Ryschawy *et al.*, 2012). Several studies across SSA have reported substantial increase in grain yield from manure application (Bekunda, Bationo and Ssali, 1997; Gichangi *et al.*, 2007), although crop yield responses are variable, and conclusions are highly site- and season-specific. For example, experiments by Bationo and Mokwunye (1991) in Niger showed a doubling of pearl millet yields a year after the first application of manure at the rate of 5t ha^{-1}. In the ASAL region of Kenya, high and consistent maize yields were obtained with manure applied at 8 t ha^{-1}, which was close to that obtained by applying mineral fertilizer at the rates of 40 kg N ha^{-1} and 17 kg P ha^{-1} (Ikombo, 1984); while Gibberd (1995) reported 58 and 75 per cent yield gains in sole crops and intercrops, respectively, from manure application at the rates of 5 and 10 t ha^{-1}. Livestock also play a critical role in the process of agricultural intensification by providing draft power. In ASAL where labour is scarce, livestock is widely used to support crop expansion by supplying draft power in land preparation and weeding for crops.

Adaptation strategies to increase productivity

In the face of climate change, adaptation is essential for sustainable increased agricultural productivity. The adoption of appropriate agricultural technologies and innovations offers numerous opportunities to mitigate and increase land productivity under smallholder systems. This contributes to agricultural production and income, food and nutritional security and poverty reduction including crop diversification, using an innovative farmer discovery process to bring about agricultural intensification and improvement in livelihoods. To illustrate the success of the intensification, three crop–livestock systems are presented below to show that this strategy has been used successfully by farmers in SSA to adapt to increasing demographic pressures, climate change, low soil fertility and decreased land size by pursuing the strategy of integration and intensification of livestock.

Box 8.1 Mixed system in the Bambey area of Senegal

Low rainfall, short growing periods coupled with inherently poor soil fertility, principally dune sands, resulting in low crop yield and sometimes crop failure are the driving factors of the system for the Serer community in Bambey areas of Senegal. The area receives about 400 mm of annual rainfall with uneven distribution during the growing period. Millet, cowpea and groundnuts are cultivated in the sand dunes and a limited area of sorghum is also cultivated, mainly on the heavier soils. Groundnuts must be planted immediately after the first heavy rains. Any delay in planting can cause a severe reduction in yield. Therefore, land preparation is carried out by a light hoeing during the dry season. The crops are quickly planted as soon as sufficient rains have fallen, usually by horse or donkey-drawn implements. Farmers cultivate crop varieties that have been commercially bred for short periods of maturity. Groundnuts are grown in pure stand, although occasionally a few cowpea seeds may be planted as an intercrop. Weeding is carried out by hand hoes or sometimes light horse or donkey-drawn weeders are used. To maintain soil fertility the Serer farming community preserves the indigenous trees, *Faidherbia albida*. The trees shed their leaves during the rainy season, and grow new leaves during the dry season, thus crops grown under the trees are not shaded. Because it is a leguminous tree, the leaf fall provides a substantial quantity of nutrients for the crops growing under the trees and assists in maintenance of soil fertility as well as in providing livestock feed. Manure is collected in the kraals where the livestock are penned at night, for maintaining soil fertility. Some organic fertilizer is sourced from outside for application to the groundnut. However, high usage is limited by affordability due to high cost. In return, livestock are fed on crop residues which are good quality fodders.

Source: Adapted from Ker (1995).

These cases from western (Box 8.1), east (Box 8.2) and central and southern Africa (Box 8.3) provide some useful lessons that can help to address current and anticipate future challenges to sustainable intensification of smallholder livestock systems. The agricultural productivity varies among these three case studies in terms of resources, demographic, inputs and capital. In all of them, there is diversification with 4–5 crops integrated with different livestock species to mitigate risks. Yields are relatively stable due to intensification.

Box 8.2 Crop–livestock systems in semi-arid eastern Kenya

Small-scale crop–livestock farms represent a large fraction of the rural population of Kenya and the level and pace of intensification vary among the regions, villages and farms. It is worth noting that agriculture is one of the most important sectors of the economy in Kenya, where about 83 per cent of land area is in the ASALs. Most parts have experienced rapid land use changes in the past three decades which has been attributed to increased pressure on land, caused by the ever-increasing human population leading to land sub-division with significant changes in per capita income and consumption patterns (FAO, 2011). These demographic and economic changes have led to higher demand for high-value livestock and crop products. The dairy production systems in Kenya vary considerably and range from grazing systems with indigenous or cross-bred cattle to more intensified mixed farming systems in the highlands mainly fed on fodder (Napier grass, crop residues and natural grasses) and supplements. In Machakos County in the semi-arid area of eastern Kenya for example, crop residues, mainly maize stovers, are the main livestock feeds during the peak of the long dry season, either grazed in situ or gathered for stall feeding with over 60 per cent of farmers using them during the months of September and October (Njarui *et al.*, 2016). Average herd sizes vary from 1–5 cows in the most intensive systems and up to 10 in the extensive systems. A high percentage of the milk production is sold in intensified systems and manure is highly valued and used in crop production (FAO, 2011). Maize is the major crop and staple food that has replaced the traditionally grown small grains such as finger and pearl millets, and sorghum as well as cassava. Amid declining yields of maize and other crops, productivity gains have come largely through land expansion into marginal areas that receive lower and more variable rainfall. The use of animal power in the region improves the timeliness of planting and therefore increases yields in areas where growing seasons are short and time of planting is crucial. Integration of crop–livestock in this case clearly shows the synergies and improved efficiency of resource use.

Source: Authors' own interpretation.

Box 8.3 Semi-intensive farming system of Zimbabwe

Chibi is a communal 'smallholding' farming area about 370 km south of Harare, the capital city of Zimbabwe. They practice a semi-extensive farming system under unimodal rainfall, averaging 450–650 mm per

(continued)

(continued)

annum which is highly variable in distribution (Collinson, 1989). The length of the growing season is variable, varying from fewer than 120 days to more than 165 days. Thus, farmers must manage both the uncertainty of the length of season and the chance of a mid-season drought at the peak of the growing season. Maize, groundnuts, finger millet, sorghum and pearl millet are the major crops cultivated. The low fertility is managed by crop rotation and use of animal manure. High population pressure has caused a downward trend in livestock holdings and consequently a decline in manure, which has undermined farmers' strategies to maintain soil fertility. The loss of draft power, has made them more vulnerable to rainfall crises as this has affected timely cultivation. To adjust, farmers have had to choose to keep cows and use them for ploughing and milk production. They employ various strategies to counter increasing animal nutrition problems; the use of cows for ploughing in winter when the animals are in dire condition, the use of maize stover for dry season feeding, and very late planting of maize and bulrush millet to provide standing fodder. Marketing facilities have been greatly improved, so that farmers could get their surplus crop to the market and make a profit. Cattle owners had a much lower crop failure rate. The higher incomes and yields reported for cattle owners demonstrate the power of higher resource endowments in managing the environment.

Source: Adapted from Ker (1995).

Challenges and opportunities of crop–livestock systems

Smallholder farmers face considerable challenges and opportunities that influence the nature of crop–livestock systems. The extents of these challenges are unique in different regions but are generally similar across the SSA. The main challenges include insecure land tenure and use, population growth, climate change, poor infrastructure and weak institutional frameworks.

Land tenure

Land is one of the key factors for production and access to land promotes agricultural productivity. The constraints related to the tenure system, such as insecurity of land tenure, unequal access to land, and lack of a mechanism to transfer rights and consolidate plots, have resulted in underdeveloped agriculture, food insecurity and degraded natural resources in SSA (Salami *et al.*, 2010). In several West African countries, land tenure and ownership is based on a traditional common property system. For example, in Mali, land is not owned, bought or sold and the rights to use the land are based on traditional authorities (Scoones and Wolmer, 2000). In Benin, where all land belongs to the state, the traditional ownership system is also respected (Deng, 2007). Similarly, in Ethiopia, land is state owned and therefore famers capitalize on maximization of production with little or no investment into the improvement of soil fertility and the conservation of ecosystems. According to Kebede (2002), privatization of land would be the most effective way to reduce insecurity associated with the tenure schemes and uncertainties created by state ownership. In Kenya, sub-division of communally owned land to private ownership has been advocated to ensure sustainable utilisation (Asienga, Perman and Kibet, 2015).

Population

The SSA has one of the highest rates of population growth in the world, at an average of 2.3 per cent, with about 70 per cent of the population living in rural areas engaged in agriculture. The population of SSA was 800 million in 2007 and is estimated to reach between 1.5 and 2 billion by 2050 and further to nearly 4 billion by 2100 (UN Department of Economic and Social Affairs, 2017). This will lead to further sub-division of already small land holdings into smaller and uneconomic units, resulting generally in fragmented production systems. The farm sizes range from as low as about one ha per household in Ethiopia, 2.0 ha in Tanzania and 2.5 ha in Uganda and Kenya (Salami *et al.*, 2010). Subsequently there has been increased pressure of cultivation resulting in degradation especially soil erosion and decline in fertility. All forms of land degradation cause a decline in the productive capacity, reducing attainable and potential yields. With limited land availability, the greatest opportunity for sustainable agricultural productivity is intensification through development of crop–livestock farming systems. These systems provide a viable method of increasing per unit productivity and income.

Climate change and variability

As pointed out *inter alia*, over 80 per cent of farmers in SSA are smallholders and agriculture is largely rain-fed and is thus vulnerable to the impacts of climate change. The Intergovernmental Panel on Climate Change (IPCC, 2007) predicted that climate change could cause potential crop yields from rain-fed agriculture to decline by 50 per cent in some African countries. Historical data have shown recurrent drought in many parts of SSA, e.g. Ethiopia in 1971–75, 1984–85, 1999–2000 and 2002–03 (Tesfai, Adugna and Nagothu, 2015). Drought events were also recorded in the 1970s in Burkina Faso and in Ghana in 1981 and 1984 (Yaro, 2004) while drought and floods simultaneously occurred in Ghana in 2007, causing widespread destruction to millet, rice and groundnuts (Kanchebe, 2010).

Rainfall variability is the single most important vulnerability imposing climatic effect. In East Africa and the Ethiopian highlands, rainfall and run-off are expected to increase with climate change, and more extensive and severe flooding is anticipated (FAO, 2010). On the other hand, monsoon winds in West Africa are likely to weaken; monsoon precipitation is likely to intensify due to the increase in atmospheric moisture (AGRA, 2014). Temperatures are expected to rise from 1.8–4°C (Sillmann *et al.*, 2013) and sea level is projected to rise between 0.4 m and 1.15 m by 2050, resulting in floods. Guinea-Bissau, Mozambique and Nigeria are projected to be the most affected by flooding (Hinkel *et al.*, 2011).

Farmers have developed coping strategies (Nyong, Adesina and Osman-Elasha, 2007; Ng'ang'a *et al.*, 2013) but have often failed (Rao *et al.*, 2011) due to poor access to information, limited inputs and low income. Concrete strategies to identify the priorities for adaptation to climate change at national, regional or continental level are lacking. Despite abundant natural water supplies, Africa has not been able to intensify its agricultural production through irrigation and improved water management. Irrigation is limited and less than 4 per cent of all agricultural output is produced under irrigation in eastern Africa (Salami *et al.*, 2010). Government measures such as establishing an efficient irrigation infrastructure may reduce the vulnerability of the farmers (Schindler, 2009). In West Africa, improvement of irrigation efficiency, the use of groundwater and promotion of community based water management (Braimoh, 2004; Sandwidi, 2007), waterways, dams, etc., to promote small agriculture (Jagtap, 1995), have been encouraged. Additionally, governments need to invest in early-warning focus systems in order to provide timely and effective information for farmers to make decisions about adaptation practices.

Infrastructure

One of the key bottlenecks facing the agricultural sector in SSA is poor infrastructure including transportation networks, access to energy, irrigation systems and stockholding facilities. In West Africa, the existing infrastructures are dilapidated due to limited or non-existing maintenance (Schindler, 2009). Poor transportation networks limit access to markets, often exacerbate high levels of post-harvest losses and inhibit efficient access to inputs such as seed and fertilizers (OECD/FAO, 2016). Because of poor road networks, smallholder farmers depend on inefficient forms of transportation, including the use of animals. The underdeveloped rural roads and other key physical infrastructure have led to high costs associated with transporting agricultural products to the market as well as farm inputs from the market, reducing farmers' competitiveness. Additionally, inadequate storage facilities in rural areas and limited processing facilities for value addition constrain marketability of perishable goods such as fish, dairy products and vegetables. Improving farmers' access to inputs such as seeds and fertilizer by upgrading infrastructure will be crucial to increase productivity.

Policies and institutional factors

Although there are efforts to address the productivity of crop–livestock intensification through research, there has been limited support for the institutional and social arrangements that enable the production system to thrive. Nonetheless, most of the countries in SSA have become aware of the need for an atmosphere of rigorous law and order and enabling policies and institutions for a viable and sustainable system. For example, the Comprehensive Africa Agriculture Development Program was prioritized in the 2003 Maputo Declaration on Agriculture and Food Security through commitments to allocate at least 10 per cent of national budgetary expenditure to agriculture in order to achieve a 6 per cent annual growth of the agricultural sector (OECD/FAO, 2016). However, only a few countries in SSA have fulfilled their commitment on agricultural spending and it is disconcerting that there is an overall decreasing trend in the share of public resources allocated to agriculture. Furthermore, it has been argued that because of the weak administrative structure in West Africa, institutions do not promote crop–livestock intensification, e.g., use of external inputs and equipment, price controls, access to markets, incentives, etc. (Jagtap, 1995). Fertilizer subsidy programmes were implemented to improve productivity in Malawi and Zambia. However, although they were successful in increasing yield, in the long term, the costs outweighed the benefits (Jayne and Rashid, 2013). Suitable policy intervention and support are necessary to provide a stable input and output market that facilitates acquisition of credit and a market for produce and support intensification. Jayne and Rashid (2013) have recommended a holistic approach to support small-scale producers that includes investment in agricultural extension programmes focused on improved farm productivity.

Conclusions and recommendations

The integrated crop–livestock system is widely practised in diverse agroecology settings and it is extensive in the ASAL; the largest agro-climatic zone that covers over 43 per cent of the land area of SSA. The close interaction of crops and livestock allows intensification with animals providing draft and manure for crop production and allows farmers to reduce the spread of risk associated with environmental and social-economic factors. The system offers opportunities for raising productivity and increasing efficiency of resources use, securing availability and access to food, increasing household incomes and reducing poverty. Further, it allows exploitation of untapped resources and more efficient use of labour.

Considerable diversity occurs in crops grown across and within agroclimatic zones and even within an individual household in an integrated crop–livestock system. Ruminants are the most important livestock in the system, which is attributable to their adaptation to the local climate, relevance to household economy and cultural aspects. The productivity may be low compared with similar systems on other continents, but nonetheless they make a vital contribution to FNS livelihood of millions of people in SSA. These systems contribute between 41 and 48 per cent of maize, rice, sorghum and millet, 90 per cent of milk and 80 per cent of meat production in SSA. The system also enhances more efficient recycling of nutrients and improves soil fertility through integration of organic manure from livestock.

There are multiple challenges to the sustainability of the crop–livestock intensification. Insecure and unequal access to land hinder investment for improved productivity. The high population growth rate has led to sub-division and land fragmentation while climate change and variability has affected the stability of the system. Poor infrastructure hinders access to inputs and markets while policies and weak institutions do not support intensification. The crop–livestock intensification systems have the potential to improve and meet growing demand for food and improve livelihoods for the rising population in SSA. The following set of recommendations proposed can help to overcome the challenges:

- i To increase productivity per unit area, it is important that farmers have secure land rights in order for them to make long-term investments. However, care is needed to avoid further land sub-division into uneconomical and unproductive units.
- ii Investment is required in early warning systems to increase farmers' access to reliable information on climate change and training to improve their adaptive capacity.
- iii Governments should invest in logistics and infrastructure, especially roads in rural areas, to facilitate acquisition of inputs and marketing of agricultural produce through faster mobility. Construction of collection and storage facilities in rural areas is vital for storage of farm produce including milk during periods of high production.
- iv Smallholder agriculture needs a consistent stable policy environment and effective institutions. Various governments in SSA should design better policies and institutions that support crop–livestock intensification as a sustainable agricultural practice.
- v Research needs to be strengthened as it plays a crucial role to help farmers identify suitable crop–livestock intensification practices, create awareness and promote best practices.
- vi Countries should ensure adequate allocation of sufficient resources for implementation of crop–livestock systems as a more sustainable option.

References

Alliance for a Green Revolution in Africa (AGRA) (2014) Africa agriculture status report: Climate change and smallholder agriculture in sub-Saharan Africa. Nairobi, Kenya.

Asienga, I.C., Perman, R. and Kibet, L.K. (2015) The role of fencing on marginal productivity of labour, land and capital in ASAL regions of Kenya. *International Journal of Development and Economic Sustainability*, 3, pp. 80–93.

Bationo, A. and Mokwunye, A.U. (1991) Role of manures and crop residues in alleviating soil fertility constraints to crop production: With special reference to the Sahelian and Sudanian zones of West Africa. *Fertilizer Research*, 29, pp. 117–125.

Bationo, A., Mokwunye, U., Vlek, P.L.G., Koala, S. and Shapiro, I. (2003) Soil fertility management for sustainable land use in the West African Sudano-Sahelian zone. *Proceedings of a Conference (8–11 October, 2002) on Grain Legumes and Green Manures for Soil Fertility in Southern Africa: Taking Stock Of Progress, Leopard Rock Hotel, Vumba, Zimbabwe*, pp. 253–292.

Bekunda, M.A., Bationo, A. and Ssali, H. (1997) Soil fertility management in Africa: A review of selected research trials. In: Buresh, R.J., Sanchez, P.A. and Calhoun, F. (eds) *Replenishing Soil Fertility in Africa*, SSSA Special publication no. 51, Madison, WI, pp. 63–79.

Braimoh, A.K. (2004) Modeling land-use change in the Volta Basin of Ghana, PhD thesis, Center for Development Research, University of Bonn. Available at: www.zef.de/publ_theses. html.

Carvalho, P.C.F., Anghinoni, I., Moraes, A., de Souza, E.D., Sulc, R.M., Lang, C.R., Flores, J.P.C., Lopes, M.L.T., Levien, R., Fontaneli, R.S., Bayer, C. (2010) Managing grazing animals to achieve nutrient cycling and soil improvement in no-till integrated systems. *Nutrient Cycling in Agro-ecosystems*, 88, pp. 259–273.

Chauvin, N.C., Mulangu, F. and Mulangu, G. (2012) Food production and consumption trends in sub-Saharan Africa: Prospects for the transformation of the agricultural sector. Working Paper, United Nations Development Programme (UNDP).

Collinson, M.P. (1989) Small farmers and technology in Eastern and Southern Africa. *Journal for International Development*, 1 (Special issue, The Green Revolution in Africa), pp. 66–83.

Debrah, S. and Sissoko, K. (1990) Sources and transfers of cash income in the rural economy: The case of smallholder mixed farmers in the semi-arid zone of Mali. African Livestock Policy Analysis Network Paper No. 25, International Livestock Centre for Africa, Addis Ababa, Ethiopia.

Deng, Z. (2007) *Vegetation Dynamics in Queme Basin, Benin, West Afrika*. Göttingen, Germany: Cuvillier Verlag.

Dixon, J., Gulliver, A. and Gibbon, D. (2001) Sub-Saharan Africa. In: *Farming Systems and Poverty: Improving farmers' livelihoods in a changing world*. Rome and Washington DC: FAO and World Bank, pp. 29–82.

Dixon, J., Tanyeri-Abur, A. and Wattenbach, H. (2003) Context and framework for approaches to assessing the impact of globalization on smallholders. In: Dixon J., Taniguchi, K. and Wattenbach H. (eds) Approaches to assessing the impact of globalization on African smallholders: Household and village economy modeling. *Proceedings of Working Session Globalization and the African Smallholder Study*, FAO and World Bank, Rome, Italy, Food and Agricultural Organization United Nations. Available at: www.fao.org/docrep/007/y5784e/y5784e02.htm.

Ellis, F. (2000) The determinants of rural livelihood diversification in developing countries. *Journal of Agricultural Economics*, 51, pp. 289–302.

Ellis, F. and Freeman, H.A. (2004) Rural livelihoods and poverty reduction strategies in four African counties. *The Journal of Development Studies*, 40 (4), pp. 1–30.

Entz, M.H., Bellotti, W.D. and Powell, J.M. (2005) Evolution of integrated crop–livestock production systems. In: McGilloway, D.A. (ed.) *Grassland: A global resource*. Wageningen: Wageningen Academic Publishers, pp. 137–148.

FAO (2002) FAOSTAT, Author's computation based on FAO statistical database. Available at: www.fao.org/faostat/en/#home.

FAO (2006) *Demands for Products of Irrigated Agriculture in Sub-Saharan Africa*, FAO Water Report No. 31, FAO, Rome, Italy.

FAO (2010) *Enhancing Crop Livestock Systems in Conservation Agriculture for Sustainable Production Intensification: A farming discovery process going to scale in Burkina Faso*. Rome: Food and Agriculture Organization, p. 42.

FAO/ILRI (2011). Global livestock production systems. Rome: FAO. Available at: www.fao.org/docrep/014/i2414e/i2414e.pdf.

Garrity, D., Dixon, J. and Boffa, J.M. (2012) Understanding African farming systems: Science and policy implications. Prepared for Food Security in Africa, Sydney, 29–30 November 2012. Sydney, Australia: Australian Centre for International Agricultural Research (ACIAR).

Gibberd, V. (1995) Yield responses of food crops to animal manure in semi-arid Kenya. *Tropical Science*, 35, pp. 418–426.

Gichangi, E. M., Njarui, D. M. G., Gatheru, M., Magiroi, K. W. N., Ghimire, S. R. (2016) Effects of Brachiaria grasses on soil microbial biomass carbon, nitrogen and phosphorus in soils of the semi-arid tropics of Kenya. *Tropical and Subtropical Agroecosystems*, 19, pp. 193–203.

Gichangi, E.M., Karanja, N.K. and Wood, C.W. (2006) Composting cattle manure from zero grazing systems with agro-organic wastes to minimize nitrogen losses in smallholder farms in Kenya. *Tropical and Subtropical Agro-ecosystems*, 6, pp. 57–64.

Gichangi, E.M., Mnkeni, P.N.S. and Brookes, P.C. (2010) Goat manure application improves phosphate fertilizer effectiveness through enhanced biological cycling of phosphorus. *Journal of Soil Science and Plant Nutrition*, 56, pp. 853–860.

Gichangi, E.M., Njiru, E.N., Itabari, J.K., Wambua, J.M., Maina, J.N. and Karuku, A. (2007) Assessment of improved soil fertility and water harvesting technologies through community based on-farm trials in the ASALs of Kenya. In: Bationo, A., Waswa, B., Kihara, J. and Kimetu, J. (eds) *Advances in Integrated Soil Fertility Management in sub-Saharan Africa: Challenges and opportunities*. Dordrecht: Springer.

Giller, K.E., Cadisch, G., Ehaliotis, C., Adams, E., Sakala, W.D. and Mafongoya, P.L. (1997) Building soil nitrogen capital in Africa. In: Buresh, R.J., Sanchez, P.J. and Calhoun, F. (eds) *Replenishing Soil Fertility in Africa*, SSSA Special Publication No. 51, Madison, WI, pp. 151–192.

Herrero, M., Thornton, P.K., Notenbaert, A.M., Wood, S., Msangi, S., Freeman, H.A., Bossio, D., Dixon, J., Peters, M., van de Steeg, J., Lynam, J., Rao, P.P.,

Macmillan, S., Gerard, B., McDermott, J., Seré, C. and Rosegrant, M. (2010) Smart investments in sustainable food production: Revisiting mixed crop–livestock systems. *Science*, 327, pp. 822–825.

Hinkel, J., Brown, S., Exner, L., Nicholls, R.J., Vafeidis, A.T. and Kebede, A.S. (2011) Sea-level rise impacts on Africa and the effects of mitigation and adaptation: an application of DIVA. *Regional Environmental Change*, 12, pp. 207–224.

IFAD (2011) *Proceedings, IFAD (International Fund for Agricultural Development) Conference on New Directions for Smallholder Agriculture, 24–25 January, 2011, IFAD, WFP, and FAO, 2012: The State of Food Insecurity in the World 2012.* Economic growth is necessary but not sufficient to accelerate reduction of hunger and malnutrition. Food and Agriculture Organization of the United Nations (FAO), Rome, Italy. Available at: www.ifad.org/asset?id =2106989.

Ikombo, B.M. (1984) Effects of farmyard manure and fertilizers on maize in semi-arid areas of Eastern Kenya. *East Africa Agriculture and Forestry Journal*, 44, pp. 266-274.

Intergovernmental Panel on Climate Change (IPCC) (2007) Summary of policy makers. In: Solomon, S., Qin, D., Manning, M., Chen, Z., Marquis, M., Averyt, K.B., Tignor, M. and Miller H.L. (eds) *Climate Change 2007: The physical science basis. Contribution of Working Group I to the Fourth Assessment Report of the Intergovernmental Panel on Climate Change.* Cambridge, UK and New York, NY: Cambridge University Press.

Jagtap, S.S. (1995) Environmental characterization of the moist lowland savanna of Africa. In: Kang, B.T. *et al.* (eds) *Moist Savannas of Africa: Potentials and constraints for crop production.* Ibadan, Nigeria: African Book Builders.

Jayne, T.S. and Rashid, S. (2013) Input subsidy programs in Sub-Saharan Africa: A synthesis of recent evidence. *Agricultural Economics*, 44, pp. 1–16.

Kanchebe, E. (2010) Local knowledge and livelihood sustainability under environmental change in Northern Ghana, PhD thesis, Center for Development Research, University of Bonn. Available at: http://hss.ulb.uni-bonn.de:90/2010/2336/2336.htm.

Karanja, N.K., Gichangi, E.M. and Wood, C.W. (2005) Simulation study to assess the potential of selected agro-organic wastes for ability to reduce N volatilization from cow manure. *Tropical and Subtropical Agro-ecosystems*, 5, pp. 25–31.

Kebede, B. (2002) Land tenure system and common pool resources in rural Ethiopia: A study based on fifteen states. *African Development Review*, 14, pp. 113–150.

Ker, A. (1995) Farming systems of the African savanna: A continent in crisis. ISBN Out of print e-ISBN 1-55250-280-5.

Kouamé, C.N., Powell, J.M., Renard, C.A. and Quesenberry, K.H. (1993) Plant yields and fodder quality related characteristics of millet-stylo intercropping systems in the Sahel. *Agronomy Journal*, 85, pp. 601–605.

Landers, J.N. (2007) Tropical crop–livestock systems in conservation agriculture: The Brazilian experience. *Integrated Crop Management*, 5, pp. 1–92.

McDermott, J.J., Staal, S.J., Freeman, H.A., Herrero, M. and Van de Steeg, J.A. (2010) Sustaining intensification of smallholder livestock systems in the tropics, *Livestock Science*, 130 (1–3), pp. 95–109.

Mohamed-Saleem, M.A. (1985) Effect of sowing time on grain and fodder potential of sorghum under sown with stylo in the subhumid zone of Nigeria, *Tropical Agriculture (Trinidad)*, 62, pp. 151–153.

Mohamed-Saleem, M.A. and Otsyina, R.M. (1986) Grain yields of maize and the nitrogen contribution following *Stylosanthes* pasture in the Nigerian subhumid zone. *Experimental Agriculture*, 22, pp. 207–214.

Ng'ang'a, S.K., Diarra, L., Notenbaert, A. and Herrero, M. (2013) Coping strategies and vulnerability to climate change of households in Mali. *CCAFS* Working Paper No. 35, CGIAR Research Program on Climate Change, Agriculture and Food Security (CCAFS), Copenhagen, Denmark.

Ngigi, S.N. (2011) *Climate Change Adaptation Strategies: Water resources management options for smallholder farming systems in sub-Saharan Africa*. The MDG Centre, East and Southern Africa, The Earth Institute at Columbia University, New York.

Njarui, D.M.G., Gichangi, E.M., Gatheru, M., Nyambati, E.M., Ondiko, C.N., Njunie, M.N., Ndungu-Magiroi, K.W., Kiiya, W.W., Kute, C.A.O. and Ayako, W. (2016) A comparative analysis of livestock farming in smallholder mixed crop–livestock systems in Kenya: 2. Feed utilization, availability and mitigation strategies to feed scarcity. *Livestock Research for Rural Development*, 28, Article 67. Available at: www.lrrd.org/lrrd28/4/njar28067.html.

Nyong, A., Adesina, F. and Osman-Elasha, B. (2007) The value of indigenous knowledge in climate change mitigation and adaptation strategies in the African Sahel: Mitigation and adaptation strategies. *Global Change*, 12, pp. 87–797.

OECD/FAO (2016) Agriculture in Sub-Saharan Africa: Prospects and challenges for the next decade. *OECD-FAO Agricultural Outlook 2016–2025*. Paris: OECD Publishing.

Palm, C.A., Myers, R.J.K. and Nandwa, S.M. (1997) Combined use of organic and inorganic nutrient sources for soil fertility maintenance and replenishment. In: Buresh, R.J., Sanchez, P.A. and Calhoun, F. (eds) *Replenishing Soil Fertility in Africa*. Madison, WI: SSSA Special Publication No. 51, pp. 193–217.

Powell, J.M., Fernandez-Rivera, S., Williams, T.O. and Renard, C. (1993) Livestock and sustainable nutrient cycling in mixed farming systems of sub-Saharan Africa, Volume 1: Conference Summary, *Proceedings of an International Conference, International Livestock Centre for Africa (ILCA), Addis Ababa, Ethiopia, 22–26 November 1993*.

Probert, M.E., Okalebo, J.R. and Jones, R.K. (1995) The use of manure on smallholders' farms in semi-arid eastern Kenya. *Experimental Agriculture*, 31, pp. 371–381.

Rao, K.P.C., Ndegwa, W.G., Kizito, K. and Oyoo, A. (2011) Climate variability and change: Farmer perceptions and understanding of intra-seasonal variability in rainfall and associated risk in semi-arid Kenya. *Experimental Agriculture*, 47, pp. 267–291.

Rockström, J. (2000) Water resources management in smallholder farms in Eastern and southern Africa: An overview. *Physics and Chemistry of the Earth*, 25, pp. 275–283.

Rockström, J. (2003) Water for food and nature in drought prone tropics: Vapour shift in rainfed agriculture. *Philosophical Transcriptions of the Royal Society of London B Biological Sciences*, 358 (1440), pp. 1997–2009.

Rockström, J., Folke, C., Gordon, L., Hatibu, N., Jewitt, G., Penning de Vries, F., Rwehumbiza, F., Sally, H., Savenije, H. and Schulze, R. (2004) A watershed approach to upgrade rainfed agriculture in water scarce regions through Water System Innovations: An integrated research initiative on water for food and rural livelihoods in balance with ecosystem functions. *Physics and Chemistry of the Earth*, 29, pp. 1109–1118.

Rufino, M.C., Dury, J., Tittonell, P., Van Wijk, M.T., Herrero, M., Zingore, S., Mapfumo, P. and Giller, K.E. (2011) Competing use of organic resources, village-level interactions between farm types and climate variability in a communal area of NE Zimbabwe. *Agricultural Systems*, 104, pp. 175–90.

Rufino, M.C., Rowe, E.C., Delve, R.J. and Giller, K.E. (2006) Nitrogen cycling efficiencies through resource-poor African crop livestock systems. *Agriculture, Ecosystems and Environment*, 112, pp. 261–282.

Rufino, M.C., Tittonell, P., van Wijk, M.T., Castellanos-Navarrete, A., Delve, R.J., Ridder, N.A. and Giller, K.E. (2007) Manure as a key resource within smallholder farming systems: Analysing farm-scale nutrient cycling efficiencies with the NUANCES framework. *Livestock Science*, 112, pp. 273–287.

Ryschawy J., Choisis, N., Choisis, J.P., Joannon, A. and Gibon, A. (2012) Mixed crop–livestock systems: an economic and environmental-friendly way of farming? *Animal*, 6 (10), pp. 1722–1730.

Salami, A., Kamara, A.B. and Brixiova, Z. (2010) *Smallholder Agriculture in East Africa: Trends, constraints and opportunities*. Working Paper No. 105, African Development Bank Group, TUNIS Belvédère, Tunisia.

Sandwidi, J.P. (2007) *Groundwater potential to supply population demand within the Kompienga dam basin in Burkina Faso*, PhD thesis, Center for Development Research, University of Bonn.

Sansoucy, R., Jabbar, M., Ehui, S. and Fitzhugh, H. (1995) Keynote paper: The contribution of livestock to food security and sustainable development, *Proceedings of the joint FAO/ILRI roundtable on livestock development strategies for low income countries, 27 February–2 March 1995, ILRI (International Livestock Research Institute), Addis Ababa, Ethiopia.*

Schiere, J.B., Baumhardt, A.L., Van Keulen, H., Whitbread, A.M., Bruinsma, A.S., Goodchild, A.V., Gregorini, P., Slingerland, M.A. and Wiedemann-Hartwell, B. (2006) Mixed crop–livestock systems in semi-arid regions. In: Peterson, G.A. (ed.) *Dryland Agriculture*, 2nd edn. Agron. Monogr. 23. Madison, WI: ASA, CSSA, and SSSA. Available at: http://ifsa.boku.ac.at/cms/fileadmin/Books/2006_Schiere.pdf.

Schindler, J. (2009) *A multi-agent system for simulating land-use and land-cover change in the Atankwidi catchment of Upper East Ghana*, PhD thesis, Center for Development Research, University of Bonn.

Schlecht, E., Fernández-Rivera, S. and Hiernaux, P. (1998) Timing, size and N-concentration of faecal and urinary excretions in cattle, sheep and goats – can they be used for better manuring of cropland? In: Renard, G., Neef, A., Becker, K. and von Oppen, M. (eds) *Soil Fertility Management in West African Land Use Systems*. Weikersheim, Germany: Margraf Verlag, pp. 361–368.

Scoones, I. and Wolmer, W. (2000) Pathways of change: Crops, livestock and livelihoods in Africa. *Lessons from Ethiopia, Mali and Zimbabwe*. Brighton: Institute of Development Studies, University of Sussex. Available at: https://assets.publishing.service.gov.uk/media/57a08d7440f0b64974001884/R6781a.pdf.

Shepherd, K. and Soule, M. (1998) Soil fertility management in west Kenya: Dynamic simulation of productivity, profitability and sustainability at different resource endowment levels. *Agriculture, Ecosystems and Environment*, 71, pp. 131–145.

Sillmann, J., Kharin,V.V., Zweiers, F.W., Zhang, X., Bronaugh, D. (2013) Climate extreme indices in the CMIP5 multi-model ensemble. Part 2: Future climate projections. *Journal of Geophysical Research: Atmospheres*, 118 (6), pp. 1–55.

Stangel, P.J. (1995) Nutrient cycling and its importance in sustaining crop–livestock systems in sub-Saharan Africa: An overview. Livestock and Sustainable Nutrient Cycling in Mixed Farming Systems of sub-Saharan Africa. Volume II: Technical Papers. *Proceedings of an International Conference held in Addis Ababa, Ethiopia, 22–26 November 1993, ILCA (International Livestock Centre for Africa), Addis Ababa, Ethiopia.*

Sumberg, J. (2003) Toward a dis-aggregated view of crop–livestock integration in Western Africa. *Land Use Policy*, 20, pp. 253–264.

Tarawali, S., Herrero, M., Descheemaeker, K., Grings, E. and Blümmel, M. (2011) Pathways for sustainable development of mixed crop livestock systems: Taking a livestock and pro-poor approach. *Livestock Science*, 139, pp. 11–21.

Tesfai, M., Adugna, A. and Nagothu, U.S. (2015) Status and trends of food security in Ethiopia. In: Nagothu, U.S. (ed.) *Food Security Development: Country case studies*. London and New York: Earthscan Routledge, pp. 147–176.

Thornton, P.K. and Herrero, M. (2001) Integrated crop–livestock simulation models for scenario analysis and impact assessment. *Agricultural Systems*, 70, pp. 581–602.

Thornton, P.K. and Herrero, M. (2015) Adapting to climate change in the mixed crop and livestock farming systems in sub-Saharan Africa. *Nature Climate Change*, 5, pp. 830–836.

UN, Department of Economic and Social Affairs (2017) World Population Prospects: The 2017 Revision, Key Findings and Advance Tables. Working Paper No. ESA/P/WP/248. Available at: https://esa.un.org/unpd/wpp/Publications/Files/WPP2017_KeyFindings.pdf.

Valbuena, D., Erenstein, O., Homann, S., Abdoulaye, T., Claessens, L. and Duncan, A.J. (2012) Conservation agriculture in mixed crop–livestock systems: Scoping crop residue trade-offs in sub-Saharan Africa and South Asia. *Field Crops Research*, 132, pp. 175–184.

Vanlauwe, B. and Giller, K.E. (2006) Popular myths around soil fertility management in sub-Saharan Africa. *Agriculture Ecosystem and Environment*, 116, pp. 34–46.

Waithaka, M., Nelson, G.C., Thomas, T.S. and Kyotalimye, M. (2013) *East African Agriculture and Climate Change: A comprehensive analysis*, International Food Policy Research Institute, 2033 K Street, NW Washington, DC 20006–1002, USA.

Wiggins, S. (2009) Can the smallholder model deliver poverty reduction and food security for a rapidly growing population in Africa?, Paper for the Expert Meeting on How to Feed the World in 2050, Rome.

Wiggins, S. and Sharada, K. (2013) Looking back, peering forward: Food prices and the food price spike of 2007/08. London: Overseas Development Institute. Available at: www.odi.org/sites/odi.org.uk/files/odi-assets/publications-opinion-files/8339.pdf.

Williams, T.O., Hiernaux, P. and Fernández-Rivera, S. (2010) Crop–livestock systems in sub-Saharan Africa: Determinants and intensification pathways. In: McCarthy, N., Swallow, B., Kirk, M. and Hazell, P. (eds) *Property Rights, Risk, and Livestock Development in Africa.* Nairobi, Kenya: International Livestock Research Institute.

World Bank (2007) *World Development Report 2008: Agriculture for Development.* The World Bank: Washington DC.

World Bank (2008) *The Growth Report: Strategies for Sustained Growth and Inclusive Development.* Washington, DC: Commission on Growth and Development, The World Bank.

World Bank (2003) *Reaching the Rural Poor: A renewed strategy for rural development.* Washington, DC: The World Bank. Available at: https://openknowledge.worldbank.org/handle/10986/14084. License: CC BY 3.0 IGO.

Yaro, J.A. (2004) *Combating Food Insecurity in Northern Ghana: Food insecurity and rural livelihood strategies in Kajelo, Chiana and Korania.* Occasional Paper No. 44, Department of Sociology and Human Geography, University of Oslo.

9 Community-driven approaches to sustainable intensification in river deltas

Lessons from the Ganges and Mekong Rivers

Douglas J. Merrey, Manoranjan K. Mondal, Chu Thai Hoanh, Elizabeth Humphreys and Nga Dao

Introduction

The livelihoods of people living in coastal deltas, especially in poor tropical countries, are being undermined by multiple insidious trends. Historically, these deltas have attracted large numbers of people, leading to high population densities, because they offer a wide range of ecosystem services. Tropical deltas are characterized by a combination of highly fertile land, multiple marine and freshwater resources, and rich biodiversity. Deltas are often the "breadbaskets" or "rice bowls" of the nations or regions where they are located: examples are the Nile, Irrawaddy, Mississippi and the Cauvery, as well as the Ganges and Mekong river deltas. However, deltas around the world are facing growing threats to their integrity and productivity. The origins of these threats are both anthropogenic and natural, and include the impacts of growing urbanization; agricultural intensification; anthropogenic alterations of flow paths and flood plains; upstream water consumption and pollution; over-extraction of groundwater; trapping of sediments; climate change; sea-level rise whose effects are amplified by sinking land levels and sedimentation of river beds; and extreme events such as river flooding and tidal surges (Renaud *et al.*, 2013).

In tropical developing countries, deltas with dense populations dependent on agriculture and marine resources and high levels of poverty and malnutrition are especially at risk. A great deal of infrastructure such as embankments, polders and irrigation schemes has been constructed in recent decades to cope with the uncertainties of deltaic agriculture and support agricultural intensification. In many cases, these structures no longer adequately protect socio-ecological systems from threats such as storm surges, salinization and rising sea levels. This is because upstream interventions such as dams for hydropower and irrigation have changed water flows; the polders themselves have seriously modified water flows and sedimentation in ways that damage local ecosystems; and they are often poorly maintained and managed. Renaud *et al.* (2013) argue that some of these deltas' socio-ecological systems are at or even beyond a "tipping point": a point where previous ecosystem services will no longer be available, leading to major changes in livelihoods, agroecosystems and mass outmigration. These trends are largely anthropogenic, which also means that in principle they can be reversed by human action.

The Mekong Delta in Vietnam and Cambodia, and the Ganges Delta in Bangladesh and India are prime examples of tropical deltas at risk, where poverty and malnutrition are high. This is why the CGIAR's Water Land and Ecosystems (WLE) Program and its predecessor, the Challenge Program on Water and Food (CPWF), have invested in identifying potential solutions that could reverse the negative trends and achieve positive future outcomes through innovative approaches to sustainably intensifying agroecosystems. A basic premise of these research programs has been that achieving sustainable agricultural intensification requires a multi-scale whole-systems landscape or agro-ecological approach.

Globally, meeting rapidly rising food demands and delivering the Sustainable Development Goals (SDGs), while remaining within planetary boundaries, will require significant transformation of global and local food systems (Dearing *et al.*, 2014; IPES-Food, 2016;; Bernard and Lux, 2016; Rockström *et al.*, 2017). In many parts of Asia and Africa, unsustainable farming practices are contributing to land degradation, water scarcity and pollution, and loss of biodiversity and soil fertility (Gray *et al.*, 2016). Some evidence suggests this is deepening rural poverty and chronic food insecurity, especially for the poor and disadvantaged, including women (Leach, Raworth and Rockström, 2013; Sugden *et al.*, 2014; de Haan and Sugden, 2014).

Reversing these negative trends is critical to delivering the SDGs, but it is also fraught with new complexities and uncertainties. However, the evidence is mounting that, if societies approach agriculture differently, all the food needed can be grown at a reasonable cost, without degrading natural systems. Sustainable intensification is a key strategy to develop more resilient agricultural systems (Pretty and Bharucha, 2014). This will also require more effective governance of agricultural ecosystems (Vasseur *et al.*, 2017). Directly linked to this is the notion that greater social equality, including gender equality, and mobilizing youth, are essential to achieve long-term sustainable and productive management of natural resources (e.g. Narayan, Saavedra-Chanduvi, and Tiwari, 2013). Resilience defines a system's capacity to withstand shocks and stresses, transform in response to changing conditions, and adapt in crisis situations. Farming systems where ecosystems and agricultural landscapes are managed in sustainable ways, i.e. sustain biodiversity and soil productivity, and safeguard freshwater resources, will be more resilient to shocks and changing conditions, and more productive in the long term.

This chapter shares experiences from projects in two related case studies aimed at reversing declines in agro-ecological productivity and promoting sustainable intensification in ways that also benefit the poorest members of communities. The research projects were carried out over a period of two decades in the Ganges and Mekong deltas. We argue that long-term systems research, carried out in partnership with community members, is the most effective way to learn how to sustainably and equitably intensify complex agro-ecological systems. The next section provides an overview of the two deltas and the challenges they face. The section following that describes the main results of the research programs. The chapter concludes with a set of recommendations for both continuing the learning process and scaling out the lessons learned.

The two deltas

The Mekong delta

The total area of the Mekong River delta, the third largest delta in the world, is 5.9 million ha, covering parts of Cambodia and Vietnam. Most of the delta area, about 3.9 million ha, is located in Vietnam. About 62 percent (2.4 million ha) of the Vietnamese portion is used for agriculture and aquaculture. The delta population in Vietnam is currently about 18.6 million people; it is growing very slowly, by only 0.42 percent in 2015, because of outmigration after the "*doi moi*" (reform) policy implemented from the end of the 1980s. The delta is Vietnam's rice bowl, producing approximately 60 percent of the country's rice, 80–90 percent of the country's rice exports, and a substantial portion of its fruit, fish and shrimp exports. Over the past three decades, Vietnam has invested heavily in water control projects: development and rehabilitation of irrigation and drainage systems; controlling brackish water and flood protection; domestic water supply; and roads. These investments, combined with increased use of fertilizer and significant policy changes, enabled many farmers to grow up to three rice crops per year (Renaud *et al.*, 2013; Tuan *et al.*, 2015).

However, cultivating three rice crops per year is becoming ecologically unsustainable. The engineered structures are causing major changes in hydrology, sediment processes and nutrient balances. Production of rice and other crops and even the quality of the domestic water supply are now threatened by several ominous trends: changes in hydrology and sediment flows as more and more dams for hydropower and irrigation are constructed upstream; pollution from urban and industrial zones and over-use of agricultural chemicals as well as intensification of aquaculture; reduced soil fertility due to agricultural intensification; intrusion of sea water; climate change and rising sea levels as a result of global climate change; sinking land levels through compaction and other processes such as groundwater exploitation; and increasing vulnerability to serious floods both from upstream and typhoon events (Syvitski *et al.*, 2009; Renaud *et al.*, 2013). Although reclamation of acid sulfate soils for agricultural production was implemented widely after the end of the Vietnam War in 1975, acidity leaching into the canal network from soil excavation is still a constraint on production in some areas, such as large depressions in the Plain of Reeds, Long Xuyen Quadrangle and Ca Mau Peninsula. Increasing amounts of fertilizer are used to maintain rice production, but the chemicals used are building up in the land and canals. Declining fish catches are a key indicator of environmental degradation: natural fish species are disappearing from the rivers, streams and canals, and along with aquatic species such as shrimps and frogs, are becoming less common in paddy fields. This reduction in freely available traditional protein sources greatly affects poorer households. As fish become harder to find, people resort to more extreme methods of catching them (Tuan *et al.*, 2015; Van *et al.*, 2016; Berg *et al.*, 2017; WLE, 2017).

Broadly, the delta can be subdivided into three sub-regions. The sub-region closest to the coastline is dedicated to farming shrimp in brackish water. Further inland is a large sub-region that is protected by a system of dikes, enabling intensive rice farming and vegetable crops, with 2–3 crops per year. Along the border of these sub-regions is a buffer zone for a mixture of rice during the rainy season and shrimp farming during the dry season. The third sub-region is at the upper part of the delta with fresh water throughout the year but impacted seriously by floodwater during the wet season. All three sub-regions are facing serious threats to their sustainability and productivity. Some researchers have argued that the Mekong delta is approaching a point where the entire socio-ecological system is at risk of tipping into an unfavorable state. Engineering solutions alone will not stabilize the system, but adaptations of the agricultural production system may constitute a viable development pathway to increase resilience (Renaud *et al.*, 2013, 2014). This is the context for WLE's recent research.

Building on the results of an earlier project supported by the United Kingdom's Department for International Development (DFID), WLE and its predecessor CPWF have carried out research in Vietnam's Bac Lieu Province in the southern tip of the Mekong Delta. From 1994 to 2000, the government installed ten large salinity control sluices along National Highway No. 1 to prevent saline water inflow during periods of low-flow in the Mekong River for an area of 260,000 ha at the central part of the Ca Mau Peninsula, while leaving the large band between the highway and coastal dike for mangrove and marine habitat (Figure 9.1). Fresh water (including floodwater) from the Mekong is still flowing into the protected area through a dense canal network. (This protection is different from "polders" in Bangladesh where dikes are built to enclose smaller areas for full protection from flood and salinity intrusion.) These investments enabled many rice farmers to produce 2–3 crops of rice or upland crops per year, but restricted the supply of saline water required by shrimp producers. Further, as rice prices fell and the price of shrimp rose, farmers were no longer able to make a decent profit. The DFID and CPWF projects supported a change in policy

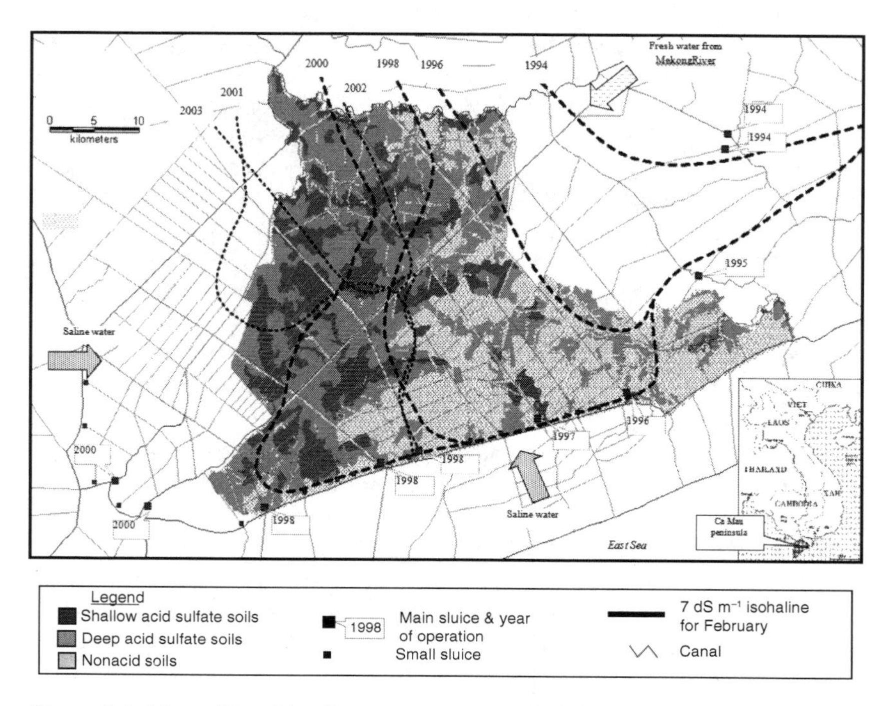

Figure 9.1 Map of Bac Lieu Province, Vietnam, with location of sluices and completion years

Source: Tuong *et al.*, 2009.

from an exclusive focus on rice to a broader land and water management policy. Modeling was used to identify how to operate the sluice gates to supply suitable water for both brackish water shrimp and rice production in the region, including the buffer zone where shrimp and rice are grown in rotation during the dry and wet seasons, respectively (Hoanh *et al.*, 2003; Wichelns *et al.*, 2010). The modeling was done in response to an urgent request by provincial authorities after shrimp farmers broke a large dam to intake saline water for shrimp culture.

However, rising sea levels, increasing intrusion of saline water, and pollution from over-use of agricultural chemicals have led to continuing deterioration of aquatic resources (fish, shrimp, frogs) and declining rice yields. As part of the WLE project *River food systems from villagers' perspectives in the Mekong Delta*, researchers have spent two years discussing what is happening to water resources with farmers and local officials, and disseminating its findings for wider discussion. Under the project, a Vietnamese NGO, The Center for Water Resources Conservation and Development (WARECOD), has joined with researchers from Can Tho

University, members of government agencies and people's organizations, and Vietnam Television to document the situation and work with local people to plan ways of reversing environmental degradation.

The Ganges delta

The Ganges is a component of the world's third largest freshwater outlet to the sea, the Ganges–Brahmaputra–Meghna river system, stretching across India, China, Nepal, Bangladesh and Bhutan. The Ganges delta is the world's largest, with an area of 105,000 km^2 in Bangladesh and India. It is a highly regulated river system with dams, barrages and extensive embankments on nearly every tributary, diminishing dry season water flow. Groundwater is also used extensively for agriculture and domestic water supply. About 250 million people depend on agriculture and aquaculture for their livelihoods (Renaud *et al.*, 2013).

Climate change is likely to have a serious impact on the reliability and intensity of the southwest monsoon, the intensity of cyclones, and river flows. Both droughts and floods have become more common, severely affecting agricultural production and people's livelihoods. Rivers are often heavily polluted, as the Ganges and its tributaries operate as receptacles for municipal and industrial wastes. Arsenic in the groundwater is a serious problem. The delta is also threatened by both the reduction in aggradation and accelerated compaction combined with rising sea levels (Syvitski *et al.*, 2009). Renaud *et al.* (2013: 5) argue that if the current status quo prevails, the socio-ecological system "could very well tip into an unfavorable configuration" – i.e., the damage would be irreversible and lead to unpredictable systemic transformations.

The coastal zone in Bangladesh is characterized by a vast network of river systems and an ever-dynamic estuary. Since the late 1960s, 139 polders have been constructed in Bangladesh's coastal zone to enable production of rice and protect people from floods. A polder is a low-lying area between the rivers, protected from tidal flooding and salinity intrusion by a peripheral earthen embankment, with internal canals (former river distributaries) linked to gated outlets in the embankment to enable management of water levels within the polder. About 54 polders are located in the Ganges coastal zone of Bangladesh (Khan *et al.*, 2015). These polders cover 1.2 million ha and are inhabited by about eight million people. In low and medium salinity regions, polders are used to cultivate rice during the rainy season, and pulses and sesame during the dry season. However, over one million ha of land lies fallow for a couple to several months during the dry season, with some livestock grazing. In the more saline regions, polders are used to raise brackish water shrimp and fish both for sale and household use, and in some areas, brackish water shrimp are grown in rotation with rice during the rainy season, as in parts of the Mekong delta.

As low-lying farmland, polders are extremely vulnerable to flooding, largely due to the high rainfall during the rainy season, lack of separation between low and high lands internally, poor drainage management, and cyclones. Saline water intrusion occurs sporadically in patches during cyclonic storm surges, or because of damaged or missing sluice gates and poor management of the gates. Despite the huge investments in polders, agricultural production is much lower than in the rest of Bangladesh and farm families face high levels of food and nutrition insecurity and poverty (Tuong *et al.*, 2014; IRRI, 2017). Most farmers grow traditional tall, photoperiod-sensitive rice varieties in the rainy season whose productivity is low. The farming systems integrate varying combinations of freshwater and brackish water prawns, rice, fish, and vegetables, which vary with the level of salinity (Dey *et al.*, 2013; Faruque *et al.*, 2016). Farmers are increasingly vulnerable to shocks from a combination of physical factors (floods, droughts, storms, saltwater intrusion), a deteriorating resource base, weak economic conditions, and limitations of existing formal and informal institutions (Huq *et al.*, 2015).

As is the case for the Mekong delta, the WLE project was built on the results achieved by the CPWF program, which in turn also had its foundation in an earlier DFID-funded research project (Poverty Elimination Through Rice Research Assistance (PETRRA); see Rural Livelihoods Evaluation Partnership, 2004). A systems approach was followed by CPWF, including field experiments, river and polder water monitoring (levels, flows, salinity) at plot, polder and regional scales, development of a comprehensive GIS database of the coastal zone, and extrapolation domains for improved cropping systems, water governance studies, modeling to assess water supply and demand to outscale the *aman-boro*[1] rice cropping pattern in Polder 30 (Figure 9.2) of Khulna District, and hydrologic modelling to predict water depth across selected polders (including Polder 30) in response to rainfall.

Polder 30 covers 6,445 ha with a net cultivable area of 4,240 ha. A 40.27 km embankment protects the area against tidal and storm surges as well as salinity intrusion. There are eight drainage sluices, ten flushing sluices, three drainage-cum-flushing sluices and six inlets to the area. There are 37 km of internal channels. Internal drainage congestion and external siltation have made some land unsuitable for crop production. The polder has a total population of 38,240 people residing in 9,490 households. The recently completed WLE project, *Community water management for improved food security, nutrition and livelihoods in the polders of the coastal zone of Bangladesh*, was led by the International Rice Research Institute (IRRI), in collaboration with Bangladesh Rural Advancement Committee (BRAC), the largest NGO in the world, the Institute of Water Modelling (Bangladesh), the International Water Management Institute (IWMI), Bangladesh Rice Research Institute (BRRI), Shushilan (another local NGO) and the BlueGold Program, a large development program.

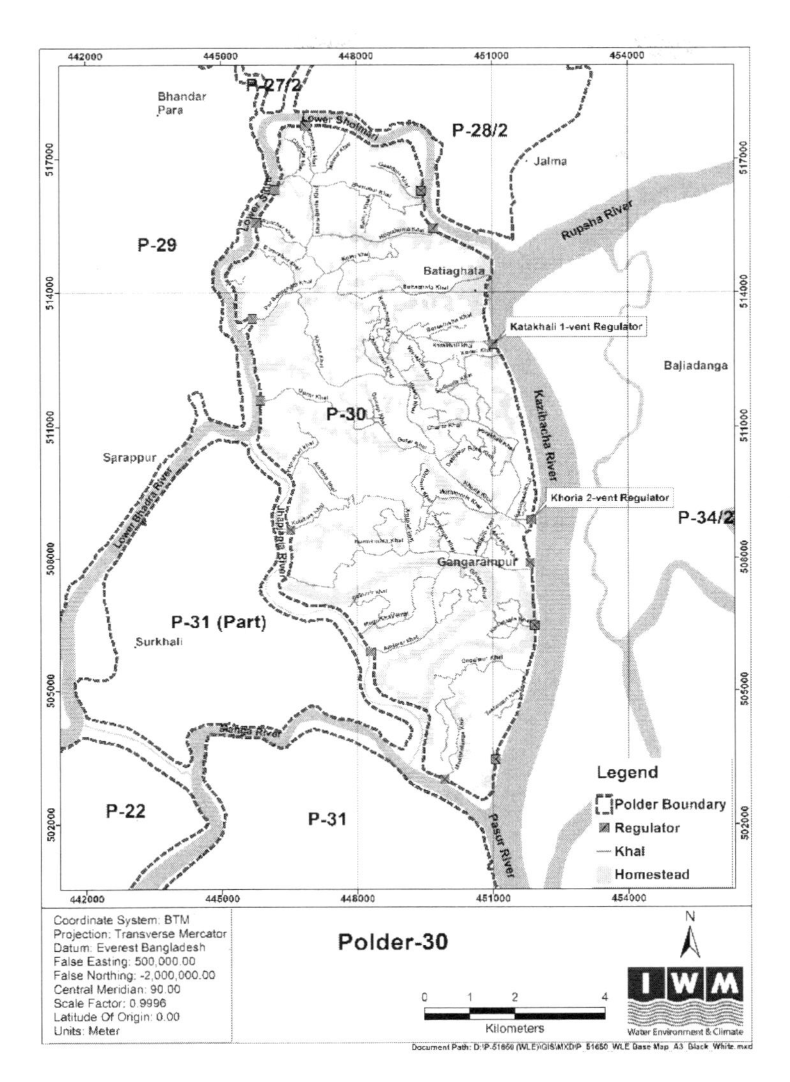

Figure 9.2 Map of Polder 30, the WLE study site

Source: CGIAR Water and Land Ecosystems Program.

Results of collaborative participatory research

In both deltas, the CPWF, WLE and other projects carried out a great deal of technical research and modeling, for example on breeding more appropriate rice varieties, cropping system (including fish) design, addressing soil quality issues, hydrology and socio-economics. This chapter does not report all of these important results in detail; rather, it focuses on key intervention strategies that can lead to more sustainable, equitable and profitable intensification of agriculture and aquaculture. The project reports listed in the reference section can be consulted for more details (see also Hoanh *et al.*, eds, 2006, 2010; Humphreys *et al.*, eds, 2015).

The Mekong delta

The WLE project was built on the foundation successfully established by a previous CPWF project, whose evaluation showed significant positive impacts on policies, capacities, and production and profitability of farms (McDonald, 2011). This project is still underway, but preliminary results combined with past results are very positive. A consortium of four partners are implementing the project: WARECOD (a Vietnamese NGO based in Hanoi), MekongNet, the College of Aquaculture and Fisheries (Can Tho University), and the Scientific and Educational Program Department of Viet Nam Television (VTV2). WARECOD is the lead organization and is conducting the *Thaibaan* component. Thaibaan is a participatory research method that is led by the farmers, not the scientists – a form of "citizen science" (Vaddhanaphuti, 2005; Mynt, 2016). It empowers local people to take responsibility for understanding and revealing knowledge about their connection with natural resources. From conception to dissemination, villagers themselves are the principal researchers. WARECOD has been supporting local communities to do research in the delta since 2012. In addition to the farmers, local officials are also involved to ensure their buy-in. An innovation of this project has been the combination of *photovoice*[2] and *Thaibaan* methods, in which the village people – especially women – are trained to use photographic techniques and storytelling.

This research has helped community members understand their own role in managing water resources, notably on how the use of pesticides and disposal of rubbish can affect their long-term livelihoods. Special efforts have been made to ensure women and ethnic minorities participate in the research so that its findings are known throughout the population. Early results show that farmers have adapted their livelihood strategies to include extensive shrimp farming, which uses the brackish water found locally and has minimum impact on the environment. Farmers grow rice when fresh water flows through the local channels and change to shrimp farming when the seawater arrives (WLE, 2017). Although not directly addressed in the WLE project, other research has suggested that the local institutional arrangements for water management need to be strengthened and improved to enable farmers to play a fully active role and to reduce the obstacles to women's active participation (Tuan *et al.*, 2015).

The results support a recent broader assessment of options for adapting to global change in Ben Tre, another coastal province of the Mekong Delta (Renaud *et al.*, 2014). That study examined historical changes in agro-ecosystems in a district that is seriously affected by salinity intrusion. The objective was to understand what enables effective adaptation and increased resilience of rice–shrimp socio-ecosystems, and the broader relevance of the findings. It concluded that the option currently favored by policymakers and even some investment banks, i.e., adding or upgrading engineered infrastructure, may provide short-term benefits such as increased rice production, but will reduce the capacity to adapt to change in the longer term. The alternative is to offer farmers more freedom of choice in how they adapt, and support their efforts to find solutions, for example by diversifying their agroecosystem. Collectively, the results suggest there is now sufficient research-based evidence to support scaling up the creation of more flexible institutions and policies for land use planning and agricultural production in the Mekong delta.

The Ganges delta

Research carried out by CPWF from 2004 to 2014 demonstrated that increasing the value of production and improving food security and livelihoods in the coastal zones of the Ganges delta is possible with existing technologies, germplasm and available water resources (Humphreys *et al.*, eds, 2015). Improved water governance and management and intensified and diversified agricultural and aquaculture systems are the keys to success. The path to shifting to high-yielding rice varieties, sustainable intensification, and crop diversification in the polders was found to be improving water management – especially drainage during and immediately after the rainy season to enable growing high yielding but shorter duration varieties of rice, which in turn enables timely establishment of *rabi* (dry season) crops. The CPWF research demonstrated the potential benefits of creating small community water management units within the polder, based on the hydrology of the landscape (grouping land that is at the same elevation) and the common aspirations of farmers (CPWF, 2014; WLE, 2014).

The recently completed WLE project was built on these results by implementing a pilot project in three villages in Katakhali sub-polder of Polder 30 (Figure 9.2). It was implemented in collaboration with the Blue Gold Development Program, a Dutch and Bangladesh government effort to reduce poverty through strengthened value chains and water management organizations, and carrying out rehabilitation and upgrading of water infrastructure (CEGIS, 2015; Buisson, Saikia and Maitra, 2017). The project implemented high yielding, diversified and climate resilient production systems, the productivity of which was 2–3 times higher than traditional farmers' practices. The WLE project highlighted the importance of organizing the community based on hydrological units within the catchment area of a sluice gate, and synchronized cropping within the units for wide-scale adoption of improved production systems in the polders of the coastal zone of Bangladesh (IRRI, 2017).

The project analyzed the short-term socio-economic impacts of both the WLE interventions in its limited area, and the broader Blue Gold interventions in Polder 30 (Buisson *et al.*, 2017). The evaluation found there were issues regarding the costs and marketing of both high yield varieties of rice and other crops, such as sunflower. Nevertheless, the results demonstrated that a community-managed drainage system significantly reduces waterlogging risks due to excessive and/or untimely rainfall in farmers' fields. Improved water management within a hydrological unit in the polder enabled the use of high yielding early maturing rice varieties, replacing traditional varieties, followed by the cultivation of high yield/value dry season crops like sunflower and maize. The project also showed that the integration of community fish culture with rice in low lying lands during the wet season can also improve food production and household incomes (Dey *et al.*, 2013; IRRI, 2017), with benefits to most households in the community. Further, these improved farming systems offer new livelihood and business opportunities for youth and women, for instance through the establishment of rice mat nurseries, use of mechanical rice transplanters, and homestead sunflower oil production (IRRI, 2017). Even the landless people can benefit: in one village, more than 60 landless and landowners were trained to collectively manage fish production in the rice land. The landless community took care of feeding the fish and guarding the rice–fish area at night and participated in harvesting of the fish. They harvested 3.5 tons of fish, of which three tons were sold and 500kg were self-consumed, bringing much needed protein into an otherwise poorly diversified diet (IRRI, 2017).

A previous project supported by CPWF on *Water governance and community-based management in coastal Bangladesh* analyzed the evolution of water management policies and their outcomes in coastal Bangladesh (Dewan, Buisson and Mukherji, 2014; Dewan, Mukherji and Buisson, 2015). The current policy advocates de-politicized community-based Water Management Organizations (WMOs), with limited involvement of local government institutions. The researchers found that such organizations are neither effective nor sustainable as they tend to be initiated externally, lack transparency and accountability, are dominated by elites, and only nominally involve women. Consequently, they cannot address underlying inequities but rather often amplify them.

If participatory water management in Bangladesh is to be effective, sustainable and equitable, water policy must recognize the inherent political nature of such organizations. Donors and the government organizations working on coastal water management should focus on formalizing the role of local government and ensuring the WMOs have access to permanent maintenance funds. The researchers argue this could lead to more effective "democratic decentralization" rather the limited effectiveness observed at present. A revised water governance framework is required to clarify the conflicting roles and counter-productive actions of actors involved in water management. Specifically, a three-tier strategy has been recommended by CPWF and WLE (2014):

1. Improving the financial sustainability and incentives of WMOs at community level to prevent lack of maintenance and deterioration of water infrastructure;
2. Formally recognizing and strengthening the roles of both the local elected government (i.e., the Union Parishads in Bangladesh) and agricultural extension staff in water management; and
3. Creating a Trust Fund, initiated by the Government of Bangladesh and donors, to solve the problem of "deferred maintenance" on a long-term basis, rather than investing in new projects. Under such an arrangement, all polders would be eligible to receive funds for repair and maintenance from the interest accruing to the fund.

The Mekong water management institutions offer an interesting model that supports this view. There is much greater cooperation between local government entities and the water management organizations than is the case in Bangladesh. However, the Mekong institutional arrangements are also inherently inequitable.

More work is needed on three issues: strengthening the value chain to enable farmers' profits to be increased; improving the capacity of the farmers to undertake collective action to manage water within the polder more effectively; and addressing equity issues – ensuring the benefits are more widely shared. It is notable that the USAID Sustainable Intensification Innovation Laboratory (SIIL) is supporting a project that is extending the WLE approach to the remainder of the sub-polder and beyond to other polders.[3]

What is similar and what is different between the two deltas?

The two deltas are quite different in some important ways: their natural conditions, social and political organizations and cultural traditions are examples. The Mekong delta is a large flat plain with a dense canal and river network connecting all sub-regions, while much of the land in the Ganges delta coastal zone is enclosed in polders (lands protected by embankments). Superficially, the Mekong River basin is characterized by a higher level of transboundary cooperation, through the Mekong River Commission, whereas what little cooperation that exists on the Ganges River Basin is bilateral (Saikia and Sharma, 2015). However, the Mekong River Commission's ability to achieve real cooperation is limited, not least because the major upstream country, China, is not a member.

The similarities between the two river basins are more striking. Both are characterized by rapidly developing upstream dams and other engineered structures affecting flows into the deltas; large investments in engineered structures within the deltas for irrigation, drainage, and control of flooding and salinity; dense populations of smallholder farmers and landless people with high levels of poverty (though poverty is worse in Bangladesh); and economies based on rice, rice-aquaculture, or year-round aquaculture. Both face similar challenges: rising sea levels; salt water intrusion; vulnerability to storm surges and also droughts; changing flows as well as increasing pollution from upstream; deterioration and weak management of polders and dikes; active NGOs; and governments that are open to research results. It is notable that in both deltas, the engineered structures for irrigation and drainage, i.e. the polders in the Ganges delta and canal systems in the Mekong delta, were originally conceived and designed by European engineers. That is why the structures are so similar to each other.

In both deltas, WLE's research has confirmed the high potential for sustainable intensification and diversification of agro-ecosystems through a focus on capacity strengthening and encouraging active citizen participation, including the use of "citizen science." Quite often, this needs to be complemented by rehabilitation, upgrading or modernization of the engineered structures (Smajgl *et al.*, 2015). In both deltas, the action-oriented participatory pilot research approach followed in the two studies is demonstrably effective and ready for scaling up. However, there is a greater need for institutional reform and strengthening in Bangladesh than may be the case in Vietnam.

Conclusions and recommendations: sustainable agroecological intensification in the Mekong and Ganges river deltas

A major conclusion of the research in both deltas is that the delta communities need considerable external support to sustainably intensify their agroecological systems and improve their livelihoods. Several recommendations have emerged from the research that are common in both deltas and indeed have wider significance.

First, strengthening, and in some cases reforming the local water management institutional arrangements within each engineered structure will be critical for sustainably increasing productivity and equity in both deltas. In Bangladesh, this should involve organizing community-based drainage systems by catchment area for each sluice gate, within all 139 polders. Based on clearly delimited "sluice gate watersheds," tailored plans can be developed for an improved drainage system, sluice gate operation timing, and mobilization of farmers to dig field drainage channels. Organizing farmers by the catchment area of each sluice gate will enable more efficient water management; and when the watershed is bounded on all sides, it also helps improve the fish culture system and ensures efforts by community members are fairly shared.

Bangladesh should strongly consider the Mekong model in which local governments (including the local department of extension) play a critical role in supporting these water management organizations. Lessons learned from the Mekong delta also show that transferring functions of the regional water management alliance for the coordination of salinity control to the river basin organization imposed and controlled by the central government does not work well because the central government cannot respond fast enough to local requirements to support both agriculture and aquaculture. Moreover, the Mekong organizations need more capacity strengthening to ensure female as well as male farmers have a strong voice in implementing improvements. In both deltas, there is a tendency for wealthier and more powerful people to dominate the water management organizations, capturing most of the benefits. This is a very difficult and deep-rooted problem; the best bet in both cases is for the government to play an active role in ensuring water management organizations are transparent and accountable and as equitable as is possible.

Second, continued support for maintaining, upgrading and improving the engineered structures is very important for success. However, the research in both deltas demonstrates that a pragmatic approach is required. Physical interventions should be driven by the local communities themselves, and should be done only when the local water management organizations are strong enough, have agreed on what is needed, and are willing to make some commitment of resources to implement the improvements. The trust fund idea recommended for Bangladesh is worth considering in both deltas.

Third, it is important to work with female as well as male farmers, and to continue to promote sustainable diversification of the agroecological systems in ways that are appropriate to the specific conditions encountered by farmers. This refers to diversification of crops – for example, where appropriate, crops other than rice during the dry season – and to aquaculture, where multiple fish and shrimp options exist, including rice-plus-fish systems. The approach should be based on participatory experimentation between scientists and farmers combined with efforts to strengthen the value chain to ensure profitability.

Fourth, in both deltas, gender inequality and the disaffection of youth are major issues with complex roots. It is important to work with interested women and youth to identify attractive opportunities, including new crops, production technologies, agricultural mechanization, and skills development and business models, that would enable them to benefit and be attracted to remaining in the area. Examples from Bangladesh are mechanical rice transplanters, rice mat nurseries, homestead sunflower oil production, and production and marketing of fish. In both deltas, supporting women and youth will require additional participatory action research, training, investments and strong policy support.

Even though the research projects reported here are on a small pilot scale, and in the case of the Mekong is not yet complete, they reflect the results of nearly two decades of research. In both cases, they have demonstrated interventions and approaches that work, i.e. are effective and sustainable.

The evidence base provides a sufficient foundation to scale up the approaches in both deltas. This will require collaboration between investors and governments; and following a flexible participatory social learning-based implementation strategy will be critical for long-term success.

We conclude with two lessons from these experiences that we believe are broadly valid for sustainably intensifying complex agro-ecosystems in developing countries.

First, the Mekong and Ganges delta research programs demonstrate the value of long-term, multidisciplinary socio-ecological systems research to achieve truly sustainable intensification of agro-ecosystems. The solutions that have been produced would not have been possible with a normal short-term project. In both cases, a wide range of scientific studies was conducted, including crop breeding, soil science, hydrological modeling, extension and socio-economic studies, all within a landscape or socio-ecological paradigm.

Second, creating partnerships with communities from the beginning and strengthening their capacities for collectively identifying, testing and adapting solutions to high-priority problems, including the integration of science and local knowledge, is the most effective pathway to sustainable intensification of agro-ecosystems. In both deltas, there was a strong emphasis from the beginning to work closely with the communities, jointly learning lessons, identifying ways to increase the capacities of communities to make their own decisions and manage their affairs, promoting strategies to ensure women and youth, as well as men, and those with little or no land as well as landowners benefit, and identifying institutional innovations that empower the communities to be successful over the long term.

These lessons reflect three main development strategies articulated by the Vietnamese Prime Minister at a recent Conference on Sustainable and Climate-Resilient Development of the Mekong Delta held in Can Tho City of the Mekong Delta on September 27, 2017:[4]

1 Facilitating the region's sustainable development and prosperity by proactively adapting to and turning challenges into opportunities, thus ensuring stable lives for the local people and preserving traditional cultural values.
2 Shifting the way of thinking in agricultural development from mainly rice cultivation to an agricultural economy, from quantity to quality, and from chemical-based agricultural production to organic and hi-tech practice with more attention to processing and support industries.
3 Respecting natural rules, thus selecting development models adaptive to natural conditions with minimum interference into nature.

While a lot of lessons have been learned on how to sustainably intensify these complex delta agro-ecosystems, there remains a lot more to learn on how to empower thousands of communities to continue to identify and adapt effective technical and institutional solutions.

Acknowledgments

The authors would like to thank Alex Smajgl, of the Mekong Region Futures Institute, for helpful comments on the changing policies in the Mekong delta. In addition, Meredith Giordano, a Principle Researcher at the International Water Management Institute (IWMI) played an important role in facilitating the preparation of this chapter and providing comments on its content. This research, including preparation of this chapter, was carried out under the CGIAR Research Program on Water, Land and Ecosystems (WLE) and builds on previous research under the CGIAR Challenge Program on Water and Food (CPWF). We would like to thank all donors who supported this research through their contributions to the CGIAR Fund (www.cgiar.org/about-us/our-funders/). The authors remain responsible for the contents of the chapter.

Notes

1 *Aman* is the December–January season; *boro* is March–May; *Aus* is July–August. Aman is the most important for rice cultivation.
2 "Photovoice" involves training people in using cameras, photographic techniques, storytelling for local researchers, and is especially aimed at women researchers – see: http://wlemekong.wikispaces.com/MK26+news (accessed 7 October 2017).
3 See: www.k-state.edu/siil/whatwedo/currentprojects/bangladesh/index.html (accessed 7 October 2017).
4 See: http://en.nhandan.com.vn/business/item/5530602-pm-asks-mekong-delta-to-develop-smart-sustainable-agriculture.html (accessed 7 October 2017).

References

Berg, H., Söderholm, A.E., Söderström, A-S. and Tam, N.T. (2017) Recognizing wetland ecosystem services for sustainable rice farming in the Mekong Delta, Vietnam. *Sustainability Science*, 12 (1), pp. 137–154.

Bernard, B. and Lux, A. (2017) How to feed the world sustainably: An overview of the discourse on agroecology and sustainable intensification. *Regional Environmental Change*, 17 (5), pp. 1279–1290. Available at: https://link.springer.com/article/10.1007/s10113-016-1027-y.

Buisson, M-C., Saikia, P. and Maitra, S. (2017) *Socioeconomic Analysis and Impact Assessment: Final Report*. Community water management for improved food security, nutrition and livelihoods in the polders of the coastal zone of Bangladesh – WLE-G9 Project. Unpublished report. Colombo: IWMI.

Center for Environmental and Geographic Information Services (CEGIS) (2015) *Final Report on Environmental Impact Assessment (EIA) on Rehabilitation of Polder 30*. Government of the People's Republic of Bangladesh, Ministry of Water Resources Bangladesh Water Development Board: Blue Gold Program. Dhaka: CEGIS. Available at: http://bluegoldbd.org/wordpress/wp-content/uploads/2016/06/EIA-Report-of-Polder-30-CEGIS.pdf (accessed 7 October 2017).

Challenge Program on Water and Food (CPWF) (2014) *Summary of CPWF Research in the Ganges River Basin*. October 2014. Colombo: CPWF. Available at: https://cgspace.cgiar.org/bitstream/handle/10568/49073/CPWF%20Ganges%20Basin%20Summary%20WEB.pdf?sequence=1 (accessed 7 October 2017).

Dearing, J.A., Wanga, R., Zhang, K., Dyke, J.G., Haberl, H., Hossain, M.S., Langdon, P.G., Lenton, T.M., Raworth, K., Brown, S., Carstensen, J., Cole, M.J., Cornell, S.E., Dawson, T.P., Doncaster, C.P., Eigenbrod, F., Flörke, F., Jeffers, E., Mackay, A.W., Nykvist, B. and Poppy, G.M. (2014) Safe and just operating spaces for regional social-ecological systems. *Global Environmental Change*, 28, pp. 227–238.

de Haan, N.C. and Sugden, F. (2014) *Social inclusion*. In:van der Bliek, J., McCornick, P.G. and Clarke, J. (eds) *On Target for People and Planet: Setting and achieving water-related sustainable development goals*. Colombo, Sri Lanka: IWMI books. Available at: www.iwmi.cgiar.org/Publications/Books/PDF/setting_and_achieving_water-related_sustainable_development_goals-chapter-5-social_inclusion.pdf (accessed 7 October 2017).

Dewan, C., Buisson, M.-C. and Mukherji, A. (2014) The imposition of participation? The case of participatory water management in coastal Bangladesh. *Water Alternatives*, 7 (2), pp. 342–366.

Dewan, C., Mukherji, A. and Buisson, M-C. (2015) Evolution of water management in coastal Bangladesh: From temporary earthen embankments to depoliticized community-managed polders. *Water International*, 40 (3), pp. 401–416.

Dey, M.M., Spielman, D.J., Haque, A.B.M.M., Rahman, M.S. and Valmonte-Santos, R. (2013) Change and diversity in smallholder rice–fish systems: Recent evidence and policy lessons from Bangladesh. *Food Policy*, 43, pp. 108–117.

Faruque, G., Sarwer, R.H., Karim, M., Phillips, M., Collis, W.J., Belton, B. and Kassam, L. (2016) The evolution of aquatic agricultural systems in southwest Bangladesh in response to salinity and other drivers of change. *International Journal of Agricultural Sustainability*, 15 (2), pp. 185–207.

Gray, E., Henninger, N., Reij, C., Winterbottom, R. and Agostini, P. (2016) *Integrated Landscape Approaches for Africa's Drylands*. World Bank Group. Available at https://elibrary.worldbank.org/doi/abs/10.1596/978-1-4648-0826-5 (accessed 7 October 2017).

Hoanh, C.T., Tuong, T.P., Gallop, K.M., Gowing, J.W., Kam, S.P., Khiem, N.T., and Phong, N.D. (2003) Livelihood impacts of water policy changes: Evidence from a coastal area of the Mekong river delta. *Water Policy*, 5 (5–6), pp. 475–488.

Hoanh, C.T., Tuong, T.P., Gowing, J.W. and Hardy, B. (eds) (2006) *Environment and Livelihoods in Tropical Coastal Zones: Managing agriculture–fishery–aquaculture conflicts*. Wallingford, Oxon, UK: CABI International with IRRI and IWMI.

Hoanh, C.T., Szuster, B.W., Suan-Pheng, K., Ismail, A.M. and Noble, A.D. (eds) (2010) *Tropical Deltas and Coastal Zones: Food production, communities and environment at the land–water interface*. Wallingford, Oxon, UK: CABI International with IRRI, IWMI, World Fish, FAO, and CPWF.

Humphreys, E., Tuong, T.P., Buisson, M-C., Pukinskis, I. and Phillips, M. (eds) (2015) *Proceedings of the CPWF, GBDC, WLE Conference on Revitalizing the Ganges Coastal Zone: Turning Science into Policy and Practices*, Dhaka, Bangladesh, 21–23 October 2015. Colombo, Sri Lanka: CGIAR Challenge Program on Water and Food (CPWF). Available at: https://cgspace.cgiar.org/handle/10568/66389 (accessed 7 October 2017).

Huq, N., Hugé, J., Boon, E., and Gain, A.K. (2015) Climate change impacts in agricultural communities in rural areas of Coastal Bangladesh: A tale of many stories. *Sustainability*, 20 (7), pp. 8437–8460.

International Panel of Experts on Sustainable Food Systems (IPES-Food). (2016) *From Uniformity to Diversity*. Available at: www.ipes-food.org/images/Reports/UniformityToDiversity_FullReport.pdf (accessed 7 October 2017).

International Rice Research Institute (IRRI) (2017) Community water management for improved food security, nutrition, and livelihoods in the polders of the coastal zone of Bangladesh. WLE Technical Report, March 2017. Unpublished paper submitted by IRRI to IWMI.

Khan, Z.H., Kamal, F.A., Khan, N.A.A., Khan, S.H. and Khan, M.S.A. (2015) Present surface water resources of the Ganges coastal zone in Bangladesh. In: Humphreys, E., Tuong, T.P., Buisson, M-C., Pukinskis, I. and Phillips, M. (eds) *Revitalizing the Ganges Coastal Zone: Turning science into policy and practices*. Conference Proceedings. Colombo, Sri Lanka: CGIAR Challenge Program on Water and Food (CPWF), pp. 14–26. Available at: https://cgspace.cgiar.org/handle/10568/66389 (accessed 7 October 2017).

Leach, M., Raworth, K. and Rockström, J. (2013) Between social and planetary boundaries: Navigating pathways in the safe and just space for humanity. In: ISSC/UNESCO (eds) *World Social Science Report 2013: Changing global environments*. OECD and UNESCO, Paris, France: OECD and UNESCO, Pp 84–89. Available at: www.oecd-ilibrary.org/social-issues-migration-health/world-social-science-report-2013/between-social-and-planetary-boundaries-navigating-pathways-in-the-safe-and-just-space-for-humanity_9789264203419-10-en (accessed 8 October 2017).

McDonald, B. (2011) Managing water and land at the interface between fresh and saline environments – an impact evaluation. Colombo, Sri Lanka: CGIAR Challenge Program for Water and Food (CPWF). (CPWF Impact Assessment Series 07). Available at: https://cgspace.cgiar.org/handle/10568/5570 (accessed 7 October 2017).

Mynt, T. (2016) Citizen science in a democracy: The case of Thai Baan research. Tocqueville Lecture Series at the Vincent and Elinor Ostrom Workshop in Political Theory and Policy Analysis, 9 September 2016. Available at: https://ostromworkshop.indiana.edu/pdf/seriespapers/2016F_Tocq/Myint%20paper.pdf (accessed 7 October 2017).

Narayan, A., Saavedra-Chanduvi, J. and Tiwari, S. (2013) Shared prosperity: Links to growth, inequality and inequality of opportunity. Washington DC: World Bank Policy Research Working Paper. Available at: https://doi.org/10.1596/1813-9450-6649 (accessed 7 October 2017).

Pretty, J. and Bharucha, Z.P. (2014) Sustainable intensification in agricultural systems. *Annals of Botany*, 114 (8), pp. 1571–1596.

Renaud, F.G., Syvitski, J.P.M., Sebesvari, Z., Werners, S.E., Kremer, H., Kuenzer, C., Ramesh, R., Jeuken, A., and Friedrich, J. (2013) Tipping from the Holocene to the Anthropocene: How threatened are major world deltas? *Current Opinion in Environmental Sustainability*, 5 (6), pp. 644–654.

Renaud, F.G., Huong Le, T.T., Lindener, C., Guong, V.T., and Sebesvari, Z. (2014) Resilience and shifts in agro-ecosystems facing increasing sea-level rise and salinity intrusion in Ben Tre Province, Mekong Delta. *Climatic Change*, 133 (1): 69–84.

Rockström, J., Williams, J., Daily, G., Noble, A., Matthews, N., Gordon, L., Wetterstrand, H., DeClerck, F., Shah, M., Steduto, P., de Fraiture, C., Hatibu, N., Unver, O., Bird, J., Sibanda, L. and Smith, J. (2017) Sustainable intensification of agriculture for human prosperity and global sustainability. *Ambio* 46, pp. 4–17.

Rural Livelihoods Evaluation Partnership (2004) Poverty Elimination through Rice Research Assistance Project (PETTRA): Independent project completion review. July/August 2004. Draft Final Report. Dhaka, Bangladesh: Rural Livelihoods Evaluation Partnership. Available at: www.lcgbangladesh.org/rlep/revdocs/E4_%20PETRRA.pdf (accessed 7 October 2017).

Saikia, P. and Sharma, B. (2015) Indo-Bangladesh Ganges water interactions: From water sharing to collective water management. In: Humphreys, E., Tuong, T.P., Buisson, M-C., Pukinskis, I. and Phillips, M. 2015. *Revitalizing the Ganges Coastal Zone: Turning Science into Policy and Practices*. Conference Proceedings. Colombo, Sri Lanka: CGIAR Challenge Program on Water and Food (CPWF), pp. 98–118. Available at: https://cgspace.cgiar.org/handle/10568/66389 (accessed 7 October 2017).

Smajgl, A., Toan, T.Q., Nhan, D.K., Ward, J., Trung, N.H., Tri, L.Q., Tri, V.P.D. and Vu, P.T. (2015) Responding to rising sea levels in the Mekong Delta. *Nature Climate Change*, 5, pp. 167–174.

Sugden, F., Maskey, N., Clement, F., Ramesh, V., Philip, A. and Rai, A. (2014) Agrarian stress and climate change in the Eastern Gangetic Plains: Gendered vulnerability in a stratified social formation. *Global Environmental Change*, 29: 258–269.

Syvitski, J.P.M., Kettner, A.J., Overeem, I., Hutton, E.W.H., Hannon, M.T., Brakenridge, G.R., Day, J., Vörösmarty, C. Saito, C.Y., Giosan, L. and Nicholls, R.J. (2009) Sinking deltas due to human activities. *Nature Geoscience*, 2, pp. 681–686.

Tuan, L.A., Minh, H.V.T., Dinh, Tuan, D.A.T. and Thao, N.T.P. (2015) Baseline study for community based water management project (CWMPs). OXFAM – DRAGON – WARECOD, Inclusion Project/Mekong Water Governance Program, Vietnam. Unpublished Technical Report. Available at: www.researchgate.net/publication/281462588_Baseline_Study_for_Community_Based_Water_Management_Project_CWMPs (accessed 8 October 2017).

Tuong T.P., Hoanh C.T. and PN10 Project Team. (2009) Managing water and land resources for sustainable livelihoods at the interface between fresh and saline water environments in Vietnam and Bangladesh. Project Report submitted to the Challenge Program on Water and Food (CPWF). Los Baños, Laguna, Philippines: International Rice Research Institute and Partner Organizations. https://cgspace.cgiar.org/handle/10568/3770 (accessed 8 October 2017).

Tuong, T.P., Humphreys, E., Khan, Z.H., Nelson, A., Mondal, M., Buisson, M., and George, P. (2014) Messages from the Ganges Basin Development Challenge: Unlocking the production potential of the coastal zone of Bangladesh through water management investment and reform. CPWF Working Paper. Available at: https://cgspace.cgiar.org/bitstream/handle/10568/41708/CPWF%20Ganges%20basin%20messages%20Sept%2014.pdf?sequence=5 (accessed 7 October 2017).

Vaddhanaphuti, C. (2005) Thai Baan research: An overview. *IUCN*. Available at: https://cmsdata.iucn.org/downloads/thai_baan_research_an_overview_1.pdf.

Van, M.V., Hien, H.V., Phuong, D.T., Quyen, N.T.K., Nga, D.T.V. and Tuan, L.A. (2016) Impact of irrigation works systems on livelihoods of fishing community in Ca Mau Peninsula, Viet Nam. *International Journal of Scientific and*

Research Publications, 6 (7), pp. 460–470. Available at: www.ijsrp.org/research-paper-0716/ijsrp-p5567.pdf (accessed 7 October 2017).

Vasseur, L., Horning, D., Thornbush, M., Cohen-Shacham, E., Andrade, A., Barrow, E., Edwards, S.R., Wit, P. and Jones, M. (2017) Complex problems and unchallenged solutions: Bringing ecosystem governance to the forefront of the UN sustainable development goals. *Ambio*, 46, pp. 731–742.

Water Land and Ecosystems CGIAR Research Program (WLE) (2014) Unlocking the potential of coastal Bangladesh: Improving water governance and community-based management. WLE Briefing Series 01, August 2014. Available at: https://wle.cgiar.org/unlocking-potential-coastal-bangladeshimproving-water-governance-and-community-based-management (accessed 7 October 2017).

WLE. (2017) Knowledge brings behavior change. Blog. Available at: https://wle.cgiar.org/knowledge-brings-behavior-change (accessed 7 October 2017).

Wichelns, D., Hoanh, C.T., Dung, L.C. and Phong, N.D. (2010) Estimating the value of improvements in environmental quality due to changes in sluice gate operations in Bac Lieu Province, Vietnam. A Report submitted to the Standing Panel on Impact Assessment of the CGIAR. Colombo: IWMI. Available at: http://ispc.cgiar.org/sites/default/files/docs/IWMI_EIA_2010%20.pdf (accessed 7 October 2017).

10 Agriculture development and sustainable intensification

Innovations to strengthen extension services and improve market value chains

Udaya Sekhar Nagothu and Allison Morrill Chatrchyan

Introduction

Agriculture, food and nutrition security will be a top priority for a majority of countries in the coming years. Climate change, however, will make it difficult for countries to produce enough food to feed their populations. This is even more relevant considering the fact that the world population will increase by more than a billion by 2050, while the net arable land will be reduced (FAO, 2009). Much of this population increase will happen in Africa, a continent already facing serious problems including food insecurity, malnutrition and hunger, and impacts of climate change leading to frequent extreme weather events. Countries in the climate vulnerable zones need to adopt more sustainable and innovative agriculture systems in the future. This demands not only technologies that are smallholder relevant, but also developing effective agricultural extension services for technology dissemination, marketing infrastructure and skills. Agricultural extension or advisory services has a crucial role to play in increasing productivity and livelihoods of farmers through sustainable agricultural intensification (SAI). Often the extension services and markets in developing countries are poorly managed and unable to provide timely and accurate technology and information to smallholders. If agriculture is not productive and profitable, there will be fewer people interested in taking up farming as a career. As a consequence, it will result in large scale migration of youth from rural areas to nearby cities and other countries for security and employment leading to an unstable social and economic environment at both ends.

Sustainable agriculture intensification as discussed in Chapter 1 is seen as an alternative to the current farming systems that are highly vulnerable to climate change. Often, constraints observed in developing countries for better adoption of SAI systems are not only technology related but also weak institutional measures, in particular inefficient extension services and marketing opportunities. Improving agricultural product value chains (VCs) therefore requires a complete transformation of production and post production services. Smallholder sustainable agriculture will succeed only when it is properly aligned with the markets, and accordingly provide adequate services to farmers to earn higher incomes. This needs to go hand in hand with technology innovation (e.g. new methods to increase input use efficiency and productivity of the major food crops). Increase in productivity also could lead to an increase in the market supply of the product and fall in product prices, i.e. price elasticity as some economists term it. In case of some perishable products such as tomato, chilli and other similar products, farmers choose not to harvest the crop and or dump the product to save costs of harvesting and transport. Governments need to provide necessary storage and insurance mechanisms in such situations to protect smallholders.

This chapter justifies, why, despite the dilemma of the market supply and demand influencing product prices, institutional innovations (e.g. improving access to inputs, extension services and market infrastructure) should be simultaneously developed with technology and policy options in support of smallholders, especially in regions that are highly vulnerable to climate change and food insecurity. The first part of the chapter focuses on: (a) the need for improved extension services, a brief history, the challenges and some examples of innovative AE models and public–private partnerships. This is followed by a section showing the importance of smallholder integration into the markets and VC improvement to achieve sustainable intensification of agriculture that climate change allegedly necessitates. Finally, the chapter provides some conclusions and recommendations for increasing the capacity of smallholder farmers through improvements in extension and advisory services, and better market infrastructure.

Agricultural extension: a brief history, challenges and successes

Agricultural extension (AE) has often been based on a top-down approach in developing countries since its establishment. Traditional AE models have taken for granted that knowledge developed within the labs and experimental fields would bring benefits to farmers. The AE systems in the process did not consider the local environmental contexts, farming systems, products, knowledge needs of other end users, their local experience and indigenous knowledge. In the 1960–70s, new agricultural information and technologies were generated at research stations by scientists without involving farmers. The knowledge was then transferred to farmers through extension agents, without adjusting the new practices to local environmental and social conditions. A typical example was the Training and Visit (T&V) Extension programme introduced in some countries such as India, where government extension agents received trainings and visited villages to disseminate knowledge to farmers (Feder, Slade and Sundaram, 1986). This approach generated mixed outcomes and led to some serious challenges in knowledge dissemination and use. During the 1980s, the agricultural knowledge and information systems (AKIS) approach was dominant and had a farmer-centred concept that recognized other types of approaches such as Farmer First (Chambers, Pacey and Thrupp, 1989), Farmer Field Schools (Waddington and White, 2014), and Farmer-to-Farmer Exchange of knowledge (F2FE) (Simpson, Heinrich and Malindi, 2012). These methods took into consideration farmer interests and their knowledge needs (information needed by farmers related to climate, new technologies etc.). Extension services varied widely across countries – in terms of their structure (public vs private sector, and university-led vs ministry-led), funding and level of efficiency and methods utilized.

In the 1990s, agricultural innovation systems (AIS) were introduced that included not only farmers but also a range of other actors that interacted in the generation, transfer, exchange and use of information (World Bank, 2006a, 2006b; Spielman *et al.*, 2008). From 2000 onwards, Information and Communication Technology (ICT) tools-based AE systems were initiated in a few countries in Asia and more recently in Africa (Waddington *et al.*, 2010; Swanson and Rajalahti, 2010; Sanga *et al.*, 2014). Throughout (Figure 10.1), the level of innovation was an important criterion in whether extension services were functioning as a one-way-only-approach in the 1970s and 1980s, versus a two-way process since the 1990s (Aerni *et al.*, 2015). However, the degree of success and impact of the extension services in developing countries, especially in sub-Saharan Africa was very limited. Farmer feedback, regular evaluation and impact assessment of extension services is important to improve the efficiency of knowledge generation and transfer which was often neglected.

In recent years, we see a gradual shift from public-sector-driven AE services with uneven spatial coverage, weak financial and institutional support and delivery systems towards private and more decentralized services. Today, AE services are provided by a range of agencies including government, private sector, research institutes, universities, non-governmental organizations (NGOs) and producer organizations, using different approaches. Some of them centred around participatory models (e.g., Farmer Participatory Research (FPR) and F2FE that support active learning and involvement of farmers, sharing knowledge and innovation capacity, whereas, others on market demand driven systems. However, the suitability of these systems needs to be determined based on the natural environment or the agro-ecological

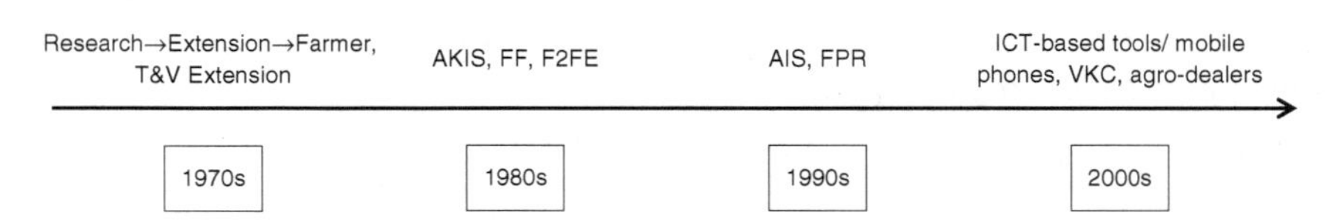

Figure 10.1 A time line illustrating the broad evolvement of AE in developing countries

Source: Authors' own compilation.

setting, target groups, specific target crops or products and end use market needs. Consequently, the impact and sustainability of the AE services is significantly influenced by their suitability to the different needs.

The transformation of agricultural extension and main challenges

The post Green Revolution period and the 1980s witnessed fewer investments in AE services and public research in a majority of the developing countries, especially in Asia, as agriculture is no longer viewed as a priority sector of the economy. Some of these countries reached self-sufficiency in food production, which was also was a driver in reducing the investments in the agriculture sector. In contrast, investments in AE were not a priority of the governments in Africa until recently. There was a total lack of innovation in the public AE services. Within the public AE system, inefficiency prevailed as extension agents became more embroiled in administrative responsibilities.

According to the World Bank (2006b) study, AE systems in many Asian countries have moved from the traditional linear model of linking research to extension to farmers, towards market-driven models. The new AE systems are more consistent with the emerging product-based agricultural system in the changing economies, for e.g., fruits and vegetables, special types of rice and potato grown for specific export market demands. Rapid economic development as observed in many countries in Asia including India, Vietnam, Thailand and China is affecting the consumption patterns and creating a demand for high-value products. Consequently, farmers also tend to shift their farming practices to meet new market demands, which in turn puts pressure on the research and AE to support specific product development. The entire product VCs from farm inputs (e.g., land, labour, access to water, seeds and fertilizer), access to storage, markets and transportation becomes important in the process. This implies that improvements at different stages of VCs are necessary as envisaged by SAI in order to improve efficiency, productivity and profitability for farmers.

All of these factors paved the way for reforms within the AE system and provided an entry point for the private sector (Umali and Schwartz, 1997). In addition, the changing economic and social context, market reforms in the 1990s, and beyond 2000, meant a need to shift from a general AE system to more specialized AE services. The shift suited the product diversification that happened in the farming sector to cater to the urban food markets and exports. Product diversification requires specialized AE services including marketing opportunities for farmers that the commercial enterprises could provide (Klerkx and Leeuwis, 2008). This can be observed in fruit and vegetable growing regions in Asian countries, where producers and buyers join together in commercial partnerships, in which the latter supply inputs and extension services and the farmers produce according to the demand of the buyers. The private sector gradually increased its investments in the AE domain that the governments could not afford. Privatization of AE services is a win–win situation for both public and private agencies. It resulted not only in decentralization and reaching target farmers, but also in commercialization at the same time (Rivera *et al.*, 2001; Andersen and Feder, 2003).

Privatization has its benefits, but also comes with accompanying risks. For example, small farmers can lose access to government extension services once AE services are privatized. Privatization also means that farmers have to operate under conditions imposed by the private enterprises who often work with commercial motives. Smallholders face constraints in implementing transactions and establishing the necessary networks in the privatization process. Individual farmers are left with fewer options for negotiation, unless the agreement is through farmer organizations or cooperatives. Only farmer organizations that are well organized have better bargaining power with the private sector actors. In China, for example, privatization has marginalized township-level extension agencies in certain provinces (Hu *et al.*, 2010).

A study by Chen and Shi (2008) provided a different view of AE reforms in China. The study presents an inclusive public agricultural technology extension system introduced in Pengzhou, Sichuan Province and in Wuchuan, Inner Mongolia Autonomous Region. In another study by Friederichsen *et al.* (2013) both extension workers as well as managers in Vietnam were observed integrating the various reform discourses received from the government, donors and academics on participation, user-orientation and private sector involvement in AE innovation into the still-dominant transfer of the technology model. Their findings demonstrated that extension workers chose selectively on the diverse discourses to foster interaction with outsiders and farmers. A key challenge according to their study was the need for improving extension workers' ability to mediate between the conflicting principles of farmers' self-organization and government control. Private AE systems use new tools of communication that help improve the overall efficiency; whereas extension personnel in state organizations lack knowledge and training in the use of ICT tools.

Knowledge development and knowledge transfer

The two facets of extension, knowledge development and knowledge transfer, are important. Knowledge is comprised of the attitudes, cumulative experiences and developed skills that enable a person (in this case a farmer) to consistently, systematically and effectively perform a function (Seidman and Michael, 2005). It is an integration of explicit and tacit knowledge. Knowledge development and compilation is the process of figuring out the action implications of the tactics for the particular situation or problem encountered. This embodies a trade-off between applicability and efficiency in that it has wide applicability across many contexts, but requires a complicated and lengthy application process to translate the declarative knowledge into a set of actions (Anderson, 1987). Previous studies have shown that declarative knowledge can apply to a variety of different situations, but is costly in terms of the time required to contextualize that knowledge to a specific problem (Nokes and Ohlsson, 2005). This is often a problem in developing countries where investments in the agriculture sector are limited. Contextualization of knowledge development and transfer is often not a strength or a priority of the Government AE system in these countries. This can be attributed to the lack of incentives for such a process, and also the lack of accountability. The variability and specific knowledge needs of farmers is not considered while developing such a knowledge base. This in fact is the reason for inefficiency and poor performance of government AE systems.

Knowledge transfer from a cognitive science point of view explains how people use and apply their prior experiential knowledge as well as acquire new knowledge to solve their problems. This largely depends on the trust created between the different entities within the knowledge transfer or extension process and the results and outcomes it generates for the end users. Previous work has identified multiple mechanisms of knowledge transfer including (but not limited to) analogy, knowledge compilation and constraint violation (Nokes, 2009). Nokes's research centres around the hypothesis that the particular profile of knowledge transfer mechanisms or processes for a given situation depends on both: (a) the type of knowledge to be transferred and how it is represented; and (b) the processing demands of the transfer. The latter largely depends on the needs and capacity of end users, the environmental and social context in which they operate, the market status and the transaction costs. For example, the extent of adoption will often depend on whether or not farmers are educated and used to new tools and knowledge, or whether they need the knowledge transferred in a simpler manner. It is also important to consider whether the knowledge transfer takes into proper consideration the specific needs of different genders. In some cases, 'seeing is believing' and farmers learn or adopt faster by seeing other farmers practising a particular technology, rather than being told. This is evident from several successful farmer-to-farmer exchange models, farmer participatory research projects, and lead farmer/or champion farmer models in many countries that have provided good results (Birner *et al.*, 2009).

According to Nokes (2009), if people have a choice or access to multiple transfer mechanisms, then it is likely that they will apply or engage those transfer mechanisms adaptively, in response to the transfer conditions, i.e., incorporating the relevant knowledge or experience they possess, how it is synthesized and disseminated, and the relation between the training and knowledge transfer problems. There will always be a trade-off between the knowledge transfer in terms of their scope of application and the amount of cognitive processing required to transfer the knowledge to the targeted end users. Any innovation in knowledge transfer would then depend on the connectivity and networking between different stakeholder groups that are developing, transferring and using the knowledge, in this case, the scientists, extension specialists, farmers, women groups and other relevant public or private agencies. Hounkonnou *et al.* (2012) state that innovation in AE is not a linear, but rather a multi-linear process and depends on all actors involved in the generation and use of the knowledge as well as the constraining and enabling institutional context. In many situations, interactive stakeholder platforms can provide a better means for new learning and innovation. The value of platforms for innovation is widely recognized, but more understanding is needed of the choices made in facilitation, to enable the platforms to perform effectively within varying value chain contexts (van Paassen *et al.*, 2013). Such platforms comprised of local stakeholders can help in building trust between the various actors and thereby effective knowledge transfer and use by farmers.

Gender, youth and agricultural extension

AE has been traditionally biased towards men and there has not been an adequate focus on developing AE services targeting women. This is a major challenge in societies where the gender divide is large, and women are not allowed to be present with men in public. The lack of AE programmes specifically catering to women's' needs and women AE agents is often a major setback to ensure the provision of accurate and timely information. This is a serious concern where out-migration of men from countries due to economic reasons leaves the burden of agriculture on women – while extension services are not organized to serve the needs of women. This is the situation in several countries, including Nepal, India, China, the Philippines and parts of Africa, where men and youth are migrating to cities, as agriculture is becoming risky and no longer profitable – leaving women behind to work.

A study by Hosenally (2011) from Mauritius shows that finding youth interested in agriculture is rare, and their first choice is business management and engineering. He attributes this to the fact that the image of agriculture does not appear 'attractive' to youth. There is a similar situation in a majority of the developing countries in Asia and Africa. His study also points to two main reasons for the phenomenon: the lack of job opportunities, in particular with the AE sector, and the fact that extension involves working in the rural areas with farmers, and requires several skills that discourage youth from pursuing AE as a career. Hosenally (2011) provides some recommendations that could make agriculture and sectors such as AE more attractive for youth, including on farm training and interaction with farmers, and showcases the potential of agri-business and new technologies. The use of ICT tools and social media in AE can be attractive to youth who use them in their daily life, and view them as modern communication channels.

Innovative AE models

A number of innovative AE systems have been developed and introduced in some Asian countries that have helped to improve farmer connectivity, provide faster and timely information and reduce transaction costs. A majority of these are private initiatives as discussed below:

Village knowledge centres (VKCs) in India: The ICT digital platform-based VKCs link farmers through smart phones and social media as a conduit for faster and effective information and knowledge to rural communities. They attempt to bridge the knowledge, gender and digital divides and empower the rural community by fostering inclusive development and participatory communication (Raj and Nagothu, 2016). VKCs located in rural areas also facilitate dissemination of knowledge through other multiple communication tools, such as notice boards, public address systems, WhatsApp, Fixed Wireless Group Audio and Video Conferencing, and Webinars to the farming community. Experience has shown that the participation of women in these VKCs is much higher as compared to conventional AE models. This can be attributed to the convenient location of the VKC in the village and the fact that the female knowledge workers who operate them are often from the same village. Timely transfer of information, in addition to providing accurate information, has a significant difference on the adoption of technologies by farmers. VKCs can also act as meeting places where knowledge exchange between farmers can be more useful. Each VKC can connect approximately 2,000–3,000 farmers directly recruited as members and many more indirectly through mobile phone services. New models such as the VKCs should complement the existing state extension services and not compete with them. The sustainability of VKCs primarily depends on the farmers' interest in taking the ownership and operating them, rather than depending on external support. VKCs can be useful models in situations where there is some awareness and use of ICT tools and mobile phones, and markets for specific products are in demand and when villages or smallholder communities are isolated and not well connected with roads.

Innovative multi-actor platforms: A multi-actor platform (MAP) is defined as a process of interactive learning, empowerment and collaborative governance that enables stakeholders with interconnected problems and ambitions, but with often different interests, to be collectively innovative and resilient when faced with the emerging risks, crises and opportunities of a complex and changing environment (Woodhill and van Vugt, 2011). MAPs represent cross-sector collaboration and are more widely used in recent years in the agriculture sector (Brouwer *et al.*, 2013). They can range from formal roundtables aiming for formal approvals of tools and technologies, to informal coordination mechanisms to manage local natural resources including soil and water. MAPs comprised of local representatives create trust-based relations that enable the empowered and active participation of all stakeholders if implemented in the correct spirit. They can be effective agents for communication and dissemination of information in the agricultural sector, especially when major policy reforms are introduced and technologies need to be scaled up. MAPs function well and are useful in environments where relevant sectors need to integrate to promote SAI and boost agricultural production.

Farmer to farmer exchange (F2FE) of knowledge: F2FE is one of the most effective ways of developing practical, effective, profitable technologies that solve identified agricultural production constraints, and address the specific needs of smallholder farmers. Previous participatory agricultural research has demonstrated the efficacy of farmer-led experimentation in building farmer knowledge, and fostering innovation and sharing, with visible impacts on food security and sustainable production methods (Bezner Kerr *et al.*, 2012). However, research so far is limited on how to effectively link farmer-led research to AE services and strengthen formal and informal networks of knowledge dissemination and innovation through the F2FE approach, or how to scale up these small-scale initiatives. Projects such as "InnovAfrica" build on Farmer Participatory Research experience in various countries (www.innovafrica.eu). F2FE extension models work in situations where smallholder communities are not educated and government extension services are not efficient.

Integrated farm planning (IFP): The IFP approach aims at changing the mindsets of small subsistence farmers, and motivating them to plan their activities and invest in their future, so that they can provide a foundation for rural development (Kessler, Van Duivenbooden and Nsabimana, 2014). During the early planning process, it is crucial that innovative farmers are trained, followed by other farmers who undergo orientation in different stages. As such, IFP helps farmers and fosters the development of a vision towards a desired future, instead of season-by-season subsistence. Farmers can work together and form a collective that helps increase the possibility of successfully entering the value chain, for example with a particular product (cash crop) or other collaborations (Kessler *et al.*, 2015). Subsequently, by combining several IFPs, farmer groups can formulate their own collective 'business plan' and obtain technical and financial support. If successful, IFPs can become the basis for village plans, commune or district-level plans, and provide inputs to larger development plans. Sustainable agricultural activities are at the heart of the IFP – enriched with all kinds of other aspects or activities, such as health (e.g. diet diversity) and micro-credits.

Inclusive public extension system: This was introduced in Pengzhou, Sichuan Province and in Wuchuan, Inner Mongolia Autonomous Region during the mid 2000s. The main goal of this new system was to meet the diverse technology and marketing information needs of farmers, and thereby to provide targeted services. The demand-driven AE model was embraced by the government and scaled up to another 25 counties due to the initial success in the two cases. A review by Hu *et al.* (2010) showed that the reform initiatives had significant impacts on the availability of AE services to farmers, and also acceptance of the services from extension agents. It influenced the farmers' adoption rates significantly with more farmers adopting as compared to regions where reforms were not initiated. The success was attributed to the inclusiveness of all farmers as end users for public extension services, a systematic approach to identifying farmer technology and information needs, accountability and incentives for extension personnel.

Integrated markets

In order to make farming profitable for smallholders, it is necessary to improve access to markets together with technologies to improve production. Providing adequate market links and prices to smallholders can help to increase their incomes and reduce poverty and hunger in rural areas. A study by Wiggins and Keats (2013) in Africa shows that a majority of smallholders are not linked to markets due to various reasons including remoteness, low farm-gate prices and lack of information. Often a catalyst is necessary that could be a private firm or a NGO to establish linkages between farmers and markets (Wiggins and Keats, 2013). However, it is a constraint to connect individual farmers, hence, organizing farmers into collectives can make it easier and provide better bargaining power. This could be either directly through farmer associations, or indirectly through local agro-dealers. We see new partnerships emerging along these lines especially for high quality export products such as fish, avocado, mango and other fruit crops. Farmer cooperatives are not new to the agriculture sector in many countries. They were popular during the 1970s, but many of them have failed due to social and political reasons. It is important to learn from the failures of the past while promoting farmer collectives in the current context where the social and economic scenarios are different from the 1970s. There is more awareness now and better connectivity and information exchange.

Public–private partnerships

New partnerships involving research institutes, public agencies, private sector entities and farmers, with an emphasis on building private extension services and market linkages have been emerging in recent years (Singh, 2013). The involvement of NGOs, and commercial enterprises, has resulted in the emergence of new agriculture development models in many countries such as India, China, Thailand, Vietnam, Kenya, Rwanda among others. Some examples of the emerging models are listed below:

Commercial enterprise (CE) and farmer contract agreements – in this case the CE provides the technology, capital and inputs and the farmer carries out the production on his or her farm under the guidance of CE, and the produce is bought by the enterprise;

Enterprise, cooperative/association and farmer – in this model, the cooperatives negotiate with the enterprise and provide the necessary resources, information and market services to the farmers. Experiences show that this model provides better results and adoption of new technologies, and the individual transaction costs for farmers are comparatively lower; and

Commercial enterprise (CE), village committees and farmers – the village committee in this model acts as an intermediary and farmers do not have to deal directly with the CE. This provides a certain degree of security to farmers and lowers risk. Involvement of village committees can provide other benefits, such as office space, additional funds and support in scaling out technology adoption.

Improving value chains with a focus on markets in the agriculture sector

During the Green Revolution, the main emphasis was on increasing food production and not much on post-production services. Scientists working within their own disciplinary silos did not visualize a comprehensive VC approach and its advantages during this phase. There were weak linkages between disciplines and sectors related to agriculture. Only within the past few years do we see the emphasis shifting towards strengthening linkages between smallholder farmers and markets. Improving market services for smallholders located in rural areas is crucial to ensure that they are able to sell their products and make profits. Smallholders are mostly involved in informal or traditional value chains, selling their produce to intermediaries, who take a larger share of the profit. Smallholder farmers need to be actively engaged with the other VC actors so that they get the added value for their products (AFI, 2017). The success of a particular product in VC development will depend on smallholder stewardship of the programme and their involvement early on in the VC development process (CGI, 2016).

According to Norton (2014), VCs work best when the cooperation of actors leads to the production of better-quality outputs and generates profits for all participants along the chain. This will also help to reduce the risks and make them more resilient. VCs that only involve exchange of information or technology may not be of interest to smallholders. Programmes must be followed up by regularizing supply of inputs, technology training and other relevant support services such as storage facilities to improve the VCs. Lack of storage facilities in a majority of the countries in sub-Saharan Africa forces farmers to sell their produce, in this case maize to intermediaries at low profits. It helps if farmers can store their produce and get credit vouchers instead, and sell the produce when they can make profits. Multi-actor platforms represented by relevant actors from government, credit and insurance agencies, buyers, farmer associations can support such interventions that are attractive to smallholders. Functioning and accessible markets, particularly for agricultural commodities, are vital for agricultural growth to realize its potential as a powerful driver of rural poverty reduction (Kürschner *et al.*, 2016). Since farming is a risky business, planning and development of VCs should consider all possible risks including climate, economic and political, to ensure that adequate mitigation measures are in place. Often fluctuating market prices, fall in product prices due to excess production and market supply is seen as one of the most practical risks smallholders face. Hence care should be taken while making reforms, to ensure minimum support price to farmer products and insurance to cover risks.

Despite more than two decades of market liberalization, considerable efforts by government and non-governmental practitioners in developing countries to transform the smallholder agricultural sector from subsistence to commercial production remains fragmented and subsistence-oriented, lacking the ability to meet changing market demands. This can be attributed mainly to: i) asymmetry in market information, ii) lack of marketing services including proper storage, transport and information, iii) lack of farmer organization into collectives or cooperatives, iv) weak regulations that do not protect farmers, and v) increased power on the buyers' side. Addressing the missing link between smallholders and other key VC actors through vertical and horizontal coordination is vital for rural economy as it enhances livelihoods of smallholders while serving the business interests of private companies (Hellin and Meijer, 2006). A better organization of actors within the VC (producers, manufacturers, importers, distributors and retailers) is therefore needed. The most important aspects for improving VCs include: i) Profitability, the essential driver of growth and more so for smallholders; ii) Institutional and policy support for smallholders; iii) Quality of products and target markets; iv) Government support through minimum support prices and regulations; and v) Investment in the private sector. Figure 10.2 shows the different actors involved with agriculture product VCs and their respective functions.

In reality, the VCs are more complex than shown in Figure 10.2. Understanding VCs is about understanding the interaction of many actors in the process, both primary actors who undertake input supplies (seeds, fertilizers, etc.), production, storage, processing, wholesale (including export), retail and consumption, and the secondary actors who perform support service roles for the primary functions such as transportation, information exchange, brokerage and service processing. Within VCs, the potential and opportunities for improving markets also differs from with environment, end use, the type of agricultural product and market situation. Problems within the VC may be due to poor quality seeds or new varieties, soil fertility factors, quality produce or harvesting technology.

According to the International Fund for Agricultural Development (IFAD) (2012), good VC pathways that favour smallholders are essential to help them sell their products. This can create a virtuous circle by boosting productivity, increasing incomes and strengthening food security (IFAD, 2012). In rural areas, social networks can help farmers to actively

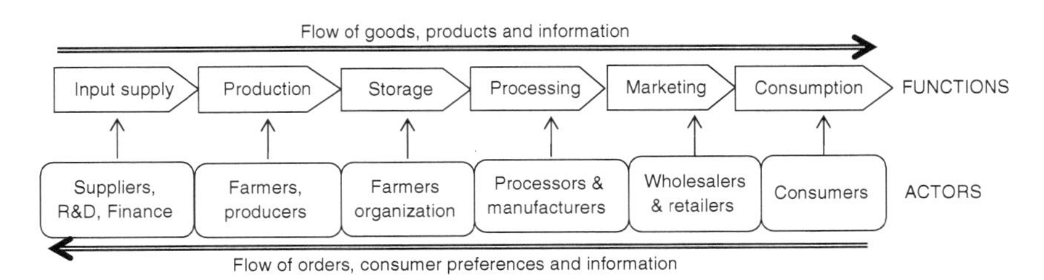

Figure 10.2 Value chain pathways for agricultural products

Source: Adapted from Hellin and Meijer, 2006.

engage with other VC actors and increase their bargaining power in the process. Successful examples of dairy farmer cooperatives in Rwanda and Kenya show that collectively farmer organizations have greater bargaining power and bring greater benefits to farmers in the VC (FAO, 2014). These cooperatives also influence governments to change policies for the benefit of smallholders, to make investments to improve infrastructure for storage, transportation and product processing and minimum support prices to products. One of the growing concepts that can help improve market VCs in many countries is the growing interest of farmer markets among producers and consumers.

A study by Asian Development Bank (ADB) (2012) in four cases, namely Bangladesh, People's Republic of China, Lao People's Democratic Republic and Nepal, has confirmed that agriculture is still a significant contributor to gross domestic product (GDP). However, the linkages to different services through VCs is varied in the four countries and has a large potential for improvement that can contribute to the overall economy. With commercialization of the agriculture sector, there is a need to connect farmers with better extension, storage, transport and processing in the farm gate stage. The study further recommends that external support to VCs is an important step for agricultural development, as this can lead to higher financial returns that will be realized through value-enhancing inputs. Overall, improvements in VCs can increase effectiveness and efficiency and thus provide benefits to all VC actors and at the same time contribute to food security and poverty reduction.

The Feed the Future programme in Kenya focuses on agriculture VC development and works with different actors to improve value chains of livestock products and grain legumes, potatoes and sorghum and orange fleshed sweet potatoes that will contribute to improved diet diversity, food security and rural incomes by improving production, productivity, supply and market access (Feed the Future, 2015). The initiative also realizes the need for actively engaging youth and women while improving market VCs, especially in developing countries, to promote rural enterprise development in recognition and motivate youth to be interested in the agriculture sector.

Farmer markets

Farmer markets that were traditionally dominant in most countries are now being revived and reintroduced in urban areas. By definition, a farmer market can be defined as 'a predominantly fresh food market that operates regularly within a community, at a focal public location that provides a suitable environment for farmers and food producers to sell farm-origin and associated value-added processed food products directly to customers' (AFMA, 2017). The drivers for the rapidly growing farmers' market movement in most countries vary and include several actors along the market VCs including local governments, farmer groups, agribusiness enterprises and consumer groups. Nevertheless, the role of local governments is quite crucial in the promotion of farmer markets, as observed in some of the provinces in Australia. Farmer markets if properly managed: i) can promote the local economy, farmer income, rural development, health and nutrition among local communities; ii) can attract tourists and provide places for experiencing local foods and rural culture; and iii) have the potential to promote local food groups, food trails and farm gate fresh food outlets and can vitally connect isolated farms with their peers and new customers (AFMA, 2017).

In India, local governments in some of the provinces organize *farmer markets*, provide market infrastructure where farmers can directly bring their products and sell to end consumers thus eliminating intermediaries. This has become a successful model initially in the state of Andhra Pradesh in India for vegetables, fish and fruit products. However, there are challenges that need to be addressed, especially the absence of proper storage and display facilities in the market place itself, problems in transporting the produce and low prices as compared to conventional markets (APAARI, 2008; Kumar, 2016). The government has subsidized construction of storage facilities through a Rural Infrastructure Development Fund, but often the facilities are not adequate. Despite the fact that prices in the farmer markets are regulated ensuring farmers a minimum profit margin of 20 per cent, local competition makes it difficult to realize the profit margins. There is a potential for generating agri-business and employment to rural youth through farmer markets. At the same time, selling local produce in the nearby markets also reduces the environmental footprint that is caused due to transport of products over long distances.

The Hub model supported by Heifer International in Kenya (HPI-K) to replace the former cooperatives that were not successful is another good example of farmer markets (AFI, 2017). A Hub, in this context, is an economic centre where farmers meet and interact to promote effective marketing of farm products. The Hub enables farmers not only to sell, but also to procure supplies and services needed by the community. The Hub model provides flexibility to farmers and uses the centre to support a network of businesses by delivering supplies and services to farmers within the network.

Farmer markets are also quite popular in Europe, the United States, Japan and Australia and preferred by a certain category of consumers that are environmentally conscious and feel the need to support local farmers. These are also popular places to sell and buy niche products grown in local regions and frequented by tourists. The local governments in these countries provide support with the necessary infrastructure and other investments needed to facilitate farmers to bring and sell their products. In Australia, the Farmer Market Association is well organized and provides some useful lessons for other countries that are in the process of improving market value chains.

As part of the improvements of market VCs it is important to align the agricultural investments with the agricultural strategy and establish formal links with the extension services (Coates *et al.*, 2011). The resources of various sub-sectors should be consolidated and focus on specific product value chain developments that have high potential. At the same time, producers could be organized into collectives to increase efficiencies. Strong public–private partnership models in the agriculture sector are an important part of the VC development process. In practice, this requires closer relationships between smallholders, government agencies and agri-business enterprises, leading to formalized partnerships and other similar models (Coates *et al.*, 2011). Rights to land and other resources have to be clear so that farmers are motivated to invest their time and resources in addressing the market needs.

One challenge is that farmers involved with public–private partnerships run the risk of incurring debt and consequent displacement from land in the event of crop failure. In such situations, farmers need to be indemnified from such a situation. The legal and institutional environment to support the agriculture sector should be strengthened, which could increase competitiveness. Interventions and new measures are needed at the macro, meso and micro level simultaneously to strengthen market VCs. There is a strong need to provide the farming community with facilities for storage and product insurance, so that wastage and produce deterioration is avoided and also to enable farmers to store produce without being compelled to sell when prices are low.

The Comprehensive Africa Agriculture Development Programme (CAADP) initiative sponsored by the New Partnership for Africa's Development (NEPAD) in Africa works together with programmes such as the Common Market for Eastern and Southern Africa (COMESA) to boost growth in the agriculture sector. One of the aims is to improve the business environment for the agriculture sector and support smallholders in different countries to enter into markets and thereby improve trade and economic growth. At the country level and/or the meso-level, the Agriculture Ministry should coordinate with different ministries concerned to consolidate resources and align the relevant policies and programmes. We often see that agriculture marketing in many countries is not within the agriculture ministry. At the micro-level, it is necessary to improve facilities for farmers, access to inputs, training, quality control, processing and marketing of products. Smallholders need to have easy access to credit and cash flows, especially during critical stages of the production. Private sector involvement is necessary for improvement of market value chains, and to provide services that government agencies may not be able to offer. While promoting marketing services, the government needs to put in place adequate safeguards to avoid any exploitation of farmers by the private trade and industries (APAARI, 2008).

Conclusions and recommendations

The chapter on a general level focused on improved extension services and increased integration of smallholders into the market that are necessary to achieve the SAI that climate change allegedly necessitates. The emphasis was on innovative extension models throughout and public–private partnerships, integrated value chains and farmer markets. The section on extension took the form of a brief history, with all its issues and challenges, and a summary of some successful models was presented at the end. The section on integration was more directly focused on possible models of improvement in extension services. Some examples of specific models of extension and integration and their associated institutions and mechanisms was also discussed. Though not discussed within the specific contextual situations, extension models must be flexible and tailored towards specific conditions, including the agroecological environments, crops or products and end users, markets and investments. It was not possible to go into such details and analyses within the scope of this chapter.

Providing accurate information at the right time, and facilitating marketing of farm produce is crucial for smallholders to realize the best use of their inputs, increase yields, adapt to climate change and make profits. If farming is not generating profits and income to farmers, it is adding to the already existing risks of climate change. In several countries, farming is not seen as a valuable economic sector preferred by youth, since it is not profitable, requires hard work, and is too risky. If on the other hand SAI increases productivity, the market supply increases and prices fall, that is in the interest of the consumers and not producers. Governments need to step in here and provide the minimum price guarantee to the producers.

As noted above, agricultural extension and advisory services vary across countries in terms of their structure (public vs private), support and capacity. A systematic cross-country analysis of extension capacity in each country, and within each agroecological environment, crops or products and end users would be a valuable first step. While some level of privatization of extension or advisory services is probably inevitable, there has not been the necessary restructuring of public extension services to work with private consultants, or to support new initiatives such as farmer-to-farmer initiatives. In many countries, government extension services still exist on paper, connected to the Ministries of Agriculture, but the staff are severely underpaid, do not have adequate research-based training, and are susceptible to corruption – leading to lack of trust. Private advisory services have not reached smallholder farmers that make up the bulk of agricultural production in developing countries. For all the talk within the international policy community of the importance of extension services in reaching farmers to help them increase SAI in light of climate change, there is very little support to achieve these lofty words. The international donor agencies should provide much more systematic funding across all countries to increase the training and capacity of innovative extension services, perhaps through the Green Climate Fund and other funding sources.

Creating awareness of the importance of markets and VC enhancement is another important step. In many developing countries, marketing is taken for granted and special efforts are not taken to promote or create value for local products. Though we see some good examples of successful VCs these days, the pace at which their awareness is growing is still slow. Conducting market analysis regularly and promoting market value chains for selected potential products, e.g. avocado, mango, coffee, or milk products that have national and international market value can be useful. It is also important to sensitize different actors on the importance of VC enhancement for cereal, legume and millet crops and value creation through simple interventions. For example, farmers need to be trained on quality sorting and packaging on the farm, and need access to packaging supplies and storage areas. This will also require capacity building of VC actors where needed. Policymakers need to be educated about the importance of value chains. And finally, there needs to be government support to build and upscale the capacity for VC analyses at different levels, through education and training in agricultural economics and business management.

References

ADB (2012) Evaluation knowledge study: Support for agricultural value chain development. Available at: www.adb.org/sites/default/files/evaluation-document/35898/files/eks-agriculturalvaluechain.pdf (accessed 18 June 2017).

Aerni, P. Nichterlein, K., Rudgard S. and Sonnino, A. (2015) Making agricultural innovation systems (AIS) work for development in tropical countries. *Sustainability*, 7, pp. 831–850.

Agriculture for Impact (AFI) (2017) Agriculture for impact: Agriculture value chains. Available at: http://ag4impact.org/sid/socio-economic-intensification/building-social-capital/agricultural-value-chains/ (accessed 15 June 2017).

AFMA (2017) Australian Farmer Market Association. Available at: www.agriculture.gov.au/ag-farm-food/food/publications/national_food_plan/issues-paper/submissions-received/australian-farmers-markets-association (accessed on 18 June 2017).

Anderson, J.R. (1987) Skill acquisition: Compilation of weak-method problem solutions. *Psychological Review*, 94, pp. 192–210.

Anderson, J.R. and Feder, G. (2003) Rural extension services. Policy Research Working Paper 2976. Washington, DC: World Bank.

APAARI (2008) Linking farmers to markets: Some success stories from Asia-Pacific region. Available at: www.apaari.org/wp-content/uploads/2009/05/ss_2008_01.pdf (accessed 23 June 2017).

Bezner Kerr, R., Msachi, R., Dakishoni, L., Shumba, L., Nkhonya, Z., Berti, P.R., Bonatsos, C., Chione, E., Mithi, M., Chitaya, A., Maona, E. and Pachanya, S. (2012) Growing healthy communities: Farmer participatory research to improve child nutrition, food security and soils in Ekwendeni, Malawi. In: Charron, D.F. (ed.) *Ecohealth Research in Practice: Innovative applications of an ecosystem approach to health*. Ottawa/New York: IDRC/ Springer, pp. 37–46.

Birner, R., Davis, K., Pender, J. and Cohen, M. (2009) From best practice to best fit: A framework for analysing pluralistic agricultural advisory services worldwide. *Journal of Agricultural Education and Extension*, 15 (4), pp. 341–355.

Brouwer, H., Hiemstra, W., van Vugt, S. and Walters, H. (2013) Analysing stakeholder power dynamics in multi-stakeholder processes: Insights of practice from Africa and Asia. *Knowledge Management for Development Journal*, 9 (3), pp. 11–31.

Clinton Global Initiative (CGI) (2016) Engaging smallholder farmers in value chains: Emerging lessons. Available at: www.clintonfoundation.org/sites/default/files/cgi_smallholder_report_final.pdf (accessed 20 June 2017).

Chambers, R., Pacey, A., and Thrupp, L.A. (1989) Farmer first: Farmer innovation and agricultural research. Available at: https://opendocs.ids.ac.uk/opendocs/handle/123456789/701 (accessed 22 September 2017).

Chen, K., and Shi, S. (2008) Mechanism and application of demand driven agricultural extension services: The role of China–Canada small farmers adapting to global markets project. Thematic Report 5. Beijing: Beijing Project Office, China–Canada Small Farmers Adapting to Global Markets Project.

Coates, M., Kitchen, R., Kebbell, G., Vignon, C., Guillemain, C. and Hofmeister, R. (2011) Financing agricultural value chains in Africa: Focus on dairy and mangos in Kenya. Available at: www2.giz.de/wbf/4tDx9kw63gma/Kenya-Financing.pdf (accessed 23 June 2017):

FAO (2009) Food security and agricultural mitigation in developing countries: Options for capturing synergies. Available at: www.fao.org/docrep/012/i1318e/i1318e00.pdf (accessed 4 November 2015).

FAO (2014) FAO supports dairy farmers to bring quality milk to the market. Available at: www.fao.org/africa/news/detail-news/en/c/238284/ (accessed 15 June 2017).

Feder, G., Slade, R.H. and Sundaram, A,K. (1986) The training and visit extension system: An analysis of operations and effects. Available at: www.sciencedirect.com/science/article/pii/0309586X86900567 (accessed 29 September 2017).

Feed the Future (2015) Feed the Future: Kenya accelerated value chain development program. Available at: https://cgspace.cgiar.org/bitstream/handle/10568/69195/AVCD_Brochure_Dec2015.pdf?sequence=6&isAllowed=y (accessed 18 June 2017).

Friederichsen, R., Minh, T.T., Neef, A. and Hoffmann, V. (2013) Adapting the innovation systems approach to agricultural development in Vietnam: Challenges to the public extension service. *Agriculture and Human Values*, 30, p. 555.

Hellin, J. and Meijer, M. (2006) Guidelines for VCA. Available at: ftp://ftp.fao.org/es/esa/lisfame/guidel_valueChain.pdf (accessed 19 May 2017).

Hounkonnou, D., Kossou, D.T, Kuyper, T.W., Leeuwis, C., Nederlof, E.S., Röling, N. Sakyi Dawson, O., Traoré M. and Van Huis, A. (2012) An innovation systems approach to institutional change: Smallholder development in West Africa. *Agricultural Systems*, 108, pp. 74–83.

Hosenally, N. (2011) Youth in agricultural extension – challenges and opportunities, Available at: http://nawsheen.com/youth-in-agricultural-Extension-challenges-and-opportunities/ (accessed 21 January 2017).

Hu, Ruifa, Cai, Y., Chen, K.Z., Cui, Y. and Huang, J. (2010) Effects of inclusive public agricultural extension service results from a policy reform experiment in western China, IFPRI Discussion Paper 01037, December 2010, Washington DC.

International Fund for Agricultural Development (IFAD) (2012) Access to markets: Making value chains work for rural people. Available at: www.ifad.org/documents/10180/650e771a-ef4a-4893-967b-2d5fd8eef313 (accessed 12 June 2017).

Kessler, A., Van Duivenbooden, N. and Nsabimana, F. (2014) The PIP Approach: A step-by-step explanation of the different phases of the creation and implementation of IFP with examples from Burundi. Project Fanning the Spark. Wageningen, Alterra.

Kürschner, E., Baumert, D., Plastrotmann, C., Poppe, A.K., Riesinger, K. and Ziesemer, S. (2016) Market access for smallholder rice producers in the Philippines. Available at: http://edoc.hu-berlin.de/series/sle/264/PDF/264.pdf (accessed 19 June 2017).

Klerkx, L., and Leeuwis, C. (2008) Balancing multiple interests: Embedding innovation intermediation in the agricultural knowledge infrastructure. *Technovation*, 28 (6), pp. 364–378.

Kumar, S.S. (2016) Problems of all huse in rythu bazaars. Available at: www.thehindu.com/news/cities/Hyderabad/problems-of-all-hues-in-rythu-bazars/article 2868129.ece (accessed 15 June 2017).

Nokes, T.J. and Ohlsson, S. (2005) Comparing multiple paths to mastery: What is learned? *Cognitive Science*, 29, pp. 769–796.

Nokes, T.J. (2009) Mechanisms of knowledge transfer. *Thinking and Reasoning*, 15 (1), pp. 1–36.

Norton, R. (2014) Agricultural value chains: A game changer for small holders, Available at: www.devex.com/news/agricultural-value-chains-a-game-changer-for-small-holders-83981 (accessed 20 June 2017).

Raj, R. and Nagothu, U.S. (2016) Gendered adaptation to climate change in canal-irrigated agro ecosystems. In: Nagothu, U.S. (ed.) *Climate Change and Agricultural Development Improving Resilience through Climate Smart Agriculture, Agroecology and Conservation*. New York: Routledge, pp. 259–278.

Rivera, W.M., Qamar, M.K. and Crowder, L.V. (2001) *Agricultural and Rural Extension Worldwide: Options for institutional reform in developing countries*. Rome: Food and Agriculture Organization of the United Nations.

Seidman, W. and Michael, M. (2005) *Optimizing Knowledge Transfer and Use*. Cerebyte, Inc.

Simpson, B.M., Heinrich, G. and Malindi, G. (2012) *Strengthening Pluralistic Agricultural Extension in Malawi: Rapid scoping mission 9–27 January 2012*. Modernizing Extension and Advisory Services Project. Washington, DC: USAID.

Singh, K.M. (2013) Public private partnership in agricultural extension management: Experiences of ATMA model in Bihar and India. Available at: https://papers.ssrn.com/sol3/papers.cfm?abstract_id=2254495 (accessed 10 October 2017).

Spielman, D., Ekboir, J., Davis, K. and Ochieng, C.M. (2008) Rural innovation systems and networks: Findings from a study of Ethiopian farmers. *Agriculture and Human Values*, 28, pp. 195–212.

Swanson, B.E. and Rajalahti, R. (2010) Strengthening agricultural extension and advisory systems: Procedures for assessing, transforming, and evaluating Extension systems. Discussion Paper 44. The IBRD/World Bank.

Sanga, C., Mussa, M., Tumbo, S, Mlozi, M.R.S, Muhiche, L. and Haug, R. (2014) On the development of mobile-based agricultural extension system in Tanzania: A technological perspective. *International Journal of Computing and ICT Research*, 8 (1), pp. 49–67.

Umali, D. L., and L. Schwartz. (1997) Public and private agricultural extension: Partners or rivals? *World Bank Research Observer*, 12 (2), pp. 203–224.

van Paassen, A., Klerkx, L., Adu-Acheampong, R., Adjei-Nsiah, S., Ouologuem, B., Zannou, E.,Vissoh, P., Soumano, L., Dembele, F. and Traore, M. (2013) Choice-making in facilitation of agricultural innovation platforms in different contexts in

West Africa: Experiences from Benin, Ghana and Mali. *Knowledge Management for Development Journal*, 9 (3), pp. 79–94.

Waddington, H. and White, H. (2014) Farmer field schools: From agricultural extension to adult education. *3ie Systematic Review Summary 1*. London: International Initiative for Impact Evaluation (3ie).

Waddington, H., Snilstveit, B., White, H. and Anderson, J. (2010) The impact of agricultural extension services. Available at: http://economia.unipv.it/naf/Working_paper/WorkingPaper/WPMWMS.pdf (accessed 08 October 2017).

Wiggins, S. and Keats, S. (2013) Leaping and learning: Linking smallholders to markets. Available at: https://workspace.imperial.ac.uk/africanagriculturaldevelopment/Public/LeapingandLearning_FINAL.pdf (Accessed 19 May 2017).

Woodhill, J. and van Vugt, S. (2011) *Guidebook for MSP Facilitation*. Chapter 3, page 41. GIZ/CCPS.

World Bank (2006a) *Enhancing Agricultural Innovation: How to go beyond the strengthening of research systems*: Agriculture and rural development, Washington DC.

World Bank (2006b) *Strengthening Agricultural Extension and Advisory Systems: Procedures for assessing, transforming, and evaluating extension systems.* Available at: http://siteresources.worldbank.org/INTARD/Resources/Stren_combined_web.pdf(accessed 24 April 2017).

11 Multi-level policy measures to support sustainable agriculture intensification for smallholders

Allison Morrill Chatrchyan, Christina Yin, Emmanuel Torquebiau, and Udaya Sekhar Nagothu

Introduction

Climate change is one of the most urgent and defining issues of our time, with the Intergovernmental Panel on Climate Change (IPCC) Fifth Assessment Report noting that "recent climate changes have had widespread impacts on human and natural systems," influencing human health, agriculture, physical systems, and biological systems (IPCC, 2014). Climate change impacts to agriculture have been increasing and are projected to become more severe by the end of the century (Hatfield *et al.*, 2014). The increased frequency and intensity of extreme weather events and climate variability will pose unprecedented challenges to agricultural production, due to the sensitivity of agricultural productivity to changing climatic conditions and the high costs of impacts and adaptations (Walthall *et al.*, 2012). Climate change will necessitate that agricultural, ecological, and social systems adapt and mitigate their greenhouse gas (GHG) emissions across multiple scales (Adger, Arnell, and Tompkins, 2005). In addition, as some regions of the world see devastating losses in crop yields from floods, increasing temperatures and droughts, other regions will have to sustainably intensify their agricultural production to meet the global demands for an adequate food supply.

In addition to biophysical challenges for the various crops outlined in this book in Chapters 2 to 8, other core challenges include the institutional dimensions of knowledge sharing and market constraints discussed in Chapter 10, and the policy frameworks that can support necessary changes at different scales. What are the policies that are needed to address the challenges of food and nutrition security in light of climate change, without repeating the environmental damage caused by mainstream, conventional agriculture? Can approaches such as Sustainable Agriculture Intensification (SAI) solve these problems? SAI is a concept that aims to increase yields without adverse environmental effects and without cultivation of more land (Petersen and Snapp, 2015). But SAI requires capacity building to help smallholder farmers learn about and adopt new techniques, and enabling policy environments at all levels of government.

In this chapter, SAI is defined as a process or system where yields are increased without adverse environmental impact and without the cultivation of more land (Royal Society, 2009).

Another closely related concept increasingly used to address the challenges of climate change for agriculture, is Climate Smart Agriculture (CSA). The Food and Agriculture Organization of the United Nations defines CSA as an "approach that helps to guide actions needed to transform and reorient agricultural systems to effectively support development and ensure food security in a changing climate" (FAO, 2013). CSA aims to tackle three main objectives, and refers to "agriculture that sustainably increases productivity, resilience (adaptation), reduces/removes GHGs (mitigation), and enhances achievement of national food security and development goals" (Lipper, 2014; FAO, 2010). CSA mainly focuses on the goals to be reached and does not prescribe the farming techniques or approaches to achieve increased agricultural adaptation, mitigation, or sustainable production. However, it does focus on sustainability, resiliency, and climate smart adaptation practices such as improving soil health that are also agroecological practices. Rather than think about SAI and CSA as distinct concepts, Campbell *et al.* (2014) argue that sustainable intensification (of agriculture) and climate smart agriculture are closely interlinked. They note that the main differences between the concepts is that CSA focuses on climate change adaptation and mitigation outcomes along with sustainable increases in yields, while SAI is critical to both increased resiliency and reduced GHG emissions.

This chapter focuses on the policy interventions that will be necessary at different levels to achieve SAI and CSA for smallholders in the future. Unlike other key economic sectors involved in adapting to climate change, adaptation in the agricultural sector ultimately comes down to the cumulative management decisions made by individual farmers – and the policies, regulations, information, and capacity that incentivize behavioral change at a personal and business level on the farm. Organizations are realizing that farmers and other stakeholders play a key role in determining whether CSA and SAI practices are actually adopted or not. Rapid change will be possible if those stakeholders who are most directly affected are included in the decision-making processes (Bartels, Furman, and Diehl, 2013; Howden *et al.*, 2007; Meinke *et al.*, 2006; Prokopy *et al.*, 2015). Farmers need the best possible information, training, support, and policy incentives to scale up climate change adaptation practices (Brugger and Crimmins, 2015).

Increased adoption of SAI and CSA practices by smallholder farmers will require incentives and policies from the global to national and local levels (Torquebiau, 2016). But a key challenge remains in the barriers to adoption and scaling up these practices, or scaling global policies down to the national and local level. Key questions considered in this chapter include: 1) What are the existing policies that support SAI and climate change and agricultural adaptation at various levels, and are they adequate to address the enormous challenge? 2) Can SAI and CSA practices be scaled up from the local, regional, and national level, or scaled down from the global level? 3) What are the linkages between these multi-level policy mechanisms and do they need strengthening or coordinating?

In the next section, the chapter provides a theoretical basis for multi-level/polycentric policies across the global, regional, and national scales for SAI and CSA, making the argument for policy coherence. It then presents an overview of agricultural policies at the international level within the United Nations Framework Convention on Climate Change (UNFCCC) and UN Sustainable Development Goals (SDGs), and through new voluntary initiatives such as the Global Alliance for Climate Smart Agriculture. This is followed by a brief account of regional mechanisms and policies to support SAI. Changes in farming practices occur at the farm-scale, with national policies to support extension, and incentives and tools to adopt SAI and CSA practices. Thus, the chapter explores cases of agricultural adaptation initiatives that are supporting SAI and CSA at the national and local level. Finally, the chapter explores the interplay between these differing levels of governance and policy, with consequences for CSA and SAI on the ground. Given the critical importance of SAI in the future agricultural development strategies; the chapter concludes with the assessment of multi-level governance mechanisms and the implications for food and nutrition security.

Multi-level and polycentric governance and policy mechanisms

Recent research on multi-level governance of climate change adaptation has questioned the issue of scaling up or scaling down policies and implementation of SAI by smallholder farmers. Currently, there are many more actors involved at different levels to address climate change adaptation, which is increasing the complexity and pace of implementation. Bulkeley *et al.* (2014) note that the responsibility to act has become increasingly diffuse across different scales for social groups, sectors, state and non-state actors, and generations, and that many new transnational climate change governance mechanisms have been established to try to bring about change quickly.

Several studies demonstrate that there has been a proliferation of actors and institutions involved in climate change governance at all levels from the local to the global, across several domains, with varying degrees of effectiveness and coordination (Andonova, 2010; Andonova, Betsill, and Bulkeley, 2009; Biermann *et al.*, 2010; Bulkeley and Newell, 2010; Jagers and Stripple, 2003; Pattberg and Stripple, 2008). This increases the complexity of climate adaptation governance with decision making at different levels (Bulkeley and Newell, 2010).

In principle, there should be a clear division of roles and responsibilities between the different policy actors. Global policy mechanisms should focus on providing the overarching frameworks and goals, set required targets for the national governments, and provide funding options where possible, while national governments should focus on providing national policy frameworks, technology development, implementation, capacity building, and support for research and education. However, it is the provincial or sub-national actors that are often engaged in the actual implementation of adaptation strategies and they are crucial for any success in the process (Bauer and Steurer, 2014).

There is another dimension, the polycentric nature of policy planning and implementation, that is important to consider. According to Jordan *et al.* (2010), the overall landscape of climate governance has become more "polycentric," that is it spans many levels (international, national, or local), and works through many modes and domains of action (such as policies, markets, networks), is more diverse and has a a greater emphasis on bottom-up initiatives. Ostrom (2010) argued that if environmental challenges are not backed up by a variety of efforts at the national, regional, and local level, they may not achieve the desired outcomes. With many independent centers of leadership, and interdependence between governing institutions and organizations at various levels, the polycentric nature of policy mechanisms further increases the complexity. Trust among these various actors is a key factor in effective governance to ensure that such mechanisms result in concrete actions (Ostrom, 2010).

Polycentric governance – with multiple levels of policies – makes it even more important to ensure better coordination across the different actors and levels both horizontal and vertical. This is termed the "policy coherence for development." Policy coherence has been defined by the OECD and others as "the systematic promotion of mutually reinforcing policies across government departments to create synergies towards achieving agreed objectives and to avoid or minimize negative spillovers in other policy areas" (OECD, 2012). Policy support for SAI across multiple levels should consider the diversity of needs across different countries and geographical regions, and analyze carefully the constraints, gaps, and options available for implementation. A recent study by Notenbaert *et al.* (2017) also describes a framework for evaluating and prioritizing potential CSA practices that can be used at multiple levels. They argue that decision making inevitably takes place at multiple scales, with multiple stakeholders and objectives, echoing the multi-level and polycentric nature of policy mechanisms in agricultural adaptation.

Frameworks to support SAI and CSA at a global level

Despite the critical nature of SAI globally in light of climate change, agriculture has received less attention than other issues such as forestry or land use under the United Nations Framework Convention on Climate Change (UNFCCC) since 1992. Caron and Treyer (2016) argue that although agriculture was central to the framework convention, it became a politically charged issue, and progress was stalemated until 2013. As a result of this lack of appropriate action, new transnational climate change governance mechanisms, such as the Global Alliance for Climate-Smart Agriculture (GACSA), and regional initiatives, such as the Comprehensive Africa Agriculture Development Programme (CAADP), have sprung up to fill this void.

Addressing SAI/ CSA under the UNFCCC

Since the establishment of the UNFCCC in 1992, agriculture has been a part of the global discussions, but progress was sidelined within the UN Negotiation processes. Article 2 of the original UNFCCC text specifies that GHGs in the atmosphere should be stabilized so as to ensure that "food production is not threatened" (UNFCCC, 1992), but does not otherwise call out agriculture, or the need for intensification of sustainable agricultural production. Agriculture is included in Article 4.1 as one of the sectors falling within the obligation of Parties to develop both mitigation plans and measures to reduce or prevent anthropogenic emissions of GHGs, as well as to cooperate in preparing for adaptation to the impacts of climate change (Richards *et al.*, 2015).

Despite that, scholars including Kalfagianni and Duyck (2015) note that agriculture has not been as much of a central issue in the multilateral negotiation process. Within the Kyoto Protocol (adopted in 1997), agriculture was incorporated into Annex I targets and was eligible under the Clean Development Mechanism (CDM). Later the Bali Action Plan (adopted in 2007) considered agriculture as a key sector to be addressed under the Ad Hoc Working Group on Long-term Cooperative Action (AWG-LCA). At the 17th Conference of the Parties (COP17) in Durban, South Africa in 2011, agriculture was moved from the LCA discussions and referred to the Subsidiary Body for Scientific and Technological Advice (SBSTA) – perhaps allowing for more of a focus on the scientific and technical aspects of agricultural mitigation and adaptation, and less on the political contentiousness of addressing agriculture between developed and developing countries. From 2011–2014, however, progress within the SBSTA was slow, as the body requested submissions from member parties on scientific and technical issues including early warning systems, assessments of agricultural risks and vulnerabilities, and organization of scientific workshops (Farming First, 2015).

Prior to COP21 in Paris in November 2015, parties to the UNFCCC were invited to submit their countries' intentions for GHG reductions and adaptation plans as Intended Nationally Determined Contributions (INDCs). Many countries included both agricultural mitigation and adaptation measures within their INDCs. There were 131 INDCs that included agriculture within both mitigation or adaptation contexts – with 103 INDCs including agricultural mitigation, and 102 INDCs including agriculture in adaptation priorities, with 30 percent of these explicitly mentioning CSA (Richards *et al.*, 2015).

While the Paris Agreement in 2015 that came out of COP21 does not specifically mention agriculture or SAI, the themes of food and nutrition security, food production, human rights, gender, ecosystems, and biodiversity are explicitly mentioned in the Agreement. The preamble of the Paris Agreement specifically notes the importance of "Recognizing the fundamental priority of safeguarding food security and ending hunger, and the particular vulnerabilities of food production systems to the adverse impacts of climate change" (UNFCCC, 2015). It is encouraging that food security is mentioned in the preamble of the agreement, as it emphasizes the importance of this key subject.

In addition, whereas gender was previously mentioned in UNFCCC texts only in the preamble, the Paris Agreement includes gender in the sections on adaptation and capacity building, thus indicating policy support for gender mainstreaming in agricultural adaptation. It states that "Parties acknowledge that adaptation action should follow a country-driven, gender responsive, participatory and fully transparent approach" (UNFCCC, 2015).

Importantly, Article 2.1 of the Paris Agreement "aims to strengthen the global response to the threat of climate change, in the context of sustainable development and efforts to eradicate poverty." This includes actions that will be needed to increase "the ability to adapt to the adverse impacts of climate change and foster climate resilience and low GHG emissions development, in a manner that does not threaten food production" (UNFCCC, 2015; see also Meadu, 2015). Still, for many developing countries, especially in Africa, that argued with the slogan at COP21 "No Agriculture, No Deal," the fact that agriculture was not *specifically* mentioned or detailed in the Paris Accord was a disappointment yet again – since a core concern for their countries was not addressed.

Since the Paris Agreement entered into force as of November 4, 2016, the main priority now is to ensure that developing countries have the adequate support and capacity to implement their Nationally Determined Contributions, 80 percent of which include a focus on agriculture. While a considerable majority of the plans include agriculture in the adaptation priorities within their NDCS, the specific agricultural adaptation practices are not detailed. Developing countries will need assistance to identify these specific practices, and funding (which is a major constraint) to help implement these strategies with farmers. This highlights the need for financial, technical, and extension support, especially for smallholder farmers, to implement SAI. As noted in Chapter 1 of this book, the ratification of the Paris Agreement, and other global initiatives, if properly implemented, can contribute to scaling up of SAI/CSA; but this will require adequate funding support in the agriculture sector both at the national and international levels.

The potential of SAI/CSA to contribute to GHG reductions through soil carbon sequestration is not properly documented or recognized either by the global or national policy actors dealing with climate change adaptation. It has been pointed out earlier in the book that one of the constraints in achieving SAI is the lack of indicators for their impact measurability. Despite the benefit of SAI practices to reduce GHG emissions, the lack of protocols and documentation has undermined their ability to contribute to mitigation. Only recently, we see that some governments are realizing the potential of the agriculture sector in mitigation. For example, the "4 per 1000" initiative of the French Government at COP 21 to build carbon sequestration in soils by 0.4 percent a year, would make it possible to offset the increases in atmospheric CO_2 (French Ministry of Agriculture, 2017; Soussana *et al.*, 2017). These efforts will be more successful if there is policy coherence between them, as well as between different multilateral initiatives, such as the new SDGs. On the other hand, countries such as India have deliberately avoided mentioning agriculture as a potential sector in their mitigation strategies. This approach by some countries is politically motivated, since they do not intend to invest in mitigation efforts in the agricultural sector, nor compromise on food production.

The agriculture focus of the UN Sustainable Development Goals (SDGs)

In January 2016, the new United Nations Sustainable Development Goals (SDGs) – part of the 2030 Agenda for Sustainable Development – officially entered into force. Within the 2030 Declaration document, agriculture is specifically mentioned four times, including in Section 24, which includes a focus on sustainable agriculture, smallholder farmers, and the importance of recognizing gender in sustainable development: "We will devote resources to developing rural areas and sustainable agriculture and fisheries, supporting smallholder farmers, especially women farmers, herders and fishers in developing countries, particularly least developed countries" (UN, 2015). This is in line with the objectives of SAI/CSA and helps national governments to draw inspiration to promote these approaches in their future strategies.

Section 27 of the 2030 Declaration also mentions sustainable agriculture as one of the sectors that needs focus and where there is a need to increase productive capacity. Agriculture is specifically mentioned in SDG 2, which sets out to "End hunger, achieve food security and improved nutrition, and promote sustainable agriculture" (UN-DESA/DSD, 2014). The SDG 2 in fact is a key goal, though broad, it provides a strong justification for national governments to invest and support the implementation of SAI.

The 2030 Agenda also acknowledges the interconnections between food, livelihoods, and natural resource management, emphasizing the need to focus on rural development and investment in agriculture (UN, 2015). Agriculture plays a major role in efforts to address climate change, and is a critical component for achieving the SDGs (FAO, 2016b). The FAO also recognizes the importance of sustainable food systems, and of finding new methods of farming and managing natural resources to realize the SDG 2 (FAO, 2016b). The SDG 2, and its sub-objectives as shown in Table 11.1 outline the targets to be achieved that relate to SAI.

In an effort to reach the SDG 2 goal and the targets against the various sub-objectives (Table 11.1), the FAO has been supporting national policies and programs supporting SAI and achievement of related SDGs in several countries. For example, one program created in 2014 by the FAO supported the launch of the Nigeria Youth Employment in Agriculture Programme, which plans to create 750,000 agricultural jobs for youth in the agricultural sector over a five-year period (FAO, 2016b). In Rwanda, the FAO is assisting the assessment of key sustainability issues, and promoting cross-sectoral policy dialogues and stakeholder engagement to devise action plans for productive landscapes (FAO, 2016b). In several Central Asian countries, the FAO is piloting sound post-harvest management practices to ensure food quality and diminish losses through the introduction of new technologies and identification of major bottlenecks in supply chains (FAO, 2016b). In

Table 11.1 SDG policies for Goal 2: achieving food security and promoting SAI (UN, 2015)

Sub-goals for SDG 2	
2.1	End hunger and ensure access by all people . . . to safe, nutritious and sufficient food
2.2	End all forms of malnutrition
2.3	Double the agricultural productivity and incomes of small-scale food producers, in particular women
2.4	Ensure sustainable food production systems and implement resilient agricultural practices that increase productivity
2.5	Maintain the genetic diversity of seeds, including through soundly managed and diversified seed and plant banks at the national, regional and international levels
Specific SDG targets or policies related to SAI and CSA	
2a	Increase investment in rural infrastructure, agricultural research and extension services, technology development
2b	Correct and prevent trade restrictions and distortions in world agricultural markets, including through the parallel elimination of all forms of agricultural export subsidies and all export measures with equivalent effect
2c	Adopt measures to ensure the proper functioning of food commodity markets and their derivatives and facilitate timely access to market information, including on food reserves, in order to help limit extreme food price volatility

Source: Adapted from UN (2015).

many other countries, the FAO is promoting SAI through integrated systems that include mixed cropping, crop diversification, conservation agriculture, and agroforestry systems, aiming to produce more food with fewer inputs and the same amount of land (FAO, 2016b). There are other SAI initiatives supported by FAO and other international and national agencies discussed in several chapters in the book.

Through these initiatives, the FAO recognizes the interconnectedness between the SDGs and the critical role that agriculture plays in achieving the goal's. FAO is working to eliminate issues such as poverty and hunger through sustainable intensification of agriculture, and building climate resilience. By supporting programs that aim to improve resource efficiency and protect natural ecosystems, the FAO also aims to protect rural livelihoods and improve the resilience of communities (FAO, 2016b).

While the SDGs go a long way towards promoting SAI and CSA, especially among smallholder and women farmers in Least Developed Countries, the success or failure of the 2030 Agenda will hinge on how they are implemented in each country with the help of the international community (Hinds, 2016). This will require that each member state incorporate the SDGs into their national plans, budgets, and strategies – including providing strong support for public institutions and coordination mechanisms to deliver the SDGs. National governments need to include stakeholders in implementation of SDG plans, strengthen data collection systems, and report their progress towards achieving goals and targets (Hinds, 2016). Work to achieve SAI within the framework of the new SDGs will clearly require coordination among international efforts and regional and national initiatives.

However, given the overall slow response of the UNFCCC process and SDGs to increase adoption of SAI and CSA practices, new transnational initiatives have arisen to hasten the pace of action, and downscale global goals to the national and local level.

New transnational mechanisms: the Global Alliance for Climate Smart Agriculture

As a follow-up to the COP 20 in Lima, Peru and leading up to the COP 21 in Paris, the Global Alliance for Climate Smart Agriculture (GACSA) was launched at the United Nations Climate Summit in September, 2014 in New York, after more than three years of initial meetings to vet the concept with state and non-state actors. GACSA's aim is to advance CSA, a concept originally developed by the FAO to address increasing global food insecurity and agricultural challenges in response to climate change (FAO, 2014a).

At its inception, GACSA had approximately 70 members (FAO 2014b), which had grown to 180 member organizations in July 2017 (FAO, 2017). GACSA membership is voluntary and does not involve a participation fee, nor does membership result in any binding obligations. Each member individually determines the extent and nature of their participation (FAO, 2014a).

Any entities interested in joining the organization, including governments, international and regional organizations, non-governmental institutions, farmers' organizations, private companies, and industry groups, should agree to the vision and objectives of the Alliance, but are not legally bound by any requirements for participation (FAO, 2014a). Members are expected, however, to provide periodic updates on progress towards their individualized goals and initiatives that pertain to GACSA's overall mission. GACSA also allows for institutions to join as observers to familiarize themselves with the program before deciding on the membership (FAO, 2014a).

GACSA primarily considers itself a "knowledge platform" to share information and to address the three main challenges to scaling up CSA: knowledge building and information sharing; mobilization of public and private funding, as well as improved efficacy; and construction of an enabling political environment. Together, members from all over the world have the collaborative space under GACSA to share and exchange information, experiences, resources, and tools to advance initiatives spanning from the local to global scale, related to food security, agriculture, and climate change. GACSA is governed by member organizations through a Strategic Committee, with a Facilitating Unit housed within the FAO (FAO, 2014b).

However, despite being active since 2015, real progress remains limited due to the challenges of establishing a new, global governance structure. Funding of the GACSA Trust Fund remains limited, which results in limited staff capacity of the Facilitation Unit, and limited funds for projects. There are also minimal funds to cover support of developing country members to take part in GACSA's international meetings, a constraint for many members with restricted funding. This is in clear differential to the UNFCCC COP meetings, where the UN provides support for developing countries representatives to participate in conferences. It remains unclear how GACSA will ensure real participation of developing country participants without supporting their participation costs.

GACSA was set up with "a light organizational structure," that is minimally funded through a five-year (2015–2019) multi-donor trust fund, with contributions from the governments of Norway, Switzerland, and the United States (FAO, 2015a). Although the Facilitation Unit is staffed by the FAO, GACSA has not been formally incorporated as a legal entity on its own. As the Framework Document (FAO, 2014a) states, "The Alliance will be a voluntary coalition, open and transparent and will build on a participatory approach."

A key criticism of GACSA and CSA in general is that it is policy neutral when considering the CSA approaches in the toolbox and neutral in terms of the agricultural practices to achieve the objectives of CSA, which can include everything from sustainable and agro-ecological methods, to conventional agricultural methods that utilize fertilizers or genetically modified organisms to increase resiliency to climate change. Researchers argue that it will be important to define the exact criteria for CSA so that it doesn't become a greenwashing concept or catch-all initiative for any agricultural practices that increase yields, but may not in fact be sustainable (Neufeldt *et al.*, 2013; Rosenstock *et al.*, 2016). Some insist that to comply with its original definition, CSA ought to be based on agroecological principles (Saj *et al.*, 2017). Another criticism of CSA is that it promotes action on the three pillars of increased production, GHG mitigation and adaptation at the same time, whereas many developing country farmers in crisis need immediate help with adaptation, and should not have to focus on reducing GHG emissions in the short term (Delvaux *et al.*, 2014; Stabinsky, 2014). Similar criticism can be observed towards SAI, which overlaps largely with CSA, that there are no set criteria for implementation or indicators to measure the impacts.

While the goals of GACSA are aspirational, it is not clear how the Alliance can realistically achieve its goal of reaching 500 million farmers and convince them to change their behaviors and practices, without adequate staffing or funding for implementation projects in developing countries. GACSA and similar global initiatives need to link with national policies necessary to scale up SAI across developing countries, and help prescribe specific policy goals for regions or countries to consider. The reality of building partnerships, and supporting agricultural producers to adopt the CSA/SAI practices and tools provided, and provide input to the process, remains incredibly challenging at the scale envisioned.

Regional policy approaches

Climate change impacts and adaptation in the agriculture sector vary by agro-ecological regions, because of differing terrain, water availability, production and cultural practices, therefore requiring different adaptation strategies by region. There will not be a one-size fits all approach that will work in varied regions, so different regions will have to support efforts to scale up specific sustainable intensification practices particular to their region. As Birner and Resnick (2010) note, an important factor in agricultural policies for smallholder farmers is the rising trend toward regional integration in Africa. Both the Economic Community of West African States (ECOWAS) and West African Economic and Monetary Union (WAEMU) have adopted agricultural policies, and both are influenced by the continent-wide Comprehensive Africa Agricultural Development Programme (CAADP). These regional governance frameworks have been established in the last few years to scale up efforts at climate change and agricultural adaptation in different parts of the world.

The Comprehensive Africa Agriculture Development Programme (CAADP)

The Comprehensive Africa Agriculture Development Programme (CAADP) funded by GIZ (Germany) is seen as one of the first African Union's (AU) continental policy frameworks for agricultural development and transformation (CAADP, 2016). The program supports the member states to achieve at least 6 percent annual growth rate in the agricultural sector. The CAADP is a positive initiative that emphasizes the importance of agriculture in adaptation and provides guidelines to national governments to follow. However, the main challenge is that a majority of the member states do not possess the funds and adequate adaptation technology support to achieve the set goals. CAADP is helping to integrate the policy and technical inputs into the National Agricultural Investment Plans (NAIPs). CAADP thus provides the necessary link to the country level programs and a common platform for cooperation and knowledge exchange among the member countries.

The CAADP framework has so far proved useful in increasing awareness, evidence based policymaking, and the ability to bring the country level policymakers into active dialogue targeting agriculture adaptation to climate change in the continent (Kibaara *et al.*, 2009). According to Ross (2013), CAADP has managed to improve awareness and country level spending on agriculture in a few countries. But so far, only two of the countries invested more than the expected 10 percent. The other challenge is to understand how the investments are used, because the investments may not be made in the best interest of certain farmer groups and regions, especially women farmers. Thus, a major drawback of the CAADP is that there is not enough focus on the local adaptation needs of smallholders or support through investments with external inputs where needed (Ross, 2013). Since women play a key role in African agriculture, gender mainstreaming should be an integral component of adaptation frameworks. Implementation of effective policy measures and strengthening partnerships between countries, sectors and stakeholders will be necessary to see positive outcomes under CAADP (FAO, 2016a).

The Africa Climate–Smart Agriculture Alliance

In recent years, new regional and country-level initiatives have been established in order to combat climate change in Africa, with support from external donors and funding agencies. New initiatives to promote SAI, CSA, agroecology, and conservation agriculture innovations suitable to smallholders have started in a few countries. Some of the more popular programs targeting agriculture include the Africa Adaptation Programme (AAP 2008–2012), the Adaptation to Climate Change in Agricultural Development in Africa (CAADP 2012–2016), and the Africa CSA Alliance.

The main focus of the African CSA Alliance is on building capacity of smallholder farmers and agriculture institutions including markets and extension approaches in various countries for improving adoption of CSA on a larger scale and farmer profits. Central to the goals of the Alliance is to improve food, nutrition, and livelihood security, resilience, and productivity, along with the equity and sustainability of agricultural systems building on proven community-based and farmer-led approaches (NEPAD, 2014).

The Alliance has set a goal of supporting the adoption of CSA practices by 6 million farming households in sub-Saharan Africa by 2021, contributing to the African Union's broader goal of engaging nearly 25 million farmers by 2025, with a focus on small shareholder farms and women and youth. This is to be achieved by promoting multi-sectoral collaboration and preparing country level CSA plans and upscaling frameworks for identified CSA measures (CCAFS, 2015). Integration with NAIPs and national funds is one of the key strategies within this framework, without which it is not possible for scaling up CSA measures. Current members in the African CSA Alliance include a mix of regional development programs, international organizations and non-governmental organizations, and farmer's unions. But without clear national commitments to this initiative, it will not be possible for the Africa CSA Alliance to help scale up CSA and ensure SAI in the countries it works with.

There are several key regional policies relevant for CSA and SAI implementation in Africa, that have been identified by the Government of Tanzania (2015). Table 11.2 provides a summary of these key regional policies and their intended focus.

Table 11.2 Examples of key regional policies to support SAI/CSA by smallholder farmers

Regional policy	*Focus*
Comprehensive Africa Agriculture Development Programme (CAADP)	Based on four reinforcing pillars for investment in agriculture to improve performance through strengthening country-focused lending programs based on coordinated sector plans, enhanced capacity for policy, analytical work, and facilitating partnerships. The four pillars are: 1) Expanding areas under sustainable land management and reliable water control systems 2) Improving rural infrastructure and trade related capacities for market access 3) Increasing food supply and reducing hunger 4) Expanding agricultural research and technology transfer and dissemination.
East Africa Community Food Security Action Plan	Developed to address food insecurity and rational agricultural production in the region. The plan will guide coordination and implementation of joint programs and projects emanating from the plan.
East Africa Community Climate Change Policy (EACS, 2011)	The purpose is to guide EAC partner states and other stakeholders on the implementation of collective measures to address climate change adaptation and mitigation actions, while assuring sustainable social and economic development.

Source: Adapted from Tanzania (2015).

Regional policies such as the ones highlighted above are particularly important to support approaches such as SAI and CSA implementation in countries with weak institutions and low capacity, and can strengthen efforts at the national and local level. It should be noted that frameworks such as the CAADP are being supported by both global agencies (United Nations), regional organizations (New Partnership for Africa's Development (NEPAD), and development initiatives (such as international development funds from EU countries or the United States). In addition, agricultural research, policy support and projects are supported by the CGIAR Consortium (formerly the Consultative Group for International Agricultural Research) – a global partnership of research carried out at 15 CGIAR centers, focused on reducing rural poverty, increasing food security, improving human health and nutrition, and ensuring sustainable management of natural resources (CGIAR, 2017a). Without support from these global and regional organizations, and research from the CGIAR and national centers, most of these regional policy initiatives would not take place.

Prioritizing SAI and CSA policy interventions at the national and local level

Ultimately, any efforts to scale up SAI have to take place with changes in farming practices of individual farmers', supported by enabling policies and incentives, research, extension support, and information and tools. Girvetz *et al.* (2017) note that what works in one location with one farmer may not work well in other regions or situations. Stakeholder involvement is key to ensuring that the right priorities for SAI are established. They suggest a straightforward approach to prioritizing initiatives, based on: 1) situational analysis (identifying climate impacts and adaptations); 2) prioritizing interventions (policy options); 3) program design and implementation; and 4) monitoring and evaluation (Girvetz *et al.*, 2017).

Recently, the CGIAR Research Program on Climate Change, Agriculture and Food Security (CCAFS) created a new CSA Guide and website to provide a step-by-step approach to developing a CSA plan for country practitioners (CGIAR, 2017b). The guide notes that widespread adoption of CSA practices can only occur if appropriate policies are developed and implemented, within a supportive enabling environment. These national policies can include agricultural components under National Determined Contributions (NDCs), National Adaptation Programmes of Action (NAPAs), National Adaptation Plans (NAPs), and Nationally Appropriate Mitigation Actions (NAMAs), all under the UNFCCC process.

The CSA Guide also notes that agriculture and food security plans are often included in national development and poverty reduction strategies; and that national trade, financial, agricultural, and environmental policies are also relevant to scaling up CSA (CGIAR, 2017b). The guidelines enacted for CSA also apply for SAI implementation in developing countries, since both concepts are similar and in some cases overlapping.

Below, we provide some case studies of national initiatives and policies for SAI and CSA in three countries that are most vulnerable to climate change in Asia, Africa, and Central America: Bangladesh, Tanzania, and Nicaragua.

Bangladesh

Agriculture is a fundamental pillar of the Bangladesh economy, contributing 16.5 percent of the country's GDP and serving as the country's largest employment sector, with 87 percent of rural inhabitants deriving at least a portion of their income from agricultural activities (CIAT, 2017). In Bangladesh, 99 percent of farms are small scale, with average areas of less than one hectare. Climate change is already causing devastating consequences for the agricultural sector, and these impacts will continue to intensify in the future. According to the Global Climate Risk Index, Bangladesh is the world's most vulnerable country to climate change. Sea level rise, temperature increases, and rainfall variability will all have significant impacts on agricultural productivity and livelihoods.

Because of the country's high vulnerability to climate change, several policies have been established to address related challenges in the agricultural sector, based on findings from the IPCC, or under the UNFCCC Framework. These include the National Environment Management Action Plan (NEMAP) for Bangladesh; the National Adaptation Programme of Action (NAPA); the Bangladesh Climate Change Strategy and Action Plan (BCCSAP); and most recently, Bangladesh's Intended Nationally Determined Contribution (INDC) under the Paris Agreement. In addition, CIAT has identified 13 key policies, strategies, and programs that relate to agriculture and climate change that are key factors enabling adoption of sustainable agriculture practices in Bangladesh.

While adaptation and productivity in the agricultural sector is a key priority for Bangladesh, effective implementation strategies and sufficient funding will be needed to realize these goals at the farm level. The government has identified four areas where improvements are needed, including: 1) providing adequate international funding to address the huge challenges of both adaptation and mitigation; 2) developing the information, communications and technology sector, to enable higher adoption of CSA practices by farmers; 3) strengthening climate and technical information services to improve farmers' capacity to adapt; and 4) improving coordination among the institutions executing CSA projects and programs (CIAT, 2017).

Tanzania

Agriculture is a key sector of the economy of Tanzania, accounting for 24.1 percent of the GDP and the livelihoods for over 75 percent of the population (Tanzania, 2015). Tanzania is highly vulnerable to climate change, especially in terms of agricultural productivity, as farming systems are highly dependent on rain-fed agriculture and there are very poor soil health conditions in the country. Increased resilience and growth in the agriculture sector are needed to help farmers and citizens adapt to the changes in weather patterns. At the same time, innovative strategies should target both women and the youth, as they produce 70 percent of the country's food requirements and 65 percent of the total labor force, respectively.

The Government of Tanzania has identified six strategic priorities as mechanisms for agricultural development in the face of climate change. These priorities actions are to be carried out within the SAI/CSA program, and include a focus on: 1) Improved productivity and incomes; 2) Building resilience and associated mitigation co-benefits; 3) Value chain integration; 4) Research for development and innovations; 5) Improving and sustaining agricultural advisory services; and 6) Improved institutional coordination. Ultimately, these strategic priorities aim to facilitate climate resilience and agricultural growth among farmers across the country (Tanzania, 2015). The Government has also developed the key national policies that are part of the national policy framework that will enable advancement of SAI and CSA (see Table 11.3).

These policy examples from Tanzania illustrate that developing countries, with a majority of smallholder farmers, are developing targeted policies to support approaches such as SAI/CSA. Notably, while introducing the CSA program for the country, the Minister for Agriculture, Steven Masatu Wassira, stated:

> I believe that this CSA Programme, which is the result of a collective, sector-wide consultation effort, places Tanzania firmly on a new and ambitious growth trajectory for the future. We recognise the vital role that agriculture must play in growing the economy and creating decent jobs . . .. The successful implementation of this programme will require multi-level partnerships between the public and private sector, civil society and citizens.
>
> (Tanzania, 2015, vii)

Nicaragua

Agriculture is a crucial sector in Nicaragua's economy, contributing about 17 percent to the country's GDP and constituting 32 percent of the national job market (CIAT, 2015). The agricultural sector contributes 12 percent to total GHG emissions in the country, and 79 percent of remaining emissions come from land-use change and forestry, such as agricultural crops and livestock systems. At the same time, droughts, floods, and variable climate patterns significantly affect agricultural productivity in Nicaragua. Coffee and beans, two of Nicaragua's most economically beneficial production systems, are seriously threatened by rising temperatures (CIAT, 2015).

Table 11.3 An example of a national policy framework of policies to support SAI and CSA

Key national policies for SAI and CSA uptake
Economic policies:
• Tanzania Development Vision
• The National Strategy for Growth and Reduction of Poverty II
Agriculture related policies:
• National Agricultural Policy
• Tanzania Agriculture and Food Security Investment Plan: maps the investments needed to achieve the CAADP target of 6 percent annual growth in agricultural GDP, to contribute to food security.
• Agricultural Sector Development Programme: to enable farmers to have better access to and use of agricultural knowledge, technologies, marketing systems, and infrastructures for higher productivity and profitability; and to promote involvement of the private sector in agricultural transformation under improved regulatory and policy frameworks.
• Livestock Sector Development Strategy
• Fisheries Sector Development Programme
• Tanzania Agriculture Climate Resilience Plan: to implement strategic adaptation and mitigation actions for crops. The plan presents a wide range of adaptation options including, but not limited to: improving agricultural land and water management, accelerating uptake of CSA, reducing impacts of climate-related shocks through risk management, and strengthening knowledge and systems to targeted climate action.
• Southern Agricultural Growth Corridor of Tanzania
• Southern Agricultural Growth Corridor of Tanzania
• Big Result Now Policy: addresses critical sector constraints and challenges to increase agriculture GDP, improve smallholder incomes, and ensure food security by 2015, mainly through smallholder aggregation models for main cereals and high potential crops contributing to import substitution, farm income and food security.
Environment, climate change, land use, and forestry policies:
• National Environmental Policy
• National Climate Change Strategy: sets out strategic interventions for climate change adaptation measures and GHG emissions reductions. Outlines objectives for all sectors and proposed strategic interventions in those sectors and themes that are highly vulnerable to climate change, such as agriculture.
• The National Strategy (and Action Plan) for Reduced Emissions from Deforestation and Forest Degradation (REDD+)

Source: Adapted from Tanzania, 2015.

CSA practices present opportunities for addressing the challenges of climate change, while also promoting economic growth and development of the agriculture sector. The Ministry of Agriculture in Nicaragua, with the support of other organizations, has agreed upon and assumed responsibility for several climate change adaptation, mitigation, and response initiatives. The Ministry is responsible for enforcing legislation that advances agro-ecological practices including management for ecological fertilizer, pesticide, and agricultural products. Key organizations are crucial for providing information and capacity-building services to help farmers respond to these climate changes. As a result of these policies, smallholder farmers have been incorporating climate-adapted seeds or grafted plant varieties on a larger scale, as well as bean, maize, and staple grain varieties that are more resistant and tolerant to erratic climate variations (CIAT, 2015).

Analysis and conclusions

Policy tools for sustainable intensification of agriculture are needed at the global, regional and national levels to concurrently address the challenges of food security, adaptation, and mitigation, and to incentivize changes in governance and financing. Table 11.4 provides an overview of initiatives that have been identified as key policies throughout the chapters of this book as means to support the sustainable intensification of agriculture

Table 11.4 Key multi-level policy mechanisms to support SAI and CSA

Local to National →	*← Regional →*	*← International*
• Supporting changes in farming practices (new varieties, cropping systems) • Adoption of policies, regulations, or tax incentives for SAI and CSA (risk management, land access, insurance, conservation and soil health programs) • Investment and capacity building for research, extension and farmer training • Passage of community and gender policies for SAI/CSA • Improving transportation, infrastructure and markets	• Facilitating and coordinating national/regional policies and efforts (market access, policies) • Supporting collection of accurate data and development of climate-based decision tools • Establishing adaptation and knowledge sharing partnerships • Supporting regional interests between national governments and international initiatives • Industry–government partnerships for investments • Providing specific financing for SAI and CSA initiatives	• Supporting development of National Plans through the UNFCCC (Nationally Determined Contributions and Adaptation Plans) or other Mechanisms (SDGs) • Coordination among national agencies through interagency working groups • Specific planning and support for SAI and CSA in internationally funded projects (e.g. Green Climate Fund) • Fostering increased investment and participation from the private sector into specific investments • Supporting Global Knowledge Sharing networks (e.g. GACSA)

Source: Authors' own compilation.

for smallholder farmers in developing countries. Interventions in policy, institutions, human capacity, and technology will be needed to bring about the transformations that will be required to achieve global food security in light of climate change.

Even though there is growing awareness of climate impacts to agriculture and the need for new SAI and CSA approaches to achieve global food security, the issue has not yet been adequately prioritized under the UN global climate change or sustainability processes. Lofty goals have not yet led to clear guidance or adequate training or funding to countries for implementation of new policies or projects. For example, there has not yet been an adequate emphasis to quantify GHG emissions reductions through soil carbon sequestration, or to quantify loss and damages in developing countries agricultural sectors.

There must be serious, ongoing efforts paid at the global level to prioritize agriculture and secure finances to support implementation of SAI. This will require continued education, coalescing of interests between developed and developing countries, and continued pressure to ensure that the needs of smallholder farmers are addressed adequately.

SAI and CSA are becoming more prevalent in the national plans countries put forward to address climate change mitigation and adaptation in agriculture, and to meet the new SDGs. But these national plans remain vague, without drilling down to the details of what will really be required for successful implementation at the national and local level. In order to adequately upscale (or downscale) adoption of SAI and CSA approaches, authors frequently highlight several priorities (and gaps from the current reality) that will take intensified effort at all levels.

These priorities for interventions must include: 1) providing increased funding and training of extension services to support smallholder farmers in changing practices; 2) providing increased support of applied-research at regional and national levels into new varieties and methods that can increase resiliency and reduce GHG emissions; 3) supporting better collection of climate and agricultural data that can drive climate smart agriculture decision tools and services; 4) making stronger investments in new adaptation and mitigation projects by the international community; 5) implementing gender policies to support changes in practices by women smallholder farmers; and 6) stronger coordination (or policy coherence) across all levels of government.

As new farming paradigms, SAI and CSA should benefit from innovative policies and incentives at all levels. The plethora of actors and mechanisms for addressing SAI is heartening, and is leading to what Ostrom envisioned with the creativity of many actors working to address issues through a greater diversity of actions and bottom-up mechanisms. But there needs to be greater attention paid to monitoring the effectiveness and achievement of goals, and accountability of results. Various policies need to be carefully analyzed to ensure that conflicts between various sectors (e.g. the agriculture sector or climate change sector), or unintended consequences, are minimized.

More effective policies should be designed to take into account the complementarity between policy instruments and implementing institutions, and should avoid tradeoffs that go against one pillar of CSA or SAI (e.g. promoting carbon sequestration, while simultaneously subsidizing mineral fertilizers; or protecting areas for developed country carbon sequestration projects in developing countries, while banning smallholder farmers from producing crops in these areas). Innovative thinking will be required in order to reconcile policy and practice along complementary lines.

As Thornton, Aggarwal, and Parsons (2017) succinctly argue, the challenges of addressing food insecurity and climate change impacts to agriculture are *urgent*. Effectively addressing these challenges will require *new approaches* that are embedded into *policies and frameworks* at the *global, regional, and national level*.

References

Adger, W., Arnell N., and Tompkins, E. (2005) Successful adaptation to climate change across scales. *Global Environmental Change*, 15, pp. 77–86.

Andonova, L. (2010) Public–private partnerships for the Earth: Politics and patterns of hybrid authority in the multilateral system. *Global Environmental Politics*, 10, pp. 25–53.

Andonova, L., Betsill, M., and Bulkeley, H. (2009) Transnational climate governance. *Global Environmental Politics*, 9(2), pp. 52–73.

Bartels, W., Furman, C., and Diehl, D. 2013. Warming up to climate change: a participatory approach to engaging with agricultural stakeholders in the Southeast US. *Regional Environmental Change*, 13(1), pp. 45–55.

Bauer, A. and Steurer, R. (2014. Multi-level governance of climate change adaptation through regional partnerships in Canada and England. *Geoforum*, 51, pp. 121–129.

Biermann F., Pattberg, P., and Zelli, F. (2010) Global climate governance beyond 2012: An introduction. In: Biermann, F., Pattberg, P., and Zelli, F. (eds) *Global Climate Governance Beyond 2010: Architecture, agency and adaptation.* Cambridge: Cambridge University Press, pp. 1–12.

Birner, R. and Resnick, D. (2010) The political economy of policies for smallholder agriculture. *World Development*, 38 (10), pp. 1442–1452.

Brugger, J. and Crimmins, M. (2015) Designing institutions to support local-level climate change adaptation: Insights from a case study of the U.S. Cooperative Extension System. *Weather, Climate, and Society*, 7, pp. 18–38.

Bulkeley, H. and Newell, P. 2010. *Governing Climate Change*. London: Routledge.

Bulkeley, H., Andonova, L., Betsill, M., Compagnon, D., Hale, T., Hoffmann, M., Newell, P., Paterson, M., Roger, C., and Vandeveer, S.D. (2014) *Transnational Climate Change Governance*. Cambridge: Cambridge University Press.

Campbell, B.M, Thornton, P., Zougmore, R., van Asten, P., and Lipper, L. (2014) Sustainable intensification: What is its role in climate smart agriculture? *Current Opinion in Environmental Sustainability*, 8, pp. 39–43.

Caron, P. and Treyer, S. (2016) Climate smart agriculture and international climate change negotiation forums. In: Torquebiau, E. (ed.) *Climate Change and Agriculture Worldwide*. Paris/Montpellier: Springer/CIRAD (Agriculture and Global Challenges Series).

CAADP. (2016) Adaptation of agriculture to climate change in Africa. Available at: www.giz.de/en/worldwide/15891.html.

CCAFS. (2015) The Africa CSA alliance: Path to implementation. Available at: https://ccafs.cgiar.org/blog/africa-csa-alliance-path-implementation#.WdT6otOGP_V.

CGIAR. (2017a). About Us. Available at: www.cgiar.org/about-us/.

CGIAR. (2017b). CSA Guide. Research program on climate change, agriculture and food security (CCAFS). Available at: https://csa.guide/.

CIAT. (2017) Climate-smart agriculture in Bangladesh. CSA country profiles for Asia series. Washington, DC: International Center for Tropical Agriculture (CIAT) and the World Bank. Available at: https://cgspace.cgiar.org/bitstream/handle/10568/83337/CSA_Profile_Bangladesh.pdf?sequence=3&isAllowed=y.

CIAT. (2015) Climate-smart agriculture in Nicaragua. CSA country profiles for Africa, Asia, and Latin America and the Caribbean Series. Washington DC: International Center for Tropical Agriculture (CIAT) and the World Bank. Available at: https://cgspace.cgiar.org/bitstream/handle/10568/68244/CSA%20in%20Nicaragua.pdf?sequence=8&isAllowed=y.

Delvaux, F., Ghani, M., Bondi, G., and Durbin, K. (2014) Climate-smart agriculture: The Emperor's new clothes? *CIDSE*, pp. 1–24.

East African Community Secretariat (EACS) (2011). East African community climate change policy. Arusha, Tanzania. Available at: www.meteorwanda.gov.rw/fileadmin/Template/Policies/EAC_Climate_Change_Policy.pdf.

Farming First. (2015) UNFCCC toolkit. Available at: www.farmingfirst.org/unfccc-toolkit-engagement-update.

Food and Agriculture Organization of the United Nations (FAO). (2017) Global alliance for climate-smart agriculture members list. Available at: www.fao.org/gacsa/members/members-list/en/.

FAO. (2016a) African countries urged to reduce challenges confronting CAADP implementation. Available at: www.fao.org/africa/news/detail-news/en/c/411237/.

FAO. (2016b) Food and agriculture: Key to achieving the 2030 Agenda for Sustainable Development. Available at: www.fao.org/3/a-i5499e.pdf.

FAO. (2015) Global alliance for climate-smart agriculture governance and structure document. GACSA Series Document 2, *pp*. 1–3.

FAO. (2014a) Global alliance for climate-smart agriculture framework document. GACSA Series Document 1, *pp*. 1–6.

FAO. (2014b) Global alliance for climate-smart agriculture framework document. Organizational website. Available online at: www.fao.org/gacsa/en/.

FAO. (2013) *Climate-smart agriculture sourcebook*. Rome, Italy: FAO.

FAO (2010). "Climate-smart" agriculture: Policies, practices and financing for food security, adaptation and mitigation. Available at: www.fao.org/docrep/013/i1881e/i1881e00.htm.

GACSA. (2014) Supporting CSA in Africa and globally. Available at: http://africacsa.org/wp-content/uploads/2015/09/White-Paper-Link-between-the-African-Global-CSA-Alliances-30.10.14.pdf.

Girvetz, E., Corner-Dolloff, C., Lamanna, C., and Rosenstock, T. (2017) "CSA-Plan": strategies to put climate-smart agriculture (CSA) into practice. *Agriculture for Development*, 30, pp. 12–16.

Hatfield, J., Takle, G., Grotjahn, R., Holden, P., Izaurralde, R.C., Mader, T., Marshall, E. and Liverman, D. (2014) Agriculture: Climate change impacts in the United States. In: Melillo, J.M., Richmond, T.C. and Yohe, G.W. (eds)

The Third National Climate Assessment. US Global Change Research Program, pp. 150–174.

Hinds, R. (2016) 5 Steps towards implementing the SDGs. Save the Children. Available online at: http://deliver2030.org/?p=6833.

Howden, S.M., Soussana, J.F., Tubiello, F.N., Chhetri, N., Dunlop, M., and Meinke, H. (2007) Adapting agriculture to climate change. *PNAS*, 104 (50), pp. 19691–19696.

IPCC. (2014) Climate Change 2014: Synthesis Report. Contribution of Working Groups I, II and III to the Fifth Assessment Report of the Intergovernmental Panel on Climate Change (Core Writing Team, R.K. Pachauri and L.A. Meyer (eds.)). IPCC, Geneva, Switzerland.

Jagers, S. and Stripple, J. (2003) Climate governance beyond the state. *Global Governance*, 9, pp. 385–399.

Jordan, A.J., Huitema, D., Hildén, M., van Asselt, H., Rayner, T., Schoenefeld, J., Tosun, J., Forster, J. and Boasson, E. (2015) Emergence of polycentric climate governance and its future prospects. *Nature Climate Change*, 5 (11), pp. 977–982.

Kalfagianni, A. and Duyck, S. (2015) The evolving role of agriculture in climate change negotiations: Progress and players. CCAFS Info Note. Available at: http://bit.ly/1YKaMI2.

Kibaara, B., Gitau, R., Kimenju, S., Nyoro, J., Bruntrup, M. and Zimmermann, R. (2009) Agricultural policy-making in sub-Saharan Africa: CAADP progress in Kenya. Available at: https://pdfs.semanticscholar.org/1d53/9a62ec18c46186ee3c5f1d9ac47fd145b2ab.pdf

Lipper, L., Thornton, P., Campbell, B.M., Baedeker, T., Braimoh, A., Bwalya, M., Caron, P., Cattaneo, D., Garrity, K., Henry, R., Hottle, L., Jackson, A., Jarvis, F., Kossam, W., Mann, N., McCarthy, A., Meybeck, H, and Neufeldt, T. (2014) Climate-smart agriculture for food security. *Nature Climate Change*, 4 (12), pp. 1068–1072.

Meadu, V. (2015) The Paris Climate Agreement: What it means for food and farming. CCAFS Info Note. Available at: https://cgspace.cgiar.org/bitstream/handle/10568/69225/CCAFS%20info%20note%20AgCop21.pdf?sequence=1&isAllowed=y.

Meinke, H., Nelson, R., Kokic, P., Stone, R., Selvaraju, R., and Baethgen, W. (2006) Actionable climate knowledge: From analysis to synthesis. *Climate Research*, 33, pp. 101–110.

Ministry of Agriculture, Food processing and Forests of the French Government (Ministère de l'Agriculture, de l'Agroalimentaire et de la Forêt). (2015–2017) 4 by 1000 Initiative. Available at: http://4p1000.org/.

New Partnership for Africa's Development NEPAD. (2014) Africa Climate Smart Agriculture Alliance launched. Available at: www.nepad.org/content/africa-climate-smart-agriculture-alliance-launched.

Neufeldt, H., Jahn, M., Campbell, B.M., Beddington, J.R., DeClerck, F., De Pinto, A., Gulledge, J., Hellin, J., Herrero, M., Jarvis, A., LeZaks, D., Meinke, H., Rosenstock, T., Scholes, M., Scholes, R., Vermeulen, S., Wollenberg, E., and Zougmoré, R. (2013) Beyond climate smart agriculture: Toward safe operating spaces for global food systems. *Agriculture and Food Security*, 2, pp. 1–6.

Notenbaert, A., Pfeifer, C., Silvestri, S., and Herrero, M. (2017) Targeting, out-scaling and prioritising climate-smart interventions in agricultural systems: Lessons from applying a generic framework to the livestock sector in sub-Saharan Africa. *Agricultural Systems*, 151, pp. 153–162.

OECD. (2012) Policy *Framework for Policy Coherence for Development*: Working Paper no 1. Available at: www.oecd.org/pcd/50461952.pdf.

Ostrom, E. (2010) Polycentric systems for coping with collective action and global environmental change. *Global Environmental Change*, 20, pp. 550–557.

Pattberg, P. and Stripple, J. (2008) Beyond the public and private divide: Remapping transnational climate governance in the 21st century. *International Environmental Agreements: Politics, Law, and Economics*, 8 (4), pp. 367–388.

Petersen, B. and Snapp, S. (2015) What is sustainable intensification? Views from experts. *Land Use Policy*, 46, pp. 1–10.

Prokopy L.S., Arbuckle, J.G., Barnes, A.P., Haden, V.R., Hogan, A., Niles, M.T., and Tyndall, J. (2015) Farmers and climate change: A cross-national comparison of beliefs and risk perceptions in high-income countries. *Environmental Management*, 56 (2), pp. 492–504.

Richards, M., Bruun, T., Campbell, B., Gregersen, L., Huyer, S., Kuntze, V., Madsen, S., Oldvig, M., and Vasileiou, I. (2015). Info Note: How countries plan to address agricultural adaptation and mitigation: an analysis of INDCs. Copenhagen, Denmark: CCAFS. Available at: https://cgspace.cgiar.org/bitstream/handle/10568/69115/CCAFS%20INDC%20info%20note-Final.pdf?sequence=3&isAllowed=y.

Rosenstock, T.S., Lamanna, C., Chesterman, S., Bell, P., Arslan, A., Richards, M., Rioux, J., Akinleye, A., Champalle, C., Cheng, Z., Corner-Dolloff, C., Dohn, J., English, W., Eyrich, A., Girvetz, E., Kerr, A., Lizarazo, M., Madalinska, A., McFatridge, S., Morris, K., Namoi, N., Poultouchidou, da Silva, R., Rayess, S., Ström, H., Tully, K. and W. Zhou, Z. (2016) The scientific basis of climate-smart agriculture: a systematic review protocol. CCAFS Working Paper No. 138. CGIAR Research Program on Climate Change, Agriculture and Food Security (CCAFS), Copenhagen, Denmark. Available at: www.ccafs.cgiar.org.

Ross, S. (2013) Fair shares: Is CAADP working? Available at: www.actionaid.org/sites/files/actionaid/fair_shares_caadp_report.pdf.

Royal Society. (2009) *Reaping the Benefits: Science and the sustainable intensification of global agriculture*. London: The Royal Society.

Saj, S., Torquebiau, E. Hainzelin, E. Pagès, J., and Maraux, F. (2017) The way forward: An agroecological perspective for Climate-Smart Agriculture. *Agriculture, Ecosystems and Environment*, 250, pp. 20–24.

Soussana, J.F., Lutfalla, S., Ehrhardt, F., Rosenstock, T., Lamanna, C., Havlík, P., Richards, M., Wollenberg, E., Chottee, J.L., Torquebiau, E., Ciaisg, P., Smith, P. and Lal, R. (2017) Matching policy and science: Rationale for the "4 per 1000 – soils for food security and climate" initiative. *Soil and Tillage Research*. https://doi.org/10.1016/j.still.2017.12.002.

Stabinsky, D. (2014) Climate-smart agriculture: Myths and problems. Heinrich Boll Stiftung. Available at: https://br.boell.org/sites/default/files/uploads/2014/09/epaper_climate_smart_eng_boll_brasil18.09.014.pdf.

Tanzania, United Republic of. (2015) Tanzania Climate Smart Agriculture Programme. Ministry of Agriculture, Food, Security and Cooperatives and Vice President's Office. Available at: http://canafrica.com/wp-content/uploads/2015/08/TANZANIA-CSA-PROGRAM-Final-version-3-August-2015.pdf.

Thornton, P., Aggarwal, P., and Parsons, D. (2017) Prioritizing climate-smart agricultural interventions at different scales. *Agricultural Systems*, 151, pp. 149–152.

Torquebiau, E. (ed.). (2016) *Climate Change and Agriculture Worldwide*. Paris/Montpellier: Springer/CIRAD (Agriculture and Global Challenges Series).

United Nations. (2015) Resolution adopted by the General Assembly on 25 September 2015, 70/1. Transforming our world: the 2030 Agenda for Sustainable Development. UN General Assembly. Available online at: www.un.org/ga/search/view_doc.asp?symbol=A/RES/70/1&Lang=E.

United Nations DESA/DSD. (2014). Sustainable Development Platform: Goal 2. Available at: https://sustainabledevelopment.un.org/?page=view&nr=164&type=230.

United Nations General Assembly. (2015) Resolution adopted by the General Assembly on 25 September 2015, 70/1. Transforming our world: The 2030 Agenda for Sustainable Development. Available at: www.un.org/ga/search/view_doc.asp?symbol=A/RES/70/1&Lang=E.

United Nations Framework Convention on Climate Change (UNFCCC). (2017) Focus on Adaptation. Available at: http://unfccc.int/focus/adaptation/items/6999.php.

UNFCCC. (2015) The Paris Agreement. Available at: https://unfccc.int/files/meetings/paris_nov_2015/application/pdf/paris_agreement_english_.pdf.

UNFCCC. (1992) Text of the UNFCCC. Available at: https://unfccc.int/resource/docs/convkp/conveng.pdf.

Walthall, C., Hatfield, J., Backlund, P., Lengnick, L., Marshall, E., Walsh, M., Adkins, S., Aillery, M., Ainsworth, E.A., Ammann, C., Anderson, C.J., and Bartomeus, I. (2012) Climate change and agriculture in the United States: Effects and adaptation. USDA Technical Bulletin 1935. Washington DC: US Department of Agriculture.

12 Summary

Sustainable intensification of agriculture, technology and policy options

Udaya Sekhar Nagothu

Introduction

The FAO estimates that farmers will have to produce twice as much food as they do today so as to feed the 9.7 billion global population by 2050 (FAO, 2017). To develop a sustainable global food system in the midst of serious challenges including climate change and population growth will be a daunting task (FAO, 2009; Nelson *et al.*, 2010). These challenges not only make it difficult to ensure food and nutrition security (FNS) for the millions, but can also become a source of serious conflicts in many countries (Beddington et al., 2012). Achieving and maintaining FNS is essential to meet the Sustainable Development Goals, most notably SDG-2 (Zero Hunger) (UN, 2015). This is highly relevant for most vulnerable regions of sub-Saharan Africa (SSA) and South Asia where suffering from hunger and malnutrition (hidden hunger) respectively is greatest in terms of number of people (FAO, 2015). As eminent scientist and Professor Swaminathan has rightly said "*If agriculture goes wrong, nothing else will go right*" (Business World, 2017). We cannot allow agriculture to fail at any cost. Mitigation of – and adaptation to – climate change will be a burden for the governments that are already resource crunched, but there is no second option. Ensuring FNS and securing rural livelihoods of millions of poor and smallholder farmers in the most climate-vulnerable regions is in the best interests of any country and even the entire global community. We are already witnessing large scale migrations across continents, displacement of families and a refugee crisis that is highly politicized. Sustainable intensification (SI) of agriculture or sustainable agricultural intensification (SAI) is one promising approach that can lead to a more stable global food system (Godfray and Garnett, 2016). SAI used in this chapter encompasses a multitude of approaches towards achieving sustainable food systems. There are a number of factors affecting FNS including: environmental, technological, socio-economic and policy related that need to be simultaneously addressed (linked to population, poverty, economic inequality, lack of appropriate policies and implementation) (EU, 2016; Nagothu *et al.*, 2015; NEPAD, 2003).

On the other hand, factors such as increase in population can increase pressure on natural resources, and limit land availability. To counter this, we need to increase the agricultural productivity in a sustainable manner. In some cases, it is the demographic patterns that are concerning, for example migration of men to urban areas for employment and an additional burden shifted to women to take up agriculture, apart from their household responsibility. Such changes in the future will have an enormous impact on the new pathways we choose for agriculture development. Besides, factors such as land tenure and related issues have been seriously responsible for under-developed agriculture, food insecurity and degraded natural resources in already vulnerable regions such as SSA (Salami, Kamara and Brixiova, 2010). Some studies suggested privatization of land as an answer to address land tenure insecurity in order to promote sustainable intensification (Asienga, Perman and Kibet, 2015; Kebede, 2002). Any SAI approach in the future must be climate-smart and allow for a multidisciplinary and multi-sectoral approach to address the wide range of challenges smallholders are facing in the process of contributing to FNS.

Early humans had already started to manipulate selected plant and animal species 12,000 years ago to meet their energy and nutrition needs. This was seen as the first Agricultural Revolution where domestication of crops started and the wave continued till 3500 BC. Interestingly, Harari (2011) writes that in fact it was the crops that domesticated humans, and not vice versa. Domestication of crops and animals was no doubt a big leap for mankind. However, the subsequent intensification of crop production and animal husbandry has also been a major cause for the degradation of the environment. Despite all the plant diversity, only a dozen or more crops, including wheat, rice, maize, potato and millets, currently provide most of the calorific needs. The Green Revolution of the 1960s was another major turn in the history of agriculture intensification that led to an increase in food production, but at a cost for environmental quality and sustainability (Tscharntke et al., 2012; IPES-Food, 2016). The 'planned biodiversity' during the Green Revolution gradually led to monocultures dominated by a few cereal crops (and varieties) such as wheat, rice and maize. Monocultures in the process have seriously influenced the composition and abundance of the associated biota such as those of the pest complex and the soil invertebrates and microorganisms, which in turn have affected plant and soil processes (Swift and Ingram, 1996). As a consequence, agricultural production in some regions has stagnated during the last decade. Agricultural scientists and development experts worldwide now need to address one intriguing question: "*How to address the challenge of FNS without repeating the environmental damage of the mid-20*th *century*?" The disparate chapters in this book authored by experienced scientists and agricultural development experts from a range of disciplines attempt to address this key question.

This chapter has started with a brief introduction. It then summarizes the relevance of SI of agriculture, and why we need to move towards a climate-smart agriculture path. This is followed by brief summaries of the various chapters in the book. Later the key technology, institutional and policy options necessary for promoting SI systems of major food crops are presented. Towards the end, the chapter provides some key messages and recommendations.

Towards sustainable agriculture intensification

All along, agriculture intensification has been a prerequisite to the development of human civilization. Humans started cultivating food crops in order to be food secure (FAO, 2004). The last hundred years witnessed a significant change in agricultural development, but at the cost of large-scale forest clearance and loss of biodiversity. Critics attribute the irreversible environmental impacts to the short-sighted agricultural policies. A 'business as usual' scenario will not help anymore if we have to feed the growing population and the growing demands of the expanding urban middle class. A transformation to SI is thus justified both by necessity (to safeguard global sustainability, a precondition for long-term agricultural viability) and by opportunity (to use sustainable practices as a vehicle for a second green revolution). The publication by Garnett *et al.* (2013) and the Montpellier Panel Report (2013) are quite relevant in this context and represent recent attempts to conceptualize 'sustainable intensification'. To simply define, SI of agriculture or SAI is an essential means of adapting to climate change, resulting in lower emissions per unit of output produced and addressing the serious challenges in the agriculture sector in order to achieve FNS in the future.

Rockstrom, Williams and Daily (2016) also argue that sustainable agriculture is the only strategy that delivers productivity enhancement to meet rising food demands and enables an earth system operating within planetary boundaries. At the global level, SAI, if well integrated into international commitments to combat climate change such as the Paris Agreement, 2015, can positively contribute to FNS and address the Sustainable Development Goals (SDGs) (UN, 2015). SAI through links to food security, nutrition, gender, health, employment, economic development and environment, can contribute significantly to achieving a wide range of SDGs, if properly put in practice. Already, the Food and Agriculture Organization (FAO, 2012), the CGIAR (Beddington *et al.*, 2012), and other international organizations are promoting SI of agriculture as a necessary approach to achieve FNS in the twenty-first century. However, it is the governments of developing countries in particular that have to take active steps to promote SAI through the necessary technological advancements, capacity building, institutional and policy innovations, in addition to investments that are necessary for scaling up the systems. In any case, since SAI systems are knowledge-intensive and require training, they need to be carefully planned for smallholders to implement (FAO, 2015; Nagothu, 2015).

SAI systems also come with certain risks that should be acknowledged and carefully addressed. The opportunity costs that smallholders practising

SAI have to bear in terms of reduced production in order to ensure ecosystem services is debatable. Some argue that SAI perhaps puts too much focus on production and pays less attention to the post-harvest aspects of food systems, including food access, storage, processing and marketing (Silici, Bias and Cavane, 2015). The FAO estimates that about 30 per cent of food produced is wasted in the entire value chains. There is also the practical challenge associated with SAI regarding the measurement of sustainability (Cleary, 2014). Challenges remain in scaling out SAI practices as they are knowledge intensive and need investments in capacity building of smallholders (Pretty, 1997). A fundamental shift therefore will be required in the institutions and policy measures for wider adoption of SAI practices that can be challenging in many developing countries where it is actually the most needed (Tilman *et al.*, 2002). The various chapters in the book attempt to provide answers to the SAI challenges and are summarized below.

Summary of chapters

Chapter 1 briefly reviewed the state of the art of literature relevant to agriculture development and SAI approach with a focus on smallholders and examined the pros and cons of the approach. The SAI concept is gaining momentum in recent years and it is being promoted by international agencies including the FAO, CGIAR and other global institutions for increasing the productivity of major crops including rice, maize, potato, wheat and others. The chapter also discussed the SAI challenges, in particular its relevance and suitability to smallholders in the most climate vulnerable regions. The chapter provided justification for focusing on the major food crops and associated sustainable farming systems that are crucial for future global FNS. This is also the main message that comes out of other chapters, e.g. Chapter 2, 4 and 5. One of the main arguments put forward is that any future investments in the agriculture (including water and livestock) sector should be linked to developing and scaling up innovations that will significantly promote FNS and at the same time adapt to climate change. Scaling up strategies require adequate investments, institutional and policy backing that can address the specific needs of smallholders. The important role of women and youth in future agricultural development and other cross-cutting issues including the policy drivers were highlighted in the chapter. The current state of the art of SAI, major gaps, and steps to be taken for improvements were finally summarized.

Chapter 2 provided a brief introduction on the need to produce more *rice* to meet future demand, highlighting the challenges that farmers will face in order to intensify rice production in a sustainable way while conserving ecosystem goods and services vital for FNS and adapting to climate change. According to the authors, improved management through more efficient use of productive and affordable production inputs and responsible management of agro-biodiversity brings a significant increase in land productivity, including rice yields, and reductions in costs and higher rural incomes. FAO is promoting the policy, concept and good practices associated with sustainable intensification of rice production under the banner of 'Save and Grow' and within the context of its Regional Rice Initiative (RRI). The chapter summarized the RRI implementation progress and results to date in various countries, and using the cases argued for increased policy support of – and investments in – agroecology-literacy and climate-smart agriculture training focusing on smallholder farmers in rice producing landscapes and regions. Farmers trained through the Farmer Field Schools (FFS) approach were found to adopt and adapt sustainable crop management practices in their own rice crops with good results, including substantial increases in land productivity and higher gross margins resulting from reduced and more efficient use of production inputs. Governments and development partners are investing in scaling out the Save and Grow FFS training. The recommendations provided in the chapter are also applicable for other rice growing regions outside Asia. Such initiatives will ensure FNS for future rural and urban populations and minimize adverse impacts on the environment through efficient use of agricultural inputs (fertilizer, water, pesticides) in smallholder rice farms.

Chapter 3 focused on *Maize*, which is one of the main staple food crops for many countries in sub-Saharan Africa and South America providing household FNS. However, its yields remain vulnerable to the vagaries of weather in rainfed systems of the region associated with climate change and variability. The chapter reviewed the major production constraints including irregular fertilizer use in the nitrogen-constrained environments, inappropriate varieties that cannot adapt to climate change and improve human nutrition, as well as pests and diseases (e.g. serious outbreak of the armyworm threatening food security in Africa), poor agronomy and unattractive markets for the crop, among many other constraints. Farmers growing maize, especially in Africa, find it difficult to make profits from maize cultivation. Experiences from the region focusing on conservation agriculture and agro-ecology as a strategy towards SI, were discussed in the chapter. Some of the recommendations to address challenges of maize production included: i) promoting SI approaches to improve soil fertility and water conservation at the farm level (e.g. CA, maize–legume crop rotations); ii) capacity building of smallholder farmers; iii) increasing access to quality inputs and mechanization; iii) developing efficient agro-advisory services; iv) setting up functional VC actor platforms; v) investing in better storage, grading, drying, fumigation facilities for smallholders in the maize growing areas; vi) introducing innovative financing schemes such as credit voucher systems; vii) mainstreaming gender and youth to actively take part in maize VCs; and viii) providing a stable and favourable policy environment for significantly improving the efficiency of maize VCs.

In *Chapter 4*, the authors shared their work experience on wheat, the epicentre of global food security because it is the most widely cultivated cereal in the world, contributing 20 per cent of global calories and proteins. The chapter vividly discussed the challenges imposed by climate change and related abiotic (heat, drought, salinity) and biotic (pathogens, insects, weeds) stresses that are putting scientists and farmers to the test to improve wheat productivity and quality for future generations. Off-the-farm, new technologies such as genomic selection, high throughput phenotyping, and genome editing are promising and will play a larger role in wheat breeding as scientists decode the complexity of the wheat genome. With these tools and others, scientists can seek higher rates of variety replacement and aggressive adoption of newly developed, more nutritious, higher yielding, and disease-resistant wheat germplasm by national wheat breeding programmes and farmers. The chapter recommended that in order to improve the rates of on-farm adoption, especially in developing countries, gender awareness and equality must be considered. Current barriers to wheat production and wheat security also involve market access, food safety, and post-harvest losses. To address the barriers, it is necessary, according to the authors, to provide better infrastructure such as roads and storage facilities, and focus on capacity-building strategies such as crop management information and training opportunities that increase farmer productivity and income, further reducing poverty and ensuring food security. The chapter finally concluded that with the right policies in place and the right commitment from policymakers, regulators, scientists, agricultural institutions, and donors, sustainable intensification of wheat production will be able to alleviate food insecurity.

Chapter 5 dealt with potato, the world's number one non-grain food commodity and third most important food crop in the world after rice and wheat in terms of human consumption. Developing countries account for more than half of the global production and its consumption is increasing each year. It is also cultivated as a cash crop in many regions and thus provides income to millions of farmers. Thus, potato-based cropping systems present increasingly important opportunities for smallholders and the rural poor, not only, in terms of food and nutrition security but also poverty alleviation. The chapter provided a literature review of major challenges associated with potato production. Some recommendations to address the challenges included cost-effective soil management measures (e.g. zero tillage), labour-saving techniques and crop management (e.g. IPM options) to increase productivity while reducing input costs and minimizing environmental impact, and improved post-harvest handling and marketing and gender integration in future strategies for sustainable intensification of potato production. The innovative case of no-tillage potato production in rice-based landscapes from Vietnam was presented to demonstrate how productive, cost-saving and profitable farming system innovations generated through Save and Grow inspired participatory technology development and education investments within the context of FFS can be successfully adopted by smallholders and scaled out with enabling government and development partner support.

In *Chapter 6*, the advantages of millet–legume crop systems were demonstrated. These could potentially be important crops for addressing FNS of the poor and are also suitable to be grown under adverse climate changes. The chapter provided a comprehensive review of literature and analysis on millets–pulses cropping systems to show their contribution to FNS, ecosystem services and the Sustainable Development Goals (SDGs) targets. The multidimensional benefits of pulses–millets cropping systems in terms of food and nutrition provision, climate change adaptation and mitigation, drought and heat tolerance, greenhouse gas emissions (GHG) reductions, soil fertility improvements and biodiversity enhancement, was described and discussed in the chapter. Moreover, possible changes needed in technology and policy that could help smallholders to adopt millets–pulses cropping systems and to realize the potentials of these future crops was suggested. These include technological improvements in improved seed varieties, improved cultivation practices including input management; and institutional and policy measures to improve market facilities and strengthen value chains, improve extension and advisory services and public food distribution systems, among others. Consolidated efforts and investments are necessary to make sure that known climate smart technologies and sustainable practices, in this case pulses and millets, get adopted by smallholders at scale through participatory technology and education efforts, such as those described in Chapter 2, 4 and 5.

In *Chapter 7* the authors argued the need to expand vegetable production in developing countries to improve the quality of diets and the incomes of smallholder farmers. As economies and incomes are growing, per capita vegetable consumption is also increasing, and consumers are willing to pay more for higher quality and off-season vegetables. However, climate change and variability in rainfall and temperature will make vegetable production in the outdoors riskier in the future. The chapter finds controlling climatic extremes through protected cultivation as a viable option that can bring large benefits to both smallholder farmers growing vegetables and consumers. Protected cultivation can help to optimize the use of increasingly expensive inputs such as land, water, labour, fertilizers and pesticides. Recent south Asian government initiatives to promote the use of protected cultivation by smallholder farmers and some success stories from the Asia region were presented in the chapter. There is a need not only to improve low cost and efficient structure designs to cultivate vegetables, but also to recommend appropriate crop sequences and cultural practices to producers, and to breed or provide seed material fit for purpose and tolerant to changing local agro-climatic conditions. A better understanding of the requirements of farmers in targeted regions would be essential to tailor such technologies to the financial and technical capacities of different socio-economic farmer groups. Formation of self-help groups and cluster farm approaches could help to tackle knowledge gaps, financial limitations and marketing issues.

In *Chapter 8*, the integrated crop–livestock systems as potential sources of livelihood to millions of smallholder farmers in arid and semi-arid lands (ASAL) was provided. The systems represent a key solution for enhancing the agricultural productivity, ensuring food and nutritional security and safeguarding the environment through sustainable intensification and prudent and efficient use of natural resources. A strong interaction between livestock and crops is necessary in ASAL. This chapter reviewed major crop–livestock farming systems covering feeds and feed systems in ASAL; forage grasses and legumes and crop-residues as well as the benefits of the integrated crop–livestock systems to smallholder farmers in ensuring food and nutritional security and sustainable livelihoods. The chapter detailed farmers' innovative practices of crop–livestock integrated systems that offer opportunities for increasing agricultural productivity and environmental sustainability in fragile ecological systems of ASAL. Specifically, the authors discussed research innovations that were successfully used in ASAL to mitigate climate change and intensify livestock-based agriculture. The factors affecting the sustainability of crop–livestock farming systems in ASAL were further examined in the context of population growth, environmental degradation, emergence of crop and animal pests and diseases and climate change. Key policy recommendations based on the best cases and experiences from ASAL also relevant for similar other regions were provided.

Chapter 9 synthesized nearly two decades of collaborative, multidisciplinary and participatory research carried out in the coastal zones of the Ganges River delta in Bangladesh and the Mekong River delta in Vietnam by the Consultative Group for the International Agricultural Research Program on Water, Land and Ecosystems and its predecessor, the Challenge Program on Water and Food. These areas are prime examples of tropical deltas at risk of flooding and droughts, where poverty and malnutrition are high and natural resources are degrading. The research projects were aimed at reversing declines in agro-ecological productivity and promoting sustainable intensification in ways that also benefit the poorest members of communities. In both deltas, a combination of in-depth scientific research, citizen science, and participatory action research identified implementable solutions that can sustainably improve productivity and livelihoods. The results demonstrate that achieving sustainable agricultural intensification in such complex systems requires a multi-scale whole-systems landscape and agro-ecological approach.

Chapter 10 argued, on a general level, that: (a) improved extension services; and (b) increased integration of smallholders into the market are necessary to achieve the SAI that climate change clearly necessitates. Elements include: (a) focus on innovative extension models; whereas in: (b) the emphasis was on public–private partnerships, integrated value chains, and farmer markets. The section on extension takes the form of a brief history, with all its issues and challenges, and a summary of some successful models at the end. The chapter takes a careful look at specific models of extension and integration and their associated institutions and mechanisms. The authors note that models must be flexible and tailored towards specific conditions, including the agro-ecological environments, crops or products and end users, markets and investments. It further reviewed the importance of market services that can help smallholders to realize better incomes. Smallholder agriculture will succeed only when it is aligned with market demands and providing opportunities for farmers to earn more income.

Chapter 11 explored the multilevel climate policy mechanisms for Sustainable Intensification of Agriculture and climate adaptation. Climate change impacts to agriculture globally have been increasing and are projected to become more severe through the end of the century. Climate change will necessitate that agricultural, ecological and social systems adapt and mitigate their greenhouse gas (GHG) emissions across multiple scales. But agricultural adaptation largely takes place on a local level and requires changes in farmer practices, increases in capacity, resources and efficient management thereof, and access to information on new innovations and technology. This chapter analysed the existing international, national and regional policies that exist to support and enable farmers to achieve adoption of SAI and CSA practices, and whether or not they are adequate. Agricultural adaptation under the UNFCCC and Paris Accord, which is placing an increasing focus on adaptation at a national level through Nationally Determined Actions, and funding from the Green Climate Fund was reviewed. The chapter analysed the adequacy of international policy mechanisms to ensure that smallholder farmers can sustainably intensify their agricultural production, without adverse effects on social and economic factors, or the environment. The chapter also reviewed the national and regional policies that exist to connect the international policies and funding mechanisms to the local level adaptation practices. Finally, the chapter explored the interplay between these differing levels of governance for agricultural adaptation and sustainable intensification.

Sustainable agriculture intensification: role of technology, capacity building, institutional and policy options

The important role of innovative technology, capacity building, institutions and policy to promote agricultural development through SAI is undisputable. The need to increase productivity without any extra burden on the environment means that future technologies, institutional and policy options have to be innovative and smallholder friendly. It will be smallholders that will continue to produce the bulk of our daily food into the twenty-first century. Hence, it is important that they have timely access to climate smart, productive, affordable and ecologically safe inputs and sources of information and educational opportunities as rightly pointed out in Chapter 1, 2, 3 and 5. Smallholder challenges are unique and differ among regions, including, insecure land tenure and water rights, lack of management skills for efficient use of resources, risks due to climate change, poor market infrastructure and weak institutional frameworks and policy environments.

Technology options

As discussed in several chapters in this book, technology in one form or the other has played an important role in agricultural development. It was the basis for the Green Revolution, where the use of new crop varieties, fertilizers and irrigation were the major factors that largely contributed to food security (Tilman *et al.*, 2002). Norman Borlaug the architect of the Green Revolution was a strong advocate for using new agricultural technology to improve food security (Ortiz *et al.*, 2007). We are at crossroads now looking for the right direction to accomplish a second Green(er) Revolution. However, this time it has be to through a sustainable intensification path without compromising on the environmental impacts. Simultaneously, we need to develop proper infrastructure and investments to take this forward.

To increase productivity without the negative environmental effects, will require innovative and improved technologies that are sustainable. For agricultural scientists, it will not be an easy task to develop such technologies, whereas, for smallholders on the other hand adoption of such technologies will be possible only if they are profitable and contributing to increased land productivity including fish harvest integrated with rice in paddy-based production systems or milk production with crops in integrated crop–livestock systems, since their risk bearing capacity is low. The advancement of SAI technologies will thus need significant efforts from various sectors and stakeholders who need to cooperate at different levels to develop and propagate the technologies. The SAI options should be sustainable and simultaneously increase the net societal benefits. In the following part of this section, technology recommendations that have the potential to contribute to higher productivity and at the same time ensure continued access to ecosystem goods and services from different chapters will be summarized. Table 12.1 lists some important technology and policy options recommended in the book chapters.

Sustainable production approaches

SAI is now more familiar and frequently appearing in the global food security and strategy documents and national governments have started following it. As demonstrated in Chapter 2, SI of rice production is feasible in Southeast Asian countries, where the Save and Grow method of rice cultivation promoted by FAO has proved to be successful in terms of facilitation of adoption and adaptation of SI at a considerable geographical scale. The results of the Save and Grow interventions in the case countries illustrated in Chapter 2 show the enormous potential of rice farmers making

Table 12.1 Most important technology, capacity building, institutional and policy options for sustainable intensification major crops

	Technology options / good agronomic practices	*Capacity building, Institutional and Policy options*
Rice	Productive and affordable production inputs; Judiciary use of agro-chemicals Responsible management of agro-biodiversity Agro-ecology and CSA approach to promote sustainable rice intensification	Regulatory policy environment Investments in education for agroecology-literacy and climate-smart agriculture (rice systems) Farmer Field Schools for training Providing innovative (dis) incentives Remove perverse subsidies on agro-chemicals
Maize	Access to good quality seed, fertilizer, mechanization and pest management technologies Conservation agriculture and agro-ecology approach Crop diversification (crop rotation with legumes)	Improving Extension Services Storage facilities close to smallholders Better roads and market information Value Chain Approach to improve maize productivity Price support to farmers Stable policy environment Investments in joint research involving several countries
Wheat	Genomic selection, high throughput phenotyping, and genome editing Sharing genetic resources for promoting advanced research Improved management practices	Better roads and storage facilities Gender equality in farming and research Capacity building and right policies Better market access, reducing post-harvest losses
Potato	Healthy seed material Soil nutrient management Integrated pest management to control late blight No tillage production	Post-harvest handling Improved marketing and storage, gender integration; Farmer education
Millets and pulses	Crop diversification (crop rotation, intercropping) Improved seed varieties and farmer driven seed systems Improved input management	Better storage and marketing Awareness about nutrition value of pulses and millets and contributions of legumes to soil health
Vegetables	Protected cultivation in control chambers	Structures that are low cost and efficient

	Improving efficiency and design of the structures Better data management	Increased investments in protected cultivation Training of smallholders Farmer collectives for easier marketing Marketing infrastructure
Crop and livestock intensification	Promoting forage grasses and legume fodder Access to land and rights to use	Land tenure security Policy support for crop–livestock integration
Sustainable water management	Promoting a multi-scale whole-systems landscape or agroecological approach.	In-depth scientific research, citizen science, and participatory action research

Source: Authors' own compilation based on summaries of the chapters in this book

more efficient use of production inputs while raising land productivity and incomes and minimizing the impact on the agro-biodiversity. In fact, this could provide incentives for the younger generation to take on a renewed interest in rice production and securing livelihoods in rural areas. Similarly, in Chapter 5, the 'no-tillage' potato cultivation with integrated pest management practices demonstrated how potato production can become more profitable and environmentally sustainable at the same time. In both cases, the FFS based learning contributed positively towards adoption, adaptation and dissemination of the sustainable crop management approaches among other farmers. Lessons learned from this case can be relevant and useful to smallholders in other regions. The emphasis of 'crop diversification' in Chapter 6 and the relevance for smallholder farming systems through crop rotations or intercropping, e.g. legumes with millets or other suitable cereals, showed how it is possible to maximize their synergetic effects on provision of food and nutrition, climate change adaptation, mitigation and soil fertility improvement.

In general, improving access to agricultural inputs, and integrated management of soil, water and pest and disease control was emphasized in several of the chapters to address the smallholder problems in growing various food crops including potato, rice, maize and legumes. According to Pretty and Bharucha (2015), Integrated Pest Management approaches that result in lower pesticide use will benefit not only farmers, but also wider environments and human health. The various chapters also emphasize that training smallholders especially women and youth on better agronomic practices leading to better input use efficiency should be the first step of any SAI strategy.

In Chapter 7, the 'protected cultivation' approach of vegetables was justified as a viable option in today's world that is not only exposed to climate change and transboundary pest proliferation, but also affected by a reduction in available arable land, water scarcity and increases in the cost of energy and agrochemicals. According to the authors, protected cultivation can form part of the solution to the challenges of increasing the efficiency of resource use and mitigating the harmful effects of biotic and abiotic factors on crop yields and quality. However, there is a need not only to improve low cost structures' designs, but also to recommend appropriate crop sequences and cultural practices to producers, and to breed or provide plant material fit for local agro-climatic conditions. Systematizing data collection on farms equipped with sensors and data loggers, experimenting with various protected agro-systems and modelling the interactions of climate, crops, pests and diseases within protected environments will help to optimize low-cost protected cropping systems and in particular soil and climatic conditions.

Integrated management and farming systems diversification, using a combination of methods, can provide better results to farmers, help ensure FNS and protect the environment and the agro-biodiversity. The case study from Laos described in Chapter 2 clearly shows the vital importance of aquatic resources harvested from rice-based landscapes and their vital importance for sustaining FNS in rural communities. The integrated crop–livestock system offers opportunities for raising productivity and increasing efficiency of resources use, securing availability and access to food, increasing household incomes and reducing poverty. Further, it allows exploitation of untapped resources and more efficient use of labour. The system also enhances more efficient recycling of nutrients and improves soil fertility through integration of organic manure from livestock. Integrated crop–livestock systems are becoming important SAI initiatives that are able to increase productivity and at the same time generate profits to smallholders as demonstrated in Chapter 8.

Seed, germplasm and biodiversity

Access to good quality seed is one of the most serious constraints to increasing productivity of major food crops, especially for smallholders, as identified in several chapters of this book. The authors in Chapter 4 clearly showed that global food security in the future will depend on the free movement and open sharing of plant genetic resources. The benefits of such sharing was clearly evident in the past; to note, the efforts of Norman Borlaug whose work depended on germplasm from various countries in order to develop the dwarf varieties of wheat and rice that contributed to the Green Revolution. These days when genetic resources and seeds are regulated by intellectual property rights and international biodiversity laws, it becomes difficult for breeders from different countries to access such resources easily, as pointed out in Chapter 4. Though the Convention on Biological Diversity (CBD) has instituted 'facilitated access' to 35 crops including wheat to the signatories, the procedures of genetic resources acquisition are still lengthy (Bjørnstad,

2016). Chapter 4 recommended that for ensuring global food security, policies and regulations should facilitate germplasm exchange for the benefit of ensuring food security. The other aspect is the availability of good quality seed at the right time and affordable by smallholders. For major food crops, timely availability of healthy seed material can make a large difference in the productivity and income of smallholders, e.g. developing potato seed material *through rapid seed multiplication or using tissue culture methods*. Public private partnerships may be necessary to ensure supply of healthy seed material, whereas in some cases farmer driven seed initiatives can be sustainable options. Chapter 3, 4, 5 and 6 emphasize investments in research involving several countries to jointly develop and produce improved seed varieties that are climate resilient and tolerant to abiotic and biotic stress through participatory plant breeding programmes.

Investment in research and technology

At this time, when most countries are growing through the financial crisis, funding towards agriculture research is not their first priority. In developing countries, investments in agricultural research and technology development are seriously constrained, and infrastructure outdated. The Comprehensive Africa Agriculture Development Program was prioritized under the 2003 Maputo Declaration on Agriculture and Food Security through commitments to allocate at least 10 per cent of national budgetary expenditure towards implementation and aimed to achieve a 6 per cent annual growth of the agricultural sector (OECD/FAO, 2016). However, only a few countries in SSA have fulfilled their commitment on agricultural spending and it is disconcerting that there is an overall decreasing trend in the share of public resources allocated to agriculture.

The recent trend is the investments from the private sector and philanthropic organizations (e.g. the Bill and Melinda Gates Foundation) into agricultural research and technology development supporting large multinational programmes involving several national and international partners (for e.g., the Sorghum breeding programme in Africa). There are other international financial organizations working across the supply value chain with a focus on increasing agricultural productivity, helping mitigate price and weather related shocks, supporting agri-based small and medium enterprises and reducing post-harvest losses (IFC, 2011).

Capacity building

One of the main constraints to implement SAI, is the lack of knowledge and management skills among smallholder farmers and fewer efforts to train them on appropriate technologies. A better understanding of the requirements of farmers in targeted regions would also help tailor such technologies to their financial and technical capacities. If SAI has to be adopted at a larger scale on smallholder farms, capacity building is crucial through farmer participatory learning (Snapp *et al.*, 2014). And governments should be willing to invest in capacity building. Chapter 2 and 4 suggested that investments in farmer education and ecosystem-literacy training, preferably through Farmers Field Schools, are vital for large-scale adoption and scaling out of the SAI options, e.g. the Save and Grow approach in Rice and 'no-tillage' cultivation of potato. Chapter 3 concluded that providing better infrastructure, crop management information, and farmer training opportunities will increase productivity and income. Chapter 8 recommended that formation of self-help groups and farmer cluster approaches could help to tackle knowledge gaps, financial limitations and marketing issues. In some cases the facilities for capacity building exist, but lack necessary technology inputs and linkages with farmers. Approaches such as participatory action research, farmer participatory research, citizen science and farmer to farmer learning are gaining importance and seen as effective ways as demonstrated in Chapter 2, 5 and 9 to improve farmer adaptive capacity. The use of ICT tools and social media can help to support online learning and follow-up of training programmes.

Institutional and policy options

Policy options currently focus mostly on production and input market improvements, e.g. the fertilizer subsidy policies. There is less focus on the post production side of the value chain. The policy, societal and regulatory challenges must be met to increase the benefits of SAI, for crops such as wheat. At the same time, it should maximize input use efficiencies and manage abiotic and biotic stresses to improve yield as demonstrated in Chapter 4. The importance of a *more enabling policy and regulatory environment* for SAI is also evident from the other chapters in this book (Chapters 2, 3, 5 and 6). In fact, Chapter 2 specifically recommends that it will be necessary to review agricultural input subsidy policies, most notably on agro-chemicals, both fertilizers and pesticides at the national level to promote sustainable production of rice. This is also relevant for other major crops. Experience has shown that as long as the policy supports subsidies for fertilizers and other inputs that are made available for free or below market prices, farmers will continue to overuse them and not base their use on the need, which is detrimental to the environment. Chapters 2, 5 and 6 recommend diversification of farming systems that are multifunctional and can achieve multiple objectives including environmental and social benefits.

If policymakers are convinced about the potential of SAI, it becomes easy to develop science based policies and implement them on a larger scale (Cook *et al.*, 2015). This would also mean putting the resources together to scale up SAI technologies. According to Odendo, Obare and Salasya (2009), awareness creation and/or raising of the benefits of a technology is a commonly acknowledged prerequisite for farmers to decide whether to adopt a technology or not. In the process, the relevant stakeholders play an important role as they can pursue the changes that are needed in agricultural policy and practice. The institutional context needs to be appropriate for delivering the necessary goods and services underlying SAI, ensuring inclusiveness across different social groups (Vanlauwe *et al.*, 2014). Success depends on the up-front investment costs that can significantly improve adoption of SAI (McCarthy, Lipper and Branca, 2011). Policies and incentives necessary for the wider adoption of SAI could both promote transitions towards greener economies as well as benefit from progress in other sectors (Pretty and Bharucha, 2014).

Gender equality

Without proper gender integration and mainstreaming, successful SAI implementation will not be possible. A number of chapters in the book pointed this out, and recommended the need for a proper gender balance, including Chapter 1, 4, 5, 6 and 10. Women play an important role in agriculture in the developing world, and any success of the smallholder farms in the future using the SAI approach will need a stronger gender integration. Gender has not been given adequate importance in agricultural research and development so far, despite the significant role of women in irrigated agriculture and the increasing burden they are facing due to climate risks (Raj and Nagothu, 2016). Several studies show that women play a significant role in ensuring household food and nutritional needs in most developing countries (Turral, Burke and Faures, 2011). At the farm level, both men and women complement each other to ensure agriculture production and income. With increased migration of men to urban areas as observed in SSA and other regions of the world, the role of women in agriculture will become even more important. Thus, while promoting SAI, the role of women, their specific vulnerabilities and capabilities for adaptation in the context of changing economic, climate and agrarian systems should be taken into consideration.

Mainstreaming the adaptation strategies with socially inclusive and gender sensitive approaches in agricultural development have the potential to reduce the gender gap (Raj and Nagothu, 2016). However, in practice it depends upon how it is being conceptualized and facilitated. In many contexts, these 'gender role changes may not be sufficient to bring uncontested positive changes in gender relations' (Tatlonghari and Paris, 2013). In their study, Dirutu, Kassie and Shiferaw (2015) find gender differences in the adoption pattern for some of the SAI practices. According to them, women farmers are less likely to adopt minimum tillage and animal manure in crop production than men, indicating the existence of certain socioeconomic inequalities and barriers. However, the study finds no gender differences in the adoption of soil and water conservation measures, improved seed varieties, chemical fertilizers, maize–legume intercropping, and maize–legume rotations. Nevertheless, a better understanding of gender differences is crucial for designing effective policies to close the gender gap while sustainably enhancing farm productivity (Theriault, Smale and Haider, 2017).

Chapter 4 emphasized the need for gender sensitive policies giving women farmers' better access to resources, improved varieties, and increasing their role in participatory variety selection, and also to bring a balance between men and women scientists' ratio in wheat research. In Chapter 5, gender mainstreaming was recommended as a necessary measure to empower women farmers and reduce gender differences through training on market choices for potato, home economics, entrepreneurship skills, decision making and leadership capacity. Oborn *et al.* (2017) emphasize integrating nutrition, gender and equity as an important component of sustainable intensification of agriculture.

Extension and Advisory Services (EASs)

The kind of Extension and Advisory Services (EASs) and their ability to communicate and share knowledge is important for promoting SAI. Proven SAI technology options once validated in real time conditions can be made accessible to smallholders for upscaling. Chapter 10 highlighted the importance of extension services and strengthening market value chains to ensure successful implementation of SAI practices by smallholders. There is a need to restructure the current extension departments in developing countries, shift towards demand based knowledge exchange, recruit more women and youth and introduce ICT based knowledge transfer mechanisms for faster communication. In Chapter 2, the authors showed the importance of approaches such as the FFS based participatory and discovery based learning process that has clearly been instrumental in allowing farmers to experiment with and learn about the FAO, Save and Grow practices, diversify farming systems and become confident in adopting location and situation specific best practices that are economically feasible as well as ecologically sound. This approach is more targeted and effective as compared to the general extension approaches adopted by state agencies in most countries. Similarly, Chapter 6 suggested that customized trainings on seed production, multiplication, storage of millets and pulses, and extension programmes particularly targeting women and youth on the health and nutrition benefits of pulses and millets is important.

Markets and investments

One of the key bottlenecks facing the agricultural sector in developing countries is poor infrastructure including transportation networks, storage, credit and insurance facilities. The middlemen take advantage of the situation and exploit the farmers, thus taking the major share of the profits. Therefore, governments should regulate the role of middlemen by establishing minimum support prices, and facilitate the connection of different value chain (VC) actors. At the same time, support should be given to public–private partnerships to improve post-harvest handling, quality control and marketing of agricultural produce. Though we see some good examples of successful product VC improvement these days, the pace at which their awareness is growing is still slow. Conducting market analysis regularly and promoting food crop VCs to promote their market value can be useful.

It is also important to sensitize different actors in the VCs on the importance of VC enhancement and value creation through simple interventions. For example, training farmers on quality sorting and packaging. This will also require capacity building of VC actors where needed. Policymakers need to be educated about the importance of value chains. And finally, there needs to be government support to build and upscale the capacity for VC analysis at different levels and strengthen the VCs, through support of education and training in agricultural economics and business management.

The Chapter 4, 5, 6 and 10 suggested that investments in logistics and infrastructure, especially rural road networks and storage facilities will facilitate acquisition of inputs and marketing of agricultural produce through faster mobility. Investment in applied agricultural research has been emphasized in Chapter 3, 4, 5, 6 and 8, as it plays a crucial role in helping farmers identify suitable crop varieties and practices, creates awareness and promotes best practices that are more sustainable. Chapter 5 specifically suggested the establishment of community owned warehouses close to farms for storage of potatoes. In Chapter 6, the authors expressed the need for developing low cost pulse/millets seed harvesting/processing machines and storage warehouses/storage units that are accessible and affordable by smallholders. Whereas, Chapter 7 demonstrated the advantage of reliable marketing networks for smallholders in protected cultivation which helps them to realize better income for higher quality produce. Woelcke (2006) in his study concluded that in order to pursue SAI, innovative credit schemes, alternative forms of labour acquisition and market reforms have to be promoted, to provide sufficient economic incentives for the adaptation.

If governments should try to develop strategic partnerships with the private sector to address SAI implementation challenges, PPPs are crucial to scale out SAI in the current situation where governments are constrained by lack of funds to invest in the agriculture sector. Partnerships among several relevant actors – public, private, NGOs and farmers will be essential to promote SAI on a wider scale. Smallholders should have a key role in such partnerships, and this is possible if farmer collectives are well organized. Good examples are the dairy cooperatives in countries such as Rwanda and Kenya that are functioning well due to the fact that smallholders are given their due share of profit (FAO, 2014). We can find similar success stories across other developing countries, e.g. the fruit growers cooperatives in the state of Maharashtra in India, and the rice value chains in Vietnam, where improvements in post-harvest activities have led to an increase in farmer income and overall value chain enhancement.

Recommendations and relevance to achieving SDGs

Agricultural scientists and policymakers have the responsibility not only to address the key goals of FNS amidst the challenges of climate change and projected increase in population, but also the SDGs. These have to be accomplished without further increasing the negative environmental impacts. Technology, capacity building, institutional and policy support is crucial for the long-term sustainability of agricultural production as highlighted in the book. The goal of SAI that is central to the book is to maximize the net societal, economic and environmental benefits simultaneously.

The book has also demonstrated that the pursuit of SAI will require not only significant increases in technology development but also adoption of the technologies effectively. An integrated VC approach should therefore be the way forward to improve sustainable productivity of major food crops. Smallholders will continue to be the major producers of food in developing countries in the years to come, and hence they should be the main target group of future agricultural development. The following are some of the key recommendations derived from the various chapters in the book and deemed necessary for future agricultural development through sustainable intensification:

- A comprehensive VC approach to promote SAI of major food crops;
- Facilitating sharing of crop genetic resources and joint research among countries for developing and making available improved seed varieties;
- Improving agricultural production input use efficiency through integrated management;
- Promoting sustainable approaches, e.g. sustainable intensification of rice, no-tillage potato cultivation, cereal–legume rotations and other CSA practices suitable for smallholders;
- Invest in capacity building, including education of smallholders for innovation, adoption and adaptation of SAI;
- Mainstreaming gender and youth (in research, education and implementation) as an integral component of future strategies to promote SAI;
- Improving post-harvest facilities with emphasis on storage, marketing, value addition, price support to smallholders;
- Increasing investments in agriculture development and research through PPPs;
- Facilitating farmer collectives for active participatory learning, research and marketing;
- Promoting SAI in complex river delta systems through a multi-scale whole-systems landscape or agroecological approach;

- Networking with relevant VC actors as part of SAI of food crops, e.g. functional Multi-Actor Platforms;
- Integrating SAI into national and international policies for agricultural development.

Finally, to conclude, the main message in the book is that a 'business as usual' approach will not work. We need to build capacity and become innovative in our future planning and implementation of sustainable agriculture intensification initiatives to address the challenges of FNS as clearly spelled out in the SDGs-2030. We need to be better prepared as we are going to face frequent climate extreme events in the coming years and decades. If agriculture is made profitable and less risky, youth will consider to take it up as the main source of livelihood. If not, we will continue to see even larger scale migrations of people to urban areas leading to a new set of problems that we are already encountering across the globe. It is up to the governments, private sector, scientists and communities at large to realize that agriculture and food security should be a top priority for investments and development in the future, especially in Africa and other regions where it is needed. It is the investments we make that matters most at this point in time in order to achieve the SDGs we have set for ourselves and for the betterment of people and the environment at large.

References

Asienga, I.C., Perman, R. and Kibet, L.K. (2015) The role of fencing on marginal productivity of labour, land and capital in ASAL regions of Kenya. *International Journal of Development and Economic Sustainability*, 3, pp. 80–93.

Beddington, J., Asaduzzaman, M., Clark, M., Fernandez, A., Guillou, M., Jahn, M., Erda, L., Mamo, T, van Bo, N., Nobre, C.A., Scholes, R., Sharma, R. and Wakhungu, J. (2012) Achieving food security in the face of climate change: Final report from the Commission on Sustainable Agriculture and Climate Change. Available at: https://cgspace.cgiar.org/bitstream/handle/10568/35589/climate_food_commission-final-mar2012.pdf (accessed 01 August 2017).

Bjørnstad, A. (2016) 'Do not privatize the giant's shoulders': Rethinking patents and plant breeding. *Cell Press*, 34, pp. 609–617.

Business World (2017) If agriculture goes wrong, nothing else will go right. Available at: http://businessworld.in/article/-If-Agriculture-Goes-Wrong-Nothing-Else-Will-Go-Right-/04-06-2017-119433/ (accessed 18 September 2017).

Cleary, D. (2014) Deconstructing sustainable intensification and issues around sustainability metrics. Available at: www.nature.org/science-in-action/science-features/david-cleary-critique.pdf (accessed 13 July 2017).

Cook, S., Silici, L., Adolph, B. and Walker, S. (2015) Sustainable agricultural intensification revisited. Available at: http://pubs.iied.org/pdfs/14651IIED.pdf (accessed 13 July 2017).

Dirutu, S.W., Kassie, M. and Shiferaw, B. (2015) Are there systematic gender differences in the adoption of sustainable agricultural intensification practices? Evidence from Kenya. Available at: www.sciencedirect.com/science/article/pii/S0306919214001109 (accessed 29 July 2017).

European Union (EU) (2016) Designing the path: A strategic approach to EU agricultural research and innovation (draft paper). Available at: https://ec.europa.eu/programmes/horizon2020/en/news/designing-path-strategic-approach-eu-agricultural-research-and-innovation (accessed 10 August 2016).

Food and Agricultural Organization (FAO) (2004), The ethics of Sustainable Agricultural Intensification. Available at: www.fao.org/docrep/007/j0902e/j0902e03.htm (accessed 13 July 2017).

Food and Agricultural Organization (FAO) (2009) 'The special challenge for sub-Saharan Africa'. High level expert forum: How to feed the world 2050. Rome: FAO. Available at: www.fao.org/fileadmin/templates/wsfs/docs/Issues_papers/HLEF2050_Africa.pdf.

Food and Agricultural Organization (FAO) (2012) Towards the future we want: End hunger and make the transition to sustainable agricultural and food systems. Available at: www.fao.org/docrep/015/an894e/an894e00.pdf (accessed 10 August 2016).

Food and Agricultural Organization (FAO) (2014) FAO supports dairy farmers to bring quality milk to the market. Available at: www.fao.org/africa/news/detail-news/en/c/238284/ (accessed 15 June 2017).

Food and Agricultural Organization (FAO), IFAD and WFP (2015) The state of food insecurity in the world 2015. Meeting the 2015 international hunger targets: taking stock of uneven progress. Available at: www.fao.org/3/a-i4646e/i4646e00.pdf (accessed 16 August 2017).

Food and Agricultural Organization (FAO) (2017) The state of food security and nutrition in the world. Available at: www.fao.org/state-of-food-security-nutrition/en/ (accessed 27 October 2017).

Garnett, T., Appleby, M.C., Balmford, A., Bateman, I.J., Benton, T.G., Bloomer, P., Burlingame, B., Dawkins, M., Dolan, L., Fraser, D., Herrero, M., Hoffmann, I., Smith, P., Thornton, P.K., Toulmin, C., Vermeulen, S.J. and Godfray, H.C.J. (2013) Sustainable intensification in agriculture: Premises and policies, *Science*, 34, pp. 33–34.

Godfray, H.C.J. and Garnett, T. (2016) Food security and sustainable intensification. Available at: www.ncbi.nlm.nih.gov/pmc/articles/PMC3928882/ (accessed 30 September, 2017).

Harari, Y.N. (2011) *Sapiens: A brief history of humankind*. Harmondsworth: Penguin, pp. 87–91.

International Finance Corporation (IFC) (2011) The private sector and global food security. Available at: www.ifc.org/wps/wcm/connect/news_ext_content/ifc_external_corporate_site/news+and+events/news/features_food_security (accessed 06 October 2017).

IPES-Food (2016) From uniformity to diversity: A paradigm shift from industrial agriculture to diversified agro-ecological systems, International Panel of Experts on Sustainable food systems. Available at: www.ipes-food.org (accessed on 15 November 2016).

Kebede, B. (2002) Land tenure system and common pool resources in rural Ethiopia: A study based on fifteen states. *African Development Review*, 14, pp. 113–150.

McCarthy, N., Lipper, L. and Branca, G. (2011) Climate-smart agriculture: Smallholder adoption and implications for climate change adaptation and mitigation. Mitigation of Climate Change in Agriculture Series 4. Rome, Italy: Food and Agriculture Organization of the United Nations (FAO).

Nagothu, U.S. (2015) The future of food security: Summary and recommendations. In: Nagothu, U.S. (ed.) *Food Security and Development: Country case studies*. London: Routledge, pp. 252–269.

Nelson, G.C., Rosegrant, M.W., Palazzo, A., Gray, I., Ingersoll, C., Robertson, R., Tokgoz, S., Zhu, T., Sulser, T., Ringler, C., Msangi, S. and You, L., (2010) *Food Security, Farming, and Climate Change to 2050: Scenarios, results, policy options*. Washington, DC: IFPRI.

NEPAD (New Partnerships For Africa's Development) (2003) Comprehensive Africa Agriculture Development Program (CAADP), Midrand, South Africa: NEPAD.

Oborn, I, Vanlauwe, B. Phillips, M., Thomas, R., Brooijmans, W. and Atta-Krah, K. (2017) Sustainable intensification in smallholder agriculture: An integrated systems research approach. Available at: www.routledge.com/Sustainable-Intensification-in-Smallholder-Agriculture-An-integrated-systems/Oborn-Vanlauwe-Phillips-Thomas-Brooijmans-Atta-Krah/p/book/9781138668089 (accessed 21 November, 2017).

Odendo, M., Obare, G. and Salasya, B. (2009) Factors responsible for differences in uptake of integrated soil fertility management practices amongst smallholders in western Kenya. *African Journal of Agricultural Research*, 4 (11), pp. 1303–1311.

OECD/FAO (2016). Agriculture in sub-Saharan Africa: Prospects and challenges for the next decade. *OECD-FAO Agricultural Outlook 2016–2025*. Paris: OECD Publishing.

Ortiz, R., Mowbray, D., Dowswell, C., and Rajaram, S. (2007) Dedication: Norman E. Borlaug the humanitarian plant scientist who changed the world. *Plant Breeding Reviews*, 28, pp. 1–37.

Pretty, J.N. (1997) The sustainable intensification of agriculture, natural resources. *Forum*, 21 (4): 247–256.

Pretty, J. and Bharucha, Z.P. (2014) Sustainable intensification in agricultural systems, *Annals of Botany*, p. 1 of 26. Available at: www.aob.oxfordjournals.org.

Pretty, J. and Bharucha, Z.P., (2015) Integrated pest management for sustainable intensification of agriculture in Asia and Africa. *Insects*, 6, pp. 152–182.

Raj, R. and Nagothu, U.S. (2016) Gendered adaptation to climate change in canal-irrigated agro-ecosystems. In: Nagothu, U.S. (ed.) *Climate Change and Agriculture Development*. London: Routledge, pp. 259–278.

Rockstrom, J., Williams, J., Daily, G. (2016) Sustainable intensification of agriculture for human prosperity and global sustainability. *Ambio*. Available at: http://link.springer.com/article/10.1007/s13280-016-0793-6.

Salami, A., Kamara, A.B. and Brixiova, Z. (2010) Smallholder agriculture in East Africa: Trends, constraints and opportunities. Working Paper No. 105, African Development Bank Group, TUNIS Belvédère, Tunisia.

Silici, L., Bias, C. and Cavane, E. (2015) Sustainable agriculture for small-scale farmers in Mozambique – A scoping report, Available at: http://pubs.iied.org/14654IIED/ (Accessed 11 July 2017).

Snapp, S.S., Blackie, M.J., Gilbert, R.A., Bezner-Kerr, R. and Kanyama-Phiri, G.Y. (2014) Modeling and participatory, farmer-led approaches to food security in a changing world: A case study from Malawi. 2014. *Secheresse*, 24, pp. 350–358.

Swift, M.J. and Ingram, J.S.O. (1996) Global change and terrestrial ecosystems, GCTE Report No. 13 Wallingford, UK.

Tatlonghari, G.T. and Paris, T.R. (2013) Gendered adaptations to climate change: A case study from the Philippines. In: Alston, M. and Whittenbury, K. (eds)

Research, Actio and Policy: Addressing the gendered impacts of climate change. Amsterdam: Springer.

The Montpellier Panel (2013) *Sustainable Intensification: A new paradigm for African agriculture*, London.

Theriault, V., Smale, M. and Haider, H. (2017) How does gender affect sustainable intensification of cereal production in the West African Sahel? Evidence from Burkina Faso. *World Development*, 92, pp. 177–191.

Tilman, D., Cassman, K.G., Matson, P.A., Naylor, R. and Polasky, S. (2002) Agricultural sustainability and intensive production practices, Available at: www.nature.com/nature/journal/v418/n6898/full/nature01014.html (accessed 17 July 2017).

Tscharntke, T., Clough, Y., Wanger, T.C., Jackson, L., Motzke, I., Perfecto, I., Vandermeer, J. and Whitbread, A. (2012) Global food security, biodiversity conservation and the future of agricultural intensification. *Biological Conservation*, 151, pp. 53–59.

Turral, H., Burke, J. and Faures, J-M. (2011) Climate change, water and food security, FAO Water Reports 36. Rome: Food and Agriculture Organization of the United Nations. Available at: www.fao.org/docrep/014/i2096e/i2096e.pdf (accessed 16 July 2017).

UN (2015) Sustainable Development Goals. Available at: https://sustainabledevelopment.un.org/?page=view&nr=164&type=230 (accessed 21 July 2017).

Vanlauwe, B., Coyne, D., Gockowski, J., Hauser, S., Huising, J., Masso, C., Nziguheba, G., Schut, M. and Van Asten, P. (2014) Sustainable intensification and the African smallholder. Available at: http://humidtropics.cgiar.org/wp-content/uploads/downloads/2014/12/Sustainable-intensification-and-the-African-smallholder.pdf (accessed 15 August 2017).

Woelcke, J. (2006) Technological and policy options for sustainable agricultural intensification in eastern Uganda. *Agricultural Economics*, 34, pp. 129–139.

Index

Page numbers in *italics* indicate information in figures; those in **bold** indicate tables; and <u>underlined</u> page numbers refer to boxes.